QSAR and Molecular Modeling Studies in Heterocyclic Drugs II

Volume Editor: Satya Prakash Gupta

With contributions by

M. K. Gupta · S. P. Gupta · D. Hadjipavlou-Litina
S. Hannongbua · S. B. Katti · Y. S. Prabhakar · V. R. Solomon
M. Vračko · C.-G. Zhan

The series *Topics in Heterocyclic Chemistry* presents critical reviews on "Heterocyclic Compounds" within topic-related volumes dealing with all aspects such as synthesis, reaction mechanisms, structure complexity, properties, reactivity, stability, fundamental and theoretical studies, biology, biomedical studies, pharmacological aspects, applications in material sciences, etc. Metabolism will be also included which will provide information useful in designing pharmacologically active agents. Pathways involving destruction of heterocyclic rings will also be dealt with so that synthesis of specifically functionalized non-heterocyclic molecules can be designed.

The overall scope is to cover topics dealing with most of the areas of current trends in heterocyclic chemistry which will suit to a larger heterocyclic community.

As a rule contributions are specially commissioned. The editors and publishers will, however, always be pleased to receive suggestions and supplementary information. Papers are accepted for *Topics in Heterocyclic Chemistry* in English.

In references *Topics in Heterocyclic Chemistry* is abbreviated *Top Heterocycl Chem* and is cited as a journal.

Springer WWW home page: springer.com
Visit the THC content at springerlink.com

Library of Congress Control Number: 2006926508

ISSN 1861-9282
ISBN-10 3-540-33233-2 Springer Berlin Heidelberg New York
ISBN-13 978-3-540-33233-6 Springer Berlin Heidelberg New York
DOI 10.1007/11731825

Springer is a part of Springer Science+Business Media

springer.com

© Springer-Verlag Berlin Heidelberg 2006
Printed in Germany

The use of registered names, trademarks, etc. in this publication does not imply, even in the absence of a specific statement, that such names are exempt from the relevant protective laws and regulations and therefore free for general use.

Cover design: *Design & Production* GmbH, Heidelberg
Typesetting and Production: LE-TeX Jelonek, Schmidt & Vöckler GbR, Leipzig

Printed on acid-free paper   02/3100 YL – 5 4 3 2 1 0

4
Topics in Heterocyclic Chemistry

Series Editor: R. R. Gupta

Topics in Heterocyclic Chemistry
Series Editor: R. R. Gupta

Recently Published and Forthcoming Volumes

Series Editor

Prof. R. R. Gupta

10A, Vasundhara Colony
Lane No. 1, Tonk Road
Jaipur-302 018, India
rrg_vg@yahoo.co.in

Volume Editor

Prof. Dr. Satya Prakash Gupta

Department of Chemistry
Birla Institute of Technology and Science
Pilani-333 031, India
spg@bits-pilani.ac.in

Editorial Board

Prof. D. Enders

RWTH Aachen
Institut für Organische Chemie
D-52074, Aachen, Germany
enders@rwth-aachen.de

Prof. Steven V. Ley FRS

BP 1702 Professor
and Head of Organic Chemistry
University of Cambridge
Department of Chemistry
Lensfield Road
Cambridge, CB2 1EW, UK
svl1000@cam.ac.uk

Prof. G. Mehta FRS

Director
Department of Organic Chemistry
Indian Institute of Science
Bangalore- 560 012, India
gm@orgchem.iisc.ernet.in

Prof. A.I. Meyers

Emeritus Distinguished Professor of
Department of Chemistry
Colorado State University
Fort Collins, CO 80523-1872, USA
aimeyers@lamar.colostate.edu

Prof. K.C. Nicolaou

Chairman
Department of Chemistry
The Scripps Research Institute
10550 N. Torrey Pines Rd.
La Jolla, California 92037, USA
kcn@scripps.edu
and
Professor of Chemistry
Department of Chemistry and Biochemistry
University of California
San Diego, 9500 Gilman Drive
La Jolla, California 92093, USA

Topics in Heterocyclic Chemistry
Also Available Electronically

For all customers who have a standing order to Topics in Heterocyclic Chemistry, we offer the electronic version via SpringerLink free of charge. Please contact your librarian who can receive a password or free access to the full articles by registering at:

springerlink.com

If you do not have a subscription, you can still view the tables of contents of the volumes and the abstract of each article by going to the SpringerLink Homepage, clicking on "Browse by Online Libraries", then "Chemical Sciences", and finally choose Topics in Heterocyclic Chemistry.

You will find information about the

 – Editorial Board
 – Aims and Scope
 – Instructions for Authors
 – Sample Contribution

at springer.com using the search function.

Preface to the Series

Topics in Heterocyclic Chemistry presents critical accounts of heterocyclic compounds (cyclic compounds containing at least one heteroatom other than carbon in the ring) ranging from three members to supramolecules. More than 50% of billions of compounds listed in *Chemical Abstracts* are heterocyclic compounds. The branch of chemistry dealing with these heterocyclic compounds is called heterocyclic chemistry, which is the largest branch of chemistry and as such the chemical literature appearing every year as research papers and review articles is vast and can not be covered in a single volume.

This series in heterocyclic chemistry is being introduced to collectively make available critically and comprehensively reviewed literature scattered in various journals as papers and review articles. All sorts of heterocyclic compounds originating from synthesis, natural products, marine products, insects, etc. will be covered. Several heterocyclic compounds play a significant role in maintaining life. Blood constituent hemoglobin and purines as well as pyrimidines, the constituents of nucleic acid (DNA and RNA) are also heterocyclic compounds. Several amino acids, carbohydrates, vitamins, alkaloids, antibiotics, etc. are also heterocyclic compounds that are essential for life. Heterocyclic compounds are widely used in clinical practice as drugs, but all applications of heterocyclic medicines can not be discussed in detail. In addition to such applications, heterocyclic compounds also find several applications in the plastics industry, in photography as sensitizers and developers, and in dye industry as dyes, etc.

Each volume will be thematic, dealing with a specific and related subject that will cover fundamental, basic aspects including synthesis, isolation, purification, physical and chemical properties, stability and reactivity, reactions involving mechanisms, intra- and intermolecular transformations, intra- and intermolecular rearrangements, applications as medicinal agents, biological and biomedical studies, pharmacological aspects, applications in material science, and industrial and structural applications.

The synthesis of heterocyclic compounds using transition metals and using heterocyclic compounds as intermediates in the synthesis of other organic compounds will be an additional feature of each volume. Pathways involving the destruction of heterocyclic rings will also be dealt with so that the synthesis of specifically functionalized non-heterocyclic molecules can be designed. Each volume in this series will provide an overall picture of heterocyclic compounds critically and comprehensively evaluated based on five to ten years of literature. Graduates, research students and scientists in the fields of chemistry, pharmaceutical chemistry, medicinal chemistry, dyestuff chemistry, agrochemistry, etc. in universities, industry, and research organizations will find this series useful.

I express my sincere thanks to the Springer staff, especially to Dr. Marion Hertel, executive editor, chemistry, and Birgit Kollmar-Thoni, desk editor, chemistry, for their excellent collaboration during the establishment of this series and preparation of the volumes. I also thank my colleague Dr. Mahendra Kumar for providing valuable suggestions. I am also thankful to my wife Mrs. Vimla Gupta for her multifaceted cooperation.

Jaipur, 31 January 2006   R.R. Gupta

Preface

The series *Topics in Heterocyclic Chemistry* now devotes its two volumes, Vols. 3 and 4, to today's most fascinating area of medicinal chemistry: quantitative structure-activity relationships (QSAR) and molecular modeling, which has revolutionized drug discovery in the present era. These two volumes together present some very timely and important reviews on QSAR and molecular modeling studies in heterocyclic drugs and are titled *QSAR and Molecular Modeling Studies in Heterocyclic Drugs I* and *QSAR and Molecular Modeling Studies in Heterocyclic Drugs II*. Since the pioneering work of Corwin Hansch from 1962–1964 that laid the foundations of QSAR by means of three important contributions: the combination of several physicochemical parameters in one equation, the definition of the lipophilicity parameter π, and the formulation of the parabolic model for nonlinear lipophilicity–activity relationships, the area of computer-aided drug design with the development of computer technology went through a revolutionary change from two-dimensional to three-dimensional and now to multi-dimensional QSAR. The QSAR and molecular modeling studies have drastically reduced the cost and the time involved in the drug design and development. With the objective that some timely in-depth reviews on such studies in heterocyclic drugs may be of great value to those involved in drug discovery, some leaders in the field were invited to contribute and the overwhelming response led to devote two volumes on the topic. Both volumes cover the excellent and novel articles of varied interest.

Volume 3 contains five articles. The first article by Castro et al. describes the application of flexible molecular descriptors in the QSAR study of heterocyclic drugs. In this article, the various formulations of optimal descriptors introduced by different authors during the last ten years are discussed for the special case of heterocyclic drugs. The second article by Basak et al. is on predicting pharmacological and toxicological activity of heterocyclic compounds using QSAR and molecular modeling. Heterocyclic compounds are important as drugs, toxicants, and agrochemicals. In this article, the authors report the QSAR modeling of pharmacological activity, insect repellency, and environmental toxicity for a few classes of heterocyclic compounds from their structure. Pharmacological activity of drugs depends mainly on the interaction with their biological targets, which have complex three-dimensional structure, and their molecular recognitions are guided by the nature of intermolecular interactions. In the third article, therefore, Ponnuswamy et al. present conformational aspects and interaction studies of different hetero-cyclic drugs. In the next article, Khanna et al. describe, in detail, *in silico* studies on PPARγ agonistic heterocyclic systems. Several heterocyclic deriva-tives like oxazolidinedione, thiazolidinedione, tetrazole, phenoxazine, etc., are being developed for the treatment of insulin resistance and type 2 diabetes mellitus. The heterocyclic head group in these systems binds to and activates peroxisome proliferator activated receptor γ (PPARγ), a nuclear receptor that regulates the expression of several genes involved in the metabolism. In this article, therefore, various molecular modeling studies have been described that are important in understanding the drug–receptor interactions, analyzing the important pharmacophore features, identifying new scaffolds, and under-standing the electronic structure and reactivity of these heterocyclic systems. The final article in this volume, written by Garg and Bhhatarai, is on QSAR and molecular modeling studies of HIV protease inhibitors. HIV protease is one of the major viral targets for the development of new chemotherapeutics against AIDS. In this article, therefore, Garg and Bhhatarai have presented a detailed study on structure–activity relationship studies on many groups of HIV protease inhibitors, providing the excellent rationale to design potent and pharmaceutically important protease inhibitors.

There are six fascinating articles in Vol. 4. These six articles present QSAR and molecular modeling on six different classes of heterocyclic drugs. The first article by Hadjipavlou-Litina is related to thrombin and factor FXa in-hibitors. Both thrombin and factor FXa are bound to and are enzymatically active in blood clots. Thus a QSAR study on them may be of great use to investi-gate potent antithrombotics or antocoagulants. Similarly, the second article by Hannongbua has reviewed structural information and drug–enzyme interac-tion of the non-nucleoside reverse transcriptase inhibitors based on quantum chemical approaches, providing the valuable guidelines to design and develop potent anti-HIV drugs. Reverse transcriptase is an important enzymatic target to inhibit the growth of human immunodeficiency virus of type 1 (HIV-1), which is the causative agent of AIDS. In the next article, Vračko has described a QSAR approach in study of mutagenicity of aromatic and heteroaromatic amines. These compounds are highly hazardous to the environment and can be carcinogenic and thus are the subject of both theoretical and experimental studies.

Cocaine is a widely abused heterocyclic drug and there is no available anti-cocaine therapeutic, but in the fourth article Zhan describes the state of the art of molecular modeling of the reaction mechanism for the hydrolysis of cocaine and the mechanism-based design of anti-cocaine therapeutics. Amongst the heterocyclic systems, thiazolidine is a biologically important scaffold known to be associated with several biological activities. Some of the prominent biologi-cal responses attributed to this skeleton are antiviral, antibacterial, antifungal, antihistaminic, hypoglycemic, and anti-inflammatory activities. In the fifth article, therefore, Prabhakar et al. have presented a very comprehensive review on the QSAR studies of diverse biological activities of the thiazolidines published during the past decade. This study may be of importance to explore the possibility if thiazolidine nucleus can be exploited to design the drugs for some other diseases. In the final article, however, Gupta has reviewed the QSAR studies on calcium channel blockers (CCBs). CCBs have potential therapeutic uses against several cardiovascular and non-cardiovascular diseases and the article throws light on how to design more effective CCBs that may be therapeutically useful.

Thus both these volumes of *Topics in Heterocyclic Chemistry* are unique and make interesting readings for all those involved, theoretically or experimentally, in design and development of drugs. As an editor of these volumes, I have greatly enjoyed reading the articles and hope all readers will too.

Pilani, March 2006   Satya Prakash Gupta

Contents

Contents of Volume 3

QSAR and Molecular Modeling Studies in Heterocyclic Drugs I

Volume Editor: Satya Prakash Gupta
ISBN: 3-540-33378-9

Top Heterocycl Chem (2006) 4: 1–53
DOI 10.1007/7081_017
© Springer-Verlag Berlin Heidelberg 2006
Published online: 21 April 2006

QSAR and Molecular Modeling Studies of Factor Xa and Thrombin Inhibitors

Dimitra Hadjipavlou-Litina

Department of Pharmaceutical Chemistry, School of Pharmacy,
Aristotle University of Thessaloniki, 54124 Thessaloniki, Greece
hadjipav@pharm.auth.gr

Abstract Thrombotic disorders remain the major cause of death and disability in Western society. Many approaches to develop antithrombotic drugs that interfere with enzymes in the coagulation system have been pursued over the past few decades. Factor FXa and thrombin are both bound to and are enzymatically active in blood clots. In this article, a brief review is presented on the QSAR and molecular modelling studies on FXa and thrombin inhibitors reported during the last decade. The results from the widely used CoMFA and CoMSIA methods have been included. Analogues with various heterocyclic rings, isoquinoline derivatives, five- or six-membered heteroaryl derivatives, benzoxazinones, and tetrazoles are included in the discussion. Hydrophobic property of molecules is shown to play a significant role. In most of the cases, CMR/MR as well as Sterimol parameters have been shown to be important. Electronic factors, with the exception of the Hammett's constant, have not been found to be of much importance.

Keywords Serine proteases · Thrombin and Factor FXa · QSAR · Molecular modeling · Heterocycles

Abbreviations

AFMoC	Adaptation of fields for molecular comparison
Å	Angstrom

AVRES	Average of absolute values of residuals
$C \log P$	Lipophilicity values calculated theoretically according to Hansch and Leo
CoMSIA	Comparative molecular similarity index analysis
CoMFA	Comparative molecular field analysis
COMBINE	Comparative binding energy
Df	Degrees of freedom
LFER	Free energy related
LMWS	Low molecular weight heparins
GA-MLR	Genetic algorithm multiple linear regression
MLR	Multiple linear regression
PLS	Partial least squares
PDB	Protein data bank
QSAR	Quantitative structure activity relationships
QSPR	Quantitative structure property relationships
PRESS	Sum of the squared errors of prediction
r_a^2	Adjusted correlation coefficient
r_{bs}^2	Bootstrapped correlation coefficient
r_{pred}^2	(SD-PRESS)/SD
RMS	Root mean square
RMSE	Root mean square error
S_{press}	Standard deviation sum of the squared errors of prediction
SDEP	Standard deviation sum of the error of prediction
SlogP	Program for the estimation of logP using the atom – weighted solvent accessible surface areas
tPA	Tissue-type plasminogen activator
TScore	Tailored scored function
uPA	Urokinase-type plasminogen activator

1
Introduction

1.1
Serine Proteases

The four major classes of protease enzymes [1–4] (aspartic, serine, cysteine and metallo) selectively catalyze the hydrolysis of polypeptide bonds. They control protein synthesis turnover. This function enables them to regulate physiological processes such as digestion, fertilization, growth, differentiation, cell signaling, migration, immunological defense, wound healing, and apoptosis. Most proteases are sequence-specific. The size and hydrophobicity/hydrophilicity of enzyme sites define possible binding amino acid side chains of polypeptide substrates. The standard nomenclature [5] used to designate substrate/inhibitor residues (e.g., P3, P2, P1, P1′, P2′, P3′) that bind corresponding enzyme subsites (S3, S2, S1, S1′, S2′, S3′) is shown in Fig. 1. Almost one-third of all proteases can be classified as serine proteases, named for the nucleophilic Ser residue at the active site. This mechanistic class is originally distinguished by the presence of the Asp-His-Ser "charge relay" system or "catalytic triad" [5]. More recently, serine proteases with novel catalytic triads and dyads have been discovered, including Ser-His-Glu-, Ser-Lys/His, His-Ser-His and N-terminal Ser [6].

Obviously, proteases are crucial for disease propagation and inhibitors of such proteases are emerging with promising therapeutic uses [5] in the treatment of cancers, parasitic, fungal and viral infections and inflammatory, immunological respiratory, cardiovascular and neurodegenerative disorders including Alzheimer's disease. Nowadays, there are many potent and selective protease inhibitors that slow or halt disease propagation. To be effective as therapeutic tools, protease inhibitors must not only be very potent but also highly selective in binding to a particular protease, and in addition must have appropriate pharmacokinetic and pharmacodynamic properties.

Proteases inhibitors have been traditionally developed by natural products, screening for lead compounds with subsequent optimization, or by empirical substrate-based methods, replacing the cleavable amide bond by a noncleavable isostere and optimizing the potency through trial-and-error structural modifications in order to reduce the peptide nature of the molecule. This substrate-based drug design has been improved with the availability of three-dimensional structural information for proteases (using the structural information about the active site of protease and fitting into it designed molecules with the aid of computers). Combinatorial chemistry also presents opportunities for assaying and optimizing lead structures for development of protease inhibitors.

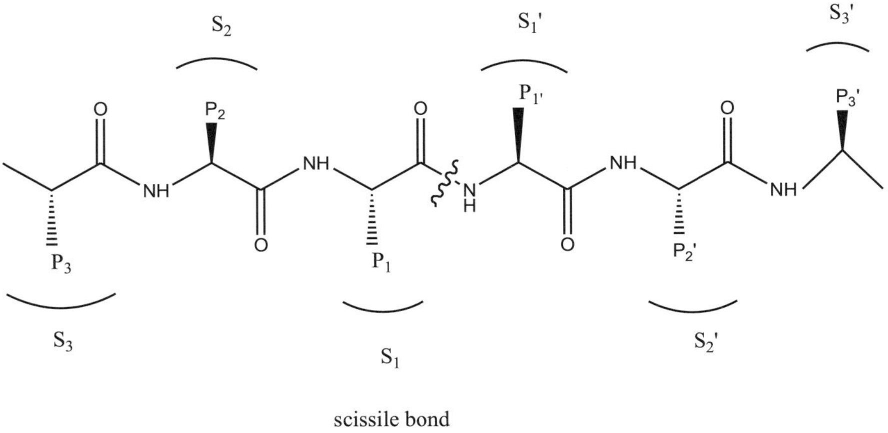

Fig. 1 Standard nomenclature for substrate residues and their corresponding binding sites

1.2
The Role of Factor Xa and of Thrombin

Serine proteases play a critical role in a variety of biologically significant processes [7, 8]. Both the two serine proteases, Factor Xa and thrombin, are bound to and are enzymatically active in blood clots [9]. This intricately controlled system of enzymatic interactions and interfacial processes maintains blood in a fluid state under physiological conditions and allows rapid clotting in response to injury [10]. The blood coagulation serine protease, Factor Xa (FXa), plays a central role in the coagulation cascade [11], linking the intrinsic pathway ("intrinsic tenase"—activated by surface contact) and extrinsic pathway ("extrinsic tenase"—activated by vessel injury-tissue Factor VIIa) to the common coagulation injury [12]. Factor Xa has a g-carboxyglutamic acid domain responsible for calcium and phospholipid binding and two epidermal growth factor (EGF)-like domains, the second of which has been suggested to mediate Factor VIIa/Va binding [13] (Fig. 2).

The mature form of Factor X consists of a 139-residue light chain and a 303-residue heavy chain linked by a sulfide. It is synthesized in the liver

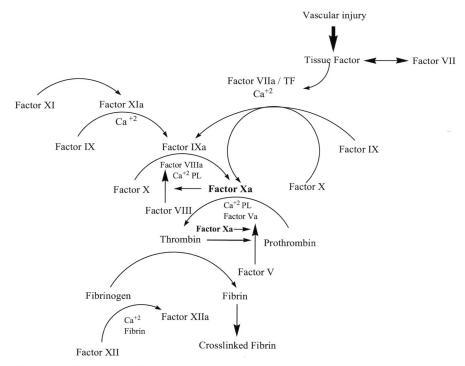

Fig. 2 Coagulation cascade

and secreted after post-traditional modifications into the blood as zymogen. Factor X is activated by the Factor VIIa-tissue factor complex in the extrinsic pathway, initiated by vascular damage, or by the Factor IXa-Factor VIIIa complex in the intrinsic pathway. The heavy chain encompasses a serine protease domain in a trypsin-like closed b-barrel fold with the active triad Ser195 – His57 – Asp102 and two neighboring protein subsites S1 and S4, which are typically explored for obtaining high-affinity inhibitors [13]. Factor Xa in combination with its cofactor (Factor Va) activates prothrombin on a phospholipid surface of membranes to form "prothrombinase complex", which activates prothrombin to thrombin. Thrombin subsequently converts fibrinogen to fibrin, inducing clot formation and platelet aggregation. Both incidents lead to serious pathological situations. The initiation of coagulation by either pathway in response to vascular injury activates Factor X to Xa, making FXa inhibition a desirable intervention point when developing novel antithrombotics [14].

Thrombin is a proteolytic enzyme and has a remarkable similarity in its overall three-dimensional structure to the digestive serine proteases, trypsin, and chymotrypsin [11–13]. Trypsin and thrombin share a common primary specificity for proteolysis next to arginine or lysine residues. Structural data of thrombin and trypsin have demonstrated strong resemblance in their substrate sites, and many small organic inhibitors are comparably active against both the enzymes [14, 15]. For this reason, no or low inhibition of trypsin is viewed as a required condition for a compound to be a successful orally bioavailable thrombin inhibitor [16].

The hydroxyl group of Serine195 acts as a nucleophile, targeting the scissile amide bond of the substrate. There are three principal binding pockets at the cleaving site: the specificity (S) pocket, the proximal (P) pocket and the distal (D) pocket modelled from the crystal structure [12, 17, 18].

Blood coagulation is the result of a cascade of enzymatic activation resulting in the production of thrombin (Factor IIa) in the penultimate step. Actually, in the coagulation cascade, thrombin is the final key enzyme that converts Factor XIII to Factor XIIIa [14]. It is possible that Factor Xa plays a central role in restenosis or in areas of vascular remodelling after plaque rupture and appears to activate mitogenesis by two independent or complementary pathways [18]. The binding site of FXa includes two main regions, named S1 and S4 pockets, and several side pockets involved in substrate recognition. The S1 pocket is a narrow cleft that extends about 8 Å deep into the core of the protein. It is bordered by planar hydrophobic walls and terminates with a negatively charged aspartate [19, 20]. On the other hand, the S4 pocket consists of a surface cleft characterized by hydrophobic aromatic (HbicArom) floor and walls. It usually accommodates hydrophobic residues. However, it has been suggested that it might participate in favorable interactions with a positive charge [20–23]. A typical direct FXa inhibitor includes three molecular fragments: (i) a positively charged group (P1) intended to fit

into the S1 pocket [23–26]; (ii) an aromatic fragment (P4) intended to interact with the S4 pocket [24, 27]; and (iii) a central linker designed to project
the substituents appropriately into their corresponding pockets [23, 28–30].
The extensive structural diversity of known direct FXa inhibitors [25–30]
suggests the existence of discrete multiple binding modes within the FXa
binding site to provide complementary pockets for such diverse ligands. Furthermore, a recently published crystallographic study revealed the existence
of two binding modes adopted by four different ligands within the S4 pocket
of FXa [24]. The apparent flexibility of FXa binding site has prompted an
attempt to identify corresponding accessible binding modes employing molecular dynamics simulation [31]. Obviously, using a single protein conformation for designing new ligands ignores important dynamic aspects of protein
ligand binding and the "induced fit" effects are ignored [32, 33]. Furthermore,
binding site flexibility estimates may be of direct use in ligand design since
they indicate which parts of the ligand may tolerate variations in size, substituents, and conformations [34]. Moreover, conformational flexibility and
strain energy of the ligand play critical roles in affinity interactions. It has
been shown that on average each freely rotating bond in a ligand reduces
binding free energy by 0.7 kcal/mol [35]. Many examples are known where
rigidifying a flexible ligand causes a substantial boost in affinity. However,
rigidifying a flexible ligand into the wrong conformation will produce a substantially less active or inactive compound [36].

The inhibition of FXa compared to thrombin may allow the effective control of thrombogenesis with minimal effect upon bleeding, because FXa inhibitors should affect coagulation specifically. Furthermore, inhibition of FXa
seems to be more efficacious because one molecule of FXa generates many
thrombin molecules. In addition, inhibition of FXa should prevent new production of thrombin, whereas a basal thrombin level might be necessary
for primary haemostasis [1]. It is suggested that Factor X activity can be
suppressed markedly without affecting haemostasis, a characteristic that is
desirable for antithrombotic agents [15, 37].

FXa and thrombin are both enzymatically active when associated with
the intravascular thrombus, so it is vitally important that an antithrombotic
agent be able to inhibit the "clot-bound" activity. Indirect inhibitors act by
antithrombin-III, a large molecule that is not able to penetrate the thrombus. Consequently, low molecular weight heparins (LMWHs) or the synthetic
pentasaccharides are not effective inhibitors of thrombus-associated FXa or
thrombin [14]. Direct FXa and thrombin inhibitors are capable of inhibiting
thrombus-associated FXa and thrombin, making these agents more attractive for medical interventions in thrombotic diseases. Ample evidence exists
for the role of FXa inhibitors as anticoagulants. Antistatin, a potent inhibitor
of blood coagulation FXa isolated from the Mexican leech (*Haementeria of-
ficinalis*) displayed antithrombotic activity in various models of arterial and
venous thrombosis [38].

DX-9065a (Fig. 3), an amidinonaphthalene derivative, is an inhibitor of Factor Xa [39], orally active and highly selective over other serine proteases. The naphthamidine group was predicted by modelling to be fixed in the S1 pocket via a salt bridge to Asp 189, which was confirmed by the crystal structure. Hydrophobic interactions around the pocket also contribute to the high binding affinity, whereas the pyrrolidine ring binds to the other major interaction site S4. DX-9065a is a promising candidate in the treatment and prevention of thrombotic diseases [40]. In both arteriovenous shunt and venous stasis models, inhibition of thrombus formation was achieved at doses that had little effect on activated partial thromboplastin time (aPTT), indicating that DX-9065a is effective in preventing thrombosis and hence has therapeutic antithrombotic potential [41]. A similar naphthamidine deriva-

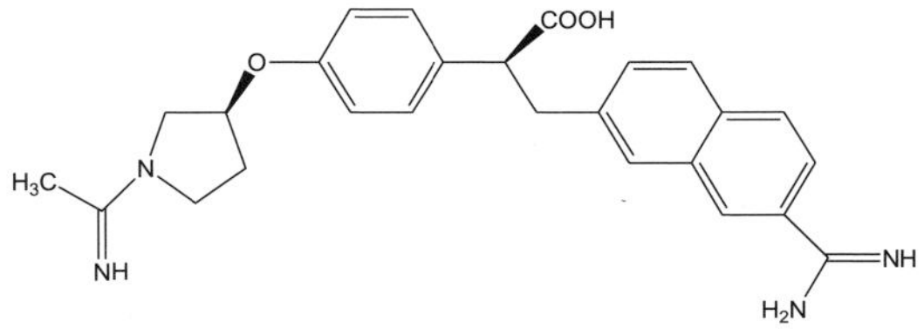

DX-9065a

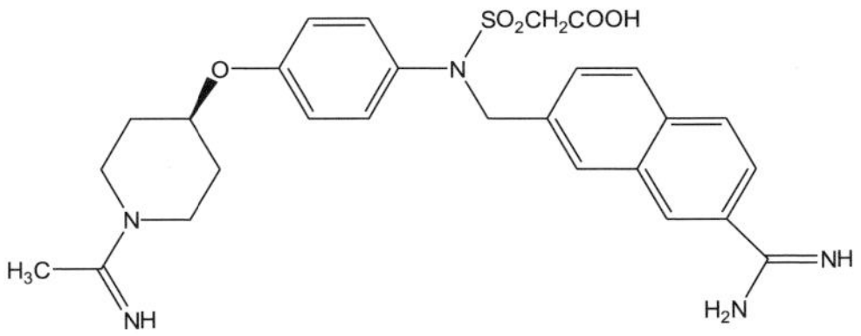

YM-60828

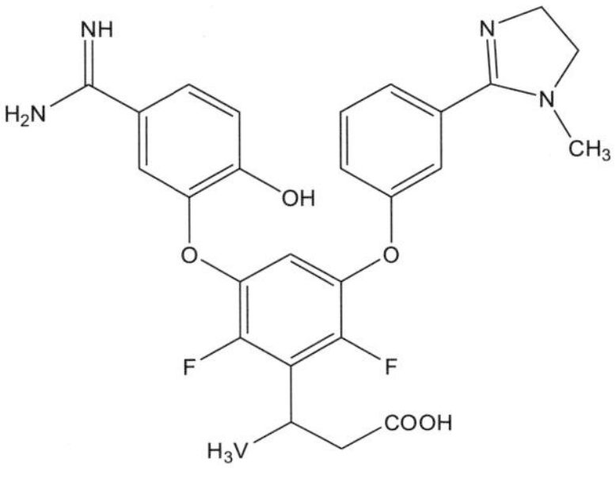

ZK-807191

Fig. 3 Significant known FXa inhibitors

tive is YM-6082841 (Fig. 3), which is selective, potent, and bioavailable. ZK-80719141 (Fig. 3) is a selective potent inhibitor of Factor Xa and is orally active. Substitutions at the 4-position of a phenol group opposite to the amidine function contributes to the selectivity and potency of this molecule.

2
Factor Xa Inhibitors with Heterocyclic Moieties

2.1
Survey of Published 3D-QSAR
and Molecular Modeling Studies on Factor Xa Inhibitors

The structures of various trypsin-like serine proteases available as potential therapeutic targets including human a-thrombin [18], FXa [19], tPA [42], and uPA [43] opened a new doorway in the structure-based design of serine protease inhibitors. The members of the trypsin-like serine protease family possess a catalytic site with high sequence homology and similar substrate specificity. Therefore, their inhibitors need to be highly potent and selective among the closely related enzymes with adequate pharmacodynamic properties. The published structures offer the possibility of understanding the specificity over related counterparts/enzymes with high sequence homology and similar substrate specificity.

Most of the known small molecule inhibitors of serine proteases bear an amidino/guanidino group as the pharmacophoric group in correlation to an heterocyclic moiety, which may help to produce high binding affinity with the protease enzymes due to its interaction with Asp189 in the S1 specificity pocket. The differences in the S1 subsite for each enzyme can be exploited for specificity [44, 45]. The potency and selectivity of the target of interest over the related enzymes is mainly responsible for the interactions of the ligand with the neighboring subsites differing from their counterparts.

Vaz et al. [13] presented an early study on a series of DX-9065a analogues (Fig. 3), using comparative molecular field analysis (CoMFA). The compounds had a hydrophobic group as well as basic and dibasic functionalities in a series of proposed alternative binding modes. The CoMFA was primarily a way of "translating" 3D information into 1D information and then back into 3D information. Based on a published model [46], the amidino group was assumed to form a salt bridge with the carboxylic group in the side chain of Asp189 of Factor Xa. The best model was derived when the conformation of the Glu 97 side chain was modified such that an H-bond interaction was maintained with the inhibitor. The model presented a tightening of the S1 pocket similar to that proposed by crystallographic study [47–49]. The highest statistical correlation occurred with the second basic group accommodated in the vicinity of Glu 97 and a hydrophobic group accommodated in the pocket defined by Phe 174, Tyr 99, and Trp 215. The CoMFA model with the best statistical significance suggested that there was a strong coulombic/hydrogen bond interaction between the side chain of Glu 97 and the acetimidoyl group of DX-9065a. The Ala 190 made the S1 pocket of Factor Xa more lipophilic allowing for the collapse of the enzyme around the naphthalene ring of DX-9065a. The number of optimal components was found to be equal to four (standard error of prediction 0.611), q^2 (cross-validated r^2) = 0.654, F-values (cross-validated run) = 6.422, Prob. ($q^2 = 0$), $r^2 = 0.973$, standard error of estimate 0.172, F-values = 159.196, Prob. ($r^2 = 0$). Contribution steric: 0.597, electrostatic = 0.403.

The next step was made by Klebe et al. [50]. Two 3D-QSAR methods were applied to get three-dimensional quantitative structure–activity relationships using a training set of 72 inhibitors of the benzamidine type with respect to their binding affinities toward Factor Xa to yield statistically reliable models of good predictive power [51–54]: the widely used CoMFA method (for steric and electrostatic properties) and the comparative molecular similarity index analysis (CoMSIA) method (for steric, electrostatic, hydrophobic, hydrogen bond donor, and hydrogen bond acceptor properties). These methods allowed the consideration of various physicochemical properties, and the resulting contribution maps could be intuitively interpreted.

The correlation results obtained by CoMSIA were graphically interpreted in terms of field contribution maps, allowing physicochemical properties relevant for binding to be easily mapped back onto molecular structures. The advantage of this feature was demonstrated using the maps to design molecules. This data set allowed the derivation of QSAR models of statistical significance. The steric properties that were presented showed that steric occupancy with bulky groups will increase affinity but other areas-isopleths should be avoided to maintain the affinity. The maps of electrostatic properties showed fewer features in space. The maps of hydrophobic properties showed areas where increasing hydrophobicity enhanced affinity for Factor Xa in the distal S3 pocket. The map for acceptor properties showed no features at all for Factor Xa.

In both the cases, the molecular property fields were evaluated between a probe atom and each molecule of the data set at the intersections of a regularly spaced grid. Since the affinity data were inter-correlated between 30 and 70%, the derivation of three significant models was not obvious or trivial. In all the cases, the CoMSIA analyses revealed significantly better correlations expressed in terms of higher q^2 values. Three different grid spacings were evaluated in both CoMFA as well as CoMSIA.

CoMFA analysis:

- grid 2 Å, $q^2 = 0.374$, $S_{press} = 0.515$, $r^2 = 0.680$, $s = 0.368$, $F = 48.3$, components 3, fraction: steric 0.701, electrostatic 0.299

- grid 1.5 Å, $q^2 = 0.415$, $S_{press} = 0.501$, $r^2 = 0.784$, $s = 0.305$, $F = 60.6$, components 4, fraction: steric 0.709, electrostatic 0.291
- grid 1 Å, $q^2 = 0.429$, $S_{press} = 0.495$, $r^2 = 0.798$, $S = 0.294$, $F = 66.2$, components 4, fraction: steric 0.701, electrostatic 0.299.

CoMSIA analysis:

- grid 2 Å, $q^2 = 0.594$, $S_{press} = 0.424$, $r^2 = 0.915$, $S = 0.194$, $F = 117.1$, components 6, fraction: steric 0.175, electrostatic 0.160, hydrophobic 0.345, donor 0.090, acceptor 0.230
- grid 1.5 Å, $q^2 = 0.589$, $S_{press} = 0.427$, $r^2 = 0.915$, $S = 0.194$, $F = 116.3$, components 6, fraction: steric 0.177, electrostatic 0.170, hydrophobic 0.334, donor 0.090, acceptor 0.229
- grid 1 Å, $q^2 = 0.590$, $S_{press} = 0.426$, $r^2 = 0.915$, $S = 0.194$, $F = 116.5$, components 6, fraction: steric 0.176, electrostatic 0.170, hydrophobic 0.336, donor 0.090, acceptor 0.228.

The numbers of partial least squares (PLS) components were higher in CoMSIA than in CoMFA. This difference probably resulted from the significantly higher number of lattice points showing steadily varying field values (e.g., inside the molecules). The optimal numbers of components were selected on the basis of lowest Spress.

The increasing interest for Factor Xa led Labute [55] to apply the construction of QSAR/QSPR models for Factor Xa activity. The previous training set of 72 inhibitors of the benzamidine analogues was used. Three sets of molecular descriptors on atomic contributions to van der Waals surface area were used: log P (octanol/water), molar refractivity and partial charge. A principal component analysis was calculated. A 15-descriptor model was formed using EOEVSA1,2,8,9,12,14 (descriptor to capture direct electrostatic interactions), Slog P-VSA5,7,8,10 (descriptor to capture hydrophobic and hydrophilic effects in the receptor or on the way to the receptor), SMR-VSA3,4,5,6,8 (descriptor to capture direct electrostatic interactions), which resulted in a r^2 of 0.69 with an RMSE of 0.35 pK_i; the leave-one-out cross-validated r^2 was 0.52 with an RMSE of 0.45 pK_i. This linear correlation was not extraordinarily high, but it was expected, since only a small set of non-3D descriptors was analyzed, which cannot be capable of describing a given molecule in great detail in relation to a specific property.

A significant work was recently presented [1] for a series of achiral FXa inhibitors, which were designed by a combination of benzamidines directed toward the FXa S1 pocket. Their basic or hydrophobic substituents were directed toward the flexible S4 subsite of Factor Xa. The 3-amidinobenzyl-1H-indole-2-carboxamide scaffold as an interesting motif was also identified. A combination of X-ray structure analysis of Factor Xa and trypsin, flexible molecular docking, and 3D-QSAR analyses was performed to understand relevant protein–ligand interactions for activity and selectivity of these analogues. The reported 3D-QSAR models allowed reliable predictions of novel candidates and uncovered binding features that were responsible for affinity and selectivity.

The CoMFA analysis was the basis for the formulation of 3D-QSAR models, which were based on the alignment rule derived from the FXa active site topology that showed a high degree of internal consistency. In this study, steric field descriptors (2,156 grid-based variables) explained 59% of the variance, whereas the electrostatic field accounted for only 41%. Similar statistical results were obtained analyzing this data set using CoMSIA steric, electrostatic, and hydrophobic fields. When a 2 Å grid spacing was used, a CoMSIA model with an r^2 (cross validated) value of 0.655 for five PLS components and a controversial r^2 of 0.863 was obtained. The steric field descriptors (2,156 variables) explained only 26% of the variance, whereas the proportion of the electrostatic descriptors remained the same (43%). The additional hydrophobic field explained the remaining 31% of the variance. As a result, the CoMFA steric field contribution could be seen as a balance between pure steric plus hydrophobic effects.

Moreover, a final 3D-QSAR model validation was done using a prospective study with an external test set. The 82 compounds from the data set were used in a lead optimization project. A CoMFA model gave an r^2 (cross validated) value of 0.698 for four relevant PLS components and a conventional r^2 of 0.938 were obtained for those 82 compounds. The steric descriptors contributed 54% to the total variance, whereas the electrostatic field explained 46%. The CoMSIA model led to an r^2 (cross validated) value of 0.660 for five PLS components and a conventional r^2 of 0.933. The contributions for steric, electrostatic, and hydrophobic fields were 25, 44, and 31%. As a result, it was proved that the basic S4-directed substituents should be replaced against more hydrophobic building blocks to improve pharmacokinetic properties. The structural and chemical interpretation of CoMFA and CoMSIA contour maps directly pointed to those regions in the Factor Xa binding site, where steric, electronic, or hydrophobic effects play a dominant role in ligand–receptor interactions.

Murcia and Ortiz [56], using the above biological results of Matter performed, a two-step, fully automatic virtual screening procedure consisting of flexible docking followed by activity prediction by comparative binding energy (COMBINE) analysis. This novel approach was applied to the above recently reported series of 133 Factor Xa (FXa) inhibitors whose activities encompassed four orders of magnitude. The docking algorithm was linked to the COMBINE analysis program and used to derive independent regression models of the 133 inhibitors docked within three different FXa structures (PDB entries 1fjs, 1f0r, and 1xka), so as to explore the effect of receptor conformation on the overall results. Reliable docking conformations and predictive regression models requiring eight latent variables could be derived for two of the FXa structures with the best model achieving a q^2 of 0.63 and a standard deviation of errors of prediction (SDEP) of 0.51 (leave-one-out). The two-step procedure was then employed to screen a designed virtual library of 112 ligands, containing both active and inactive compounds. In the best case, a recognition rate of $\sim 80\%$ of known binders at $\sim 15\%$ false positive rate was achieved, corresponding to an enrichment factor of $\sim 450\%$ over random.

Matter et al. [57], continuing their research, reported the design and structure activity relationship of a series of nonchiral 3-oxybenzamides as inhibitors of FXa by means of X-ray crystallography, 3D-QSAR, and tailored scoring functions. Their design rationale was based on X-ray structures of FXa in uncomplexed form [58] and when complexed with inhibitors [59–62] plus knowledge of privileged motifs directed toward the S1 pocket and structure activity information on privileged substructures accumulated in the project. The CoMFA [63–65] and CoMSIA [66] were used to correlate molecular property fields to biological activities based on the X-ray structures of some analogues providing the active inhibitor conformation for alignment. The superposition of all other molecules onto these templates produced consistent models in agreement with binding site requirements. The contour maps from 3D-QSAR models enhanced the understanding of electrostatic, hydrophobic and steric requirements for ligand binding, guiding the design of inhibitors to regions where structural variations revealed a correlation to biological properties. Tailored scoring functions were also derived on the basis of two approaches to establish predictive models for structure-based optimization. The first approach was the AFMoC (adaptation of fields for molecular comparison) approach. Knowledge-based pair potentials were adapted to a binding site by considering ligand information. Atom-type specific interaction fields, capturing binding site characteristics and ligand complementarity, were correlated to biological affinities. In contrast, the in-house approach TScore captures protein–ligand interaction on an amino acid level. For each binding site of residue with ligands, terms describing hydrogen-bonding, lipophilicity and steric contacts produce a protein–ligand interaction profile. Their statistical analysis by correlating them to affinities led to relevant models. A total of six X-ray structures of FXa/inhibitor complexes led us to identify the major protein–ligand interactions. The binding mode was characterized by a lipophilic dichlorophenyl substituent interacting with Tyr 228 in the protease S1 pocket, while polar parts were accommodated in S4. This alignment in combination with docking allowed derivation of 3D-QSAR models and tailored scoring functions to rationalize biological affinity and to provide guidelines for optimization. The resulting models showed good correlation coefficients, and predictions of external test sets corresponded to binding site topologies in terms of steric, electrostatic, and hydrophobic complementarity. Good correlations to experimental affinities were obtained for both AFMoC and the novel TScore function.

A QSAR study was presented by Kunal Roy et al. [67, 68] on human Factor Xa inhibitors, N2-aroylanthranilamides, which were synthesized by Yee [69] (Fig. 4). For these derivatives, it was suggested that the phenyl ring at R_1 in-

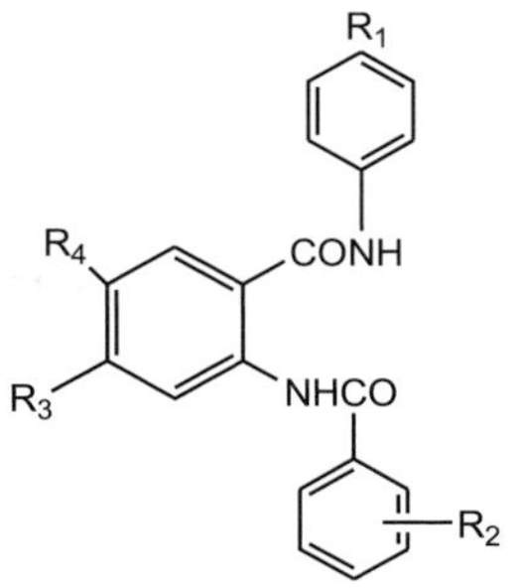

Fig. 4 General structure of FXa inhibitors *N*2-aroylanthranilamides

teracts with small, hydrophobic and less solvent-exposed S1 region of FXa, whereas the R_2-substituent interacts with relatively more solvent-exposed S4 region. In their first study [67], the factor analysis for the subsets A, B and C was done to explore the contribution pattern of R_2, R_1 and R_3/R_4 substituents, respectively, to the activity. Equations 1, 2, 3, 4 were derived:

Set A

$$\log K_{ass} = -1.469(\pm 0.639)\sigma_{R2,p} - 1.574(\pm 0.488)I_{R2,m} + 3.136(\pm 0.270)$$
$$n = 13, \ r_{a2} = 0.872, \ r = 0.945, \ r^2 = 0.893, \ s = 0.363, \ F = 41.8, \ df = 2.10,$$
$$\text{AVRES} = 0.280 \tag{1}$$

Set B

$$\log K_{ass} = -1.533(\pm 0.538)I_{R2,m} - 1.410(\pm 0.708)\sigma_{R2,p} - 2.470(\pm 1.364)\sigma R_1$$
$$- 0.508(\pm 0.279)MR_{R1,2} + 2.739(\pm 0.367)$$
$$n = 19, \ r_{a2} = 0.783, \ r = 0.912, \ r^2 = 0.832, \ s = 0.423, \ F = 17.2, \ df = 4.14,$$
$$\text{AVRES} = 0.313 \tag{2}$$

$$\log K_{ass} = -1.561(\pm 0.528)I_{R2,m} - 1.448(\pm 0.693)\sigma_{R2,p} -$$
$$1.797(\pm 1.306)\sigma R_1 - 0.458(\pm 0.238)MR_{R1,2} + 2.739(\pm 0367)$$
$$n = 19, \ r_{a2} = 0.793, \ r = 0.916, \ r^2 = 0.839, \ s = 0.413, \ F = 18.3, \ df = 4.14,$$
$$\text{AVRES} = 0.304 \tag{3}$$

Set C

$$\log K_{ass} = 0.569(\pm 0.389)\text{Ind}1 - 1.974(\pm 0.835)\sigma m\text{-}R_3 + 0.660(\pm 0.489)IR_4$$
$$+ 3.502(\pm 0.228)$$
$$n = 13, \ r_{a2} = 0.852, \ r = 0.943, \ r^2 = 0.889, \ s = 0.274, \ F = 24.1, \ df = 3,9,$$
$$\text{AVRES} = 0.182 \tag{4}$$

$\log K_{ass}$ are the FXa binding affinities of the compounds (K_{ass}), which are converted to logarithmic scale and for factor analysis $\sigma_{R2,p}$ is the electronic parameter of the *para* substituents, σR_1 is the electronic parameter of the R_1 substituents, $\sigma m\text{-}R_3$ is the electronic parameter of the R_3 substituents, MR_{R1} is the steric parameter (molar refractivity) of the R_1 substituents, IR_4 Indicator variable denoting the presence of $NHSO_2Me$ group as R_4 substituents and Ind1 is an indicator indicating presence of tert-butyl (value 0) or dimethylamino (value 1) as R_2 *para*-substituents; r_{a2} is the adjusted correlation coefficient r^2).

From the above equations, it was documented that electron-donating R_1-substituents with less bulk and optimum hydrophilic–lipophilic balance, such as methyl and methoxy groups, increase the activity, whereas the presence of electron donating para R_2-substituent with free (unsubstituted) meta position is conducive to the binding affinity. Further, electron-withdrawing R_3-substituents are detrimental to the activity, whereas bulkier R_4-substituents (particularly $NHSO_2Me$ group) increase the activity. Significant contributions of R_3- and R_4- substituents suggest that the central phenyl ring also may be involved in the interactions with the receptor proteins.

Another study [68] was reported by the same group using the free energy related (LFER) model with electrotopological state index of atom (ETSA) to explore the atoms/regions of the compounds that modulated the activity comparatively to the greater extent. Ten equations were derived with high statistical significance ($r = 0.915\text{--}0.956$). The conclusions derived from these equations were almost same as those derived from the Hansch-type analysis (Eqs. 1–4), i.e., the presence of electron-donating R_2-substituent at the para position (the affinity was decreased by a meta R_2-substituent), electron-donating R_1-substituents with less bulk and optimum hydrophilic–lipophilic balance, electron-withdrawing R_3-substituents and R_4-bulkier substituents were advantageous to the activity. The electrotopological parameter that was used showed the high significance of the size of meta R_2-substituent on binding affinity. In this way, the role of this substituent was clarified, whereas in Hansch-type analysis it was accounted for by an indicator variable.

From the studies above, it was shown that the use of 3D-QSAR models led to the identification of binding site regions, where steric, electronic, or hydrophobic effects played a dominant role. Although cationic interactions in both S1 and S4 subsites were favorable for in vitro affinity, they might be detrimental for oral bioavailability. Thus, the CoMFA steric field contribution could be seen as a balance between pure steric plus hydrophobic effects. The contributions for steric, electrostatic and hydrophobic fields from the CoMSIA studies were 25, 44, and 31%, respectively.

Gadad et al. [69] reported a series of indole/benzoimidazole-5-carboxamidines which inhibited various trypsin-like serine proteases such as uPA, tPA, Factor Xa, thrombin, plasmin, and trypsin. 3D-QSAR models were generated for indole/benzoimidazole-5-carboxamidines using in vitro inhibitory activity pK_i (lM) as a dependent variable and the CoMFA technique to study their selectivity trends toward various trypsin-like serine proteases. Molecular superimposition was carried out on the template structure using the atom-based RMS fit method. The low-energy conformer obtained from systematic search routine was used for molecular superimposition of ligands on the template structure by atom-based RMS fitting (Fig. 5).

The CoMFA models were established from the training set of 25–29 molecules and validated by predicting the activities of seven to eight test set molecules. To better evaluate, in the context of QSAR studies, new validation techniques such as bootstrapping and cross-validation have been used. Bootstrapping and cross-validation are more powerful indicators of possible chance correlation than are the classical tests based on assumed normal independent distribution of variables. Cross-validation was used to help determine whether a relationship or model found within this set was likely to be generalizable to others, most often on a term-by-term basis, while bootstrapping was used to generate confidence limits for each of the parameters within the model. The cross-validated r^2 may be defined completely analogously to the definition of the conventional r^2 whereas the bootstrapped r^2 is the mean of conventional r^2. The model obtained from only the steric field showed r^2 (cross validated) = 0.492 with three components, r^2 (cross validated) = 0.775, F-value = 24.047, r^2_{bs} = 0.855, whereas the model from the electrostatic field showed r^2 (cross validated) = 0.622 with five components, r^2 (cross validated) = 0.956, F-value = 82.442, r^2_{bs} = 0.966. The CoMFA model with both steric and electrostatic fields showed r^2 (cross validated) = 0.513 from first two components, r^2 (cross validated) = 0.811, F-value = 47.252, r^2_{bs} = 0.819, r^2 (predicted) = 0.753 with 56.1% steric and 43.9% electrostatic field contributions.

The CoMFA model generated using steric and electrostatic fields for FXa inhibition exhibited better statistical significance than the CoMFA models generated using ClogP as an additional descriptor. The nature of the substituent at C-6 position (Fig. 5) was the crucial factor in determining the selectivity trends of various serine protease inhibitors. The researchers observed that optimum low electron density substituent at C-6 may enhance the FXa activity. Replacement of C-6 proton with a halogen (chloro) markedly

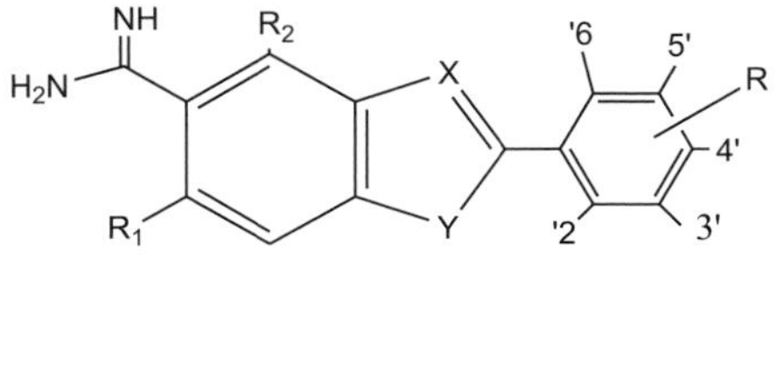

X= N, CH, NMe
Y = NH, O, N
2' = OH, H, OMe
3' =H, NO_2, OH, F, Br, Me, OMe, naphthyl
4'= H, Me, NEt_2
5'= H, F, Br, Me, OMe, NO_2,Cl
6'= H, OH
R_1=H, F, OMe,Cl, OH
R_2 = H, Cl

Fig. 5 Structure of the test set molecules

decreased the FXa inhibition. Thus, the validated CoMFA models with steric and electrostatic fields were used to generate 3D contour maps, which might provide possible modification of molecules for better selectivity/activity.

The flexibility of activated Factor X (FXa) binding site was assessed employing ligand-based pharmacophore modelling combined with genetic algorithm (GA)-based QSAR modelling. Four training subsets of wide structural diversity were selected from a total of 199 direct FXa inhibitors (Fig. 6) and were employed to generate different FXa pharmacophoric hypotheses using CATALYST software over two subsequent stages.

The corresponding bioactivities are expressed as the dissociation constants of the enzyme-inhibitor complexes (K_i values). In the first stage, high-quality binding models (hypotheses) were identified. However, in the second stage, these models were refined by applying variable feature weight analysis to assess the relative significance of their features in the ligand-target affinity. Genetic evolution yielded the following Eq. 5 (70) as optimal QSAR model after removing seven statistical outliers:

$$\log(K_i) = -0.864[\pm 0.322] + 0.237[\pm 0.164] \log(\text{Hypo4}/1) + 0.402[\pm 0.195]$$
$$\log(\text{Hypo4}/2) + 0.589[\pm 0.225] \log(\text{Hypo9}/6)$$
$$n = 192, \ r = 0.76, \ F = 86.17 \tag{5}$$

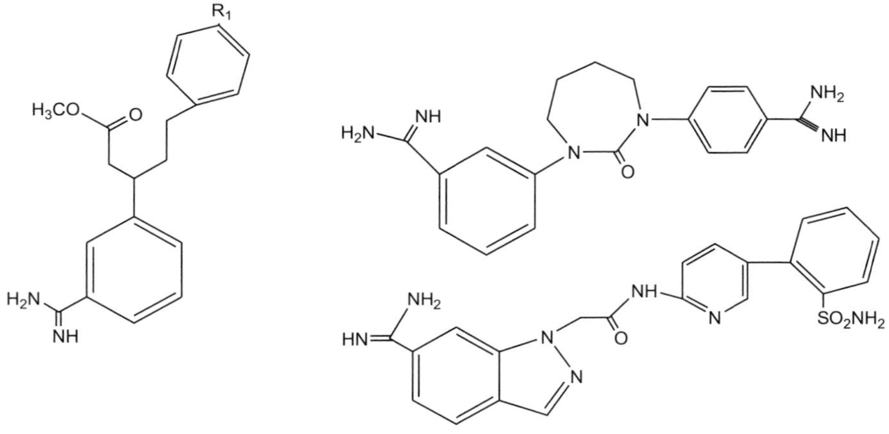

General representative structure

C=NH(NH2)-Ar-C(Z)-W

Ar =phenyl, heterocycles
Z = COOH, CONHR
W = amidine substituted heterocycles or benzamidine

Fig. 6 Mined FXa inhibitors (examples and general structure) [70]

The variables log (Hypo4/1), log (Hypo4/2) and log (Hypo9/6) represent the logarithmic transformations of bioactivity estimates determined employing Hypo4/1, Hypo4/2 and Hypo9/6, respectively.

AlDamen et al. [70] were prompted to use additional molecular descriptors, obtained by Alchemy2000 and QSARIS, as independent variables in the GA–MLR–QSAR modelling. The most significant QSAR model was achieved after 4000 iterations of GA–MLR analysis followed by excluding statistical outliers. The resulting QSAR model was cross-validated automatically using leave one-out cross-validation. The following Eq. 6 showed the resulting QSAR model of 192 FXa inhibitors (without outliers)

$\log(K_i) = -6.586[\pm 3.131] + 0.194[\pm 0.129]\log(\text{Hypo4}/1) + 0.373[\pm 0.1587]\log(\text{Hypo4}/2) + 0.389[\pm 0.177]\log(\text{Hypo9}/6) - 0.3189[\pm 0.096]\text{SdsCH} + 0.219[\pm 0.108]\text{SssssC} - 0.0530[\pm 0.035]\text{SaaN} - 0.183[\pm 0.105]\text{SsCH3\_acnt} - 13.570[\pm 4.230]\text{MaxHp} + 0.395[\pm 0.156]\text{SHBint2\_Acnt} - 0.111[\pm 0.050]\text{SHBint8\_Acnt} - 0.0827[\pm 0.049]\text{SHBint10\_Acnt} - 0.200[\pm 0.073]\text{SHHBd} + 0.537[\pm 0.197]\text{Ioniz Pot} - 0.721[\pm 0.306]\text{HOMO} + 0.322[\pm 0.155]\text{Py} + 49.520[\pm 15.125]v10\text{ch} + 1.628[\pm 0.277]\text{knotpv}$

$n = 192,\ r = 0.91,\ F = 49.64,\ \text{Multiple } q^2 = 0.79,$ \hfill (6)

where multiple q^2 is the leave-one-out correlation coefficient, SdsCH is the sum of all $= \text{CH} -$ topological E-state values in a molecule, SssssC is the sum of all E-state values of quaternary carbon atoms in a molecule, SaaN is the sum of all E-state values of aromatic nitrogens, $\text{SsCH}_3\_\text{acnt}$ is the count of all CH_3-groups in a molecule, MaxHp is the largest positive charge on a hydrogen atom, SHBint2_Acnt, SHBint8_Acnt, SHBint10_Acnt are the sums of E-state descriptors of strength for potential hydrogen bonds of path lengths 2, 8 and 10, respectively, SHHBd is the E-state indices for HBDs, Ioniz Pot is the ionization potential, HOMO is the energy of the highest occupied molecular orbital and Py is the component of the dipole moment along the inertial Y-axis, $v10$ch is the simple 10th order chain chi index and knotpv is the difference between chi valence cluster-3 and chi valence path/cluster-4. Re-emergence of Hypo4/1, Hypo4/2 and Hypo9/6 in Eq. 6, despite evolutionary competition imposed by the newly added physicochemical descriptors, strongly emphasized the statistical significance of this pharmacophoric combination. However, it was clearly evident from the above equation that the new descriptors improved the overall statistical criteria of the model.

The binding models were validated according to their coverage (capacity as a three-dimensional (3D) database search queries) and predictive potential as three-dimensional quantitative structure–activity relationship (3D-QSAR) models. The validation process identified 17 plausible pharmacophore models. Subsequently, genetic algorithm and multiple linear regression (MLR) analysis were employed to construct different QSAR models from high-quality pharmacophores and explore the statistical significance of combination models in explaining bioactivity variations across 199 FXa inhibitors. Three orthogonal pharmacophoric models emerged in the optimal QSAR equation suggesting three binding modes accessible to ligands in the binding pocket within FXa in agreement with X-ray structures and 3D-QSAR.

2.2
Survey of Published QSAR on Factor Xa Inhibitors

Diaryloxypyridines

Guilford et al. [71] designed and synthesized a series of diaryloxypyridines (1–4) as selective nanomolar Factor Xa (FXa) inhibitors. Their early lead structure contained two arylamidine groups. In continuation, they modified one of the arylamidine groups, since earlier studies had shown that one amidine was necessary for activity.Using these data, Eq. 7 was derived [72], which indicated that hydrophobicity was critical for the activity. Three indicator variables were used. I-R_{OH}, which was used with a value of 1 for the presence of a hydroxy group at position 6 of the benzyl ring was shown to be most significant. I-R_1NME_2, referred to the existence of an NMe_2 group at the 3-position of the benzyl ring, and I-DR_1-IM was used for the presence of an imidazolyl ring in the same position. The positive sign of the coefficient all the indicators exhibited the positive effects of all these substituents. The

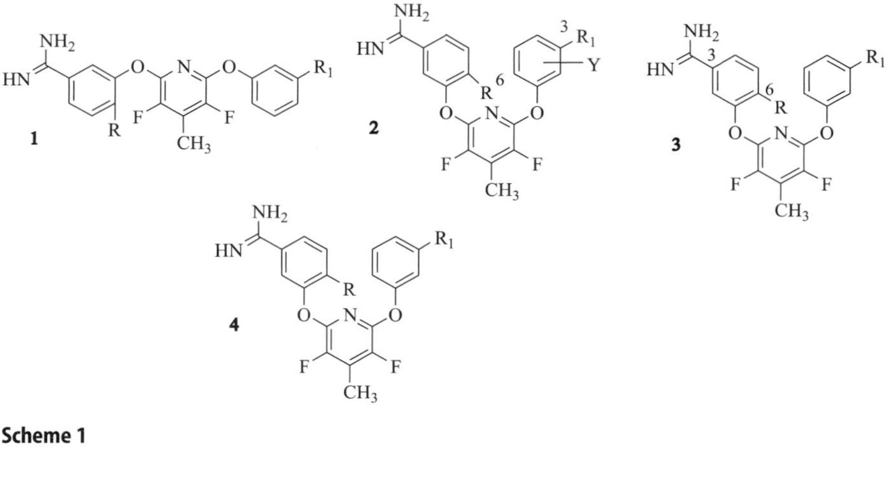

Scheme 1

N-dimethylamide and the imidazolinyl ring seemed to be good alternatives for the second amidine.

$$\log(1/K_i) = -0.246(\pm0.120)C\log P + 1.480(\pm0.337)\text{I-R}_{OH} +$$
$$0.614(\pm0.368)\text{I}_{R_1NME_2} + 0.581(\pm0.481)\text{ID}_{R1\text{-}IM} + 6.751(\pm0.392)$$
$$n = 72,\ r = 0.860,\ r^2 = 0.740,\ q^2 = 0.693,$$
$$s = 0.485,\ F_{4,72} = 4.515,\ \alpha = 0.01 \tag{7}$$

Aminophenol-Based Inhibitors

Aminophenol was used as a potent and selective scaffold [73] since it was more synthetically accessible than the benzimidazole template. For a set of aminophenol-based inhibitors (5–8), the correlation obtained was as shown by Eq. 8. In this equation,

$$\log(1/K_i) = -0.406(\pm0.243)C\log P + 0.608(\pm0.541)\text{MR-}_{R4} -$$
$$1.640(\pm0.452)\text{I}_{H\text{-}B} + 7.182(\pm2.933)$$
$$n = 18,\ r = 0.908,\ r^2 = 0.824,\ q^2 = 0.660,$$
$$s = 0.385,\ F_{3,14} = 4:287,\ \alpha = 0.05 \tag{8}$$

$I_{H\text{-}B}$ was used with a value of 1 for the possibility of existence of a hydrogen bond at the R_2 substituent and $MR\text{-}R_4$ was the molar refractivity for the R_4 group. The positive coefficient of $MR\text{-}R_4$ indicated that the larger and more polarizable is the R_4-substituent the more would be the inhibitory activity. The negative sign with $I_{H\text{-}B}$ showed that the presence of a hydrogen bond results in lower activity. Equation 8 [72] also exhibits the importance of hydrophilicity of the compounds for the inhibition of FXa.

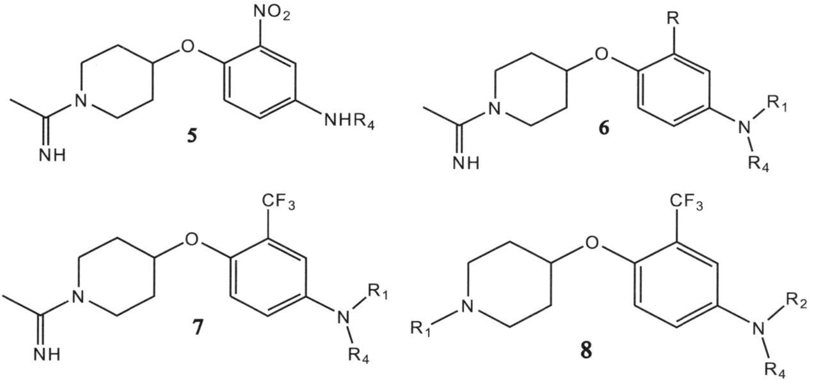

Scheme 2

Isoxazoline Derivatives

a) Bisbenzamidine Isoxazoline Derivatives

Most of the nonpeptide FXa inhibitors reported in the literature were compounds with two basic groups [74]. To date very few monobasic FXa inhibitors have been reported. Ellis and co-workers [74] have reported their effort on synthesizing a series of bisbenzamidine isoxazoline derivatives and a series of monobasic substituted biaryl isoxazoline derivatives (9–12). For these compounds, Eq. 9 was derived [72]. In Eq. 9, Ipyr was used to take values

$$\log(1/K_i) = 0.297(\pm0.260)C \log P - 0.106(\pm0.063)C \log P^2 +$$
$$0.337(\pm0.218)I_{pyr} + 0.925(\pm0.203)IR_{TETR} + 8.230(\pm0.260)$$
$$n = 36, \; r = 0.907, \; r^2 = 0.823, \; q^2 = 0.750, \; s = 0.261, \; F_{4,31} = 4.708, \; \alpha = 0.01$$
$$C \log P_o = 1.402(\pm0.705) \text{ from } 0.397 \text{ to } 1.807 \tag{9}$$

1/0 for the presence and absence of a pyridyl ring in the molecule, and IR_{TETR} was an indicator variable for the existence of a tetrazolyl ring in the molecule. The presence of the pyridyl and the tetrazolyl rings accounted for a positive effect. The fact that $\log P$ has been used to model hydrophobicity implies

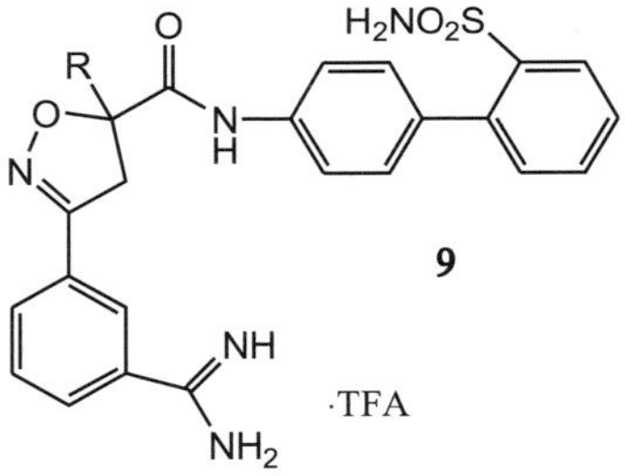

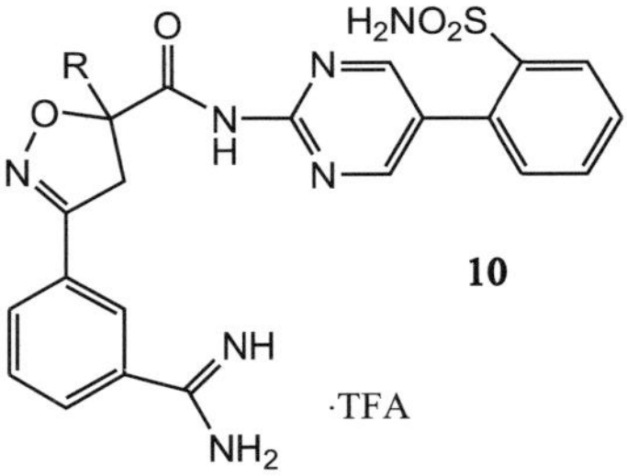

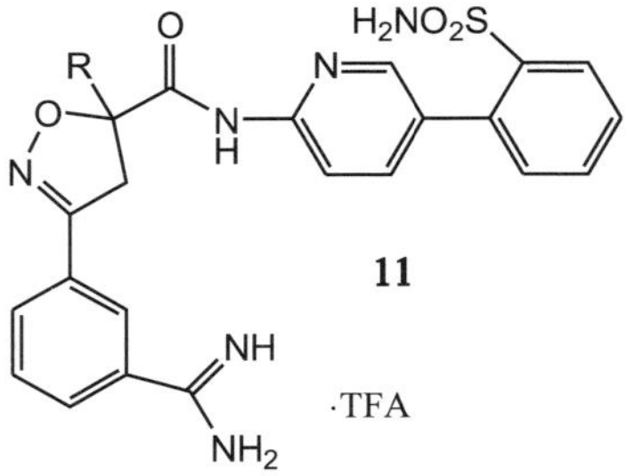

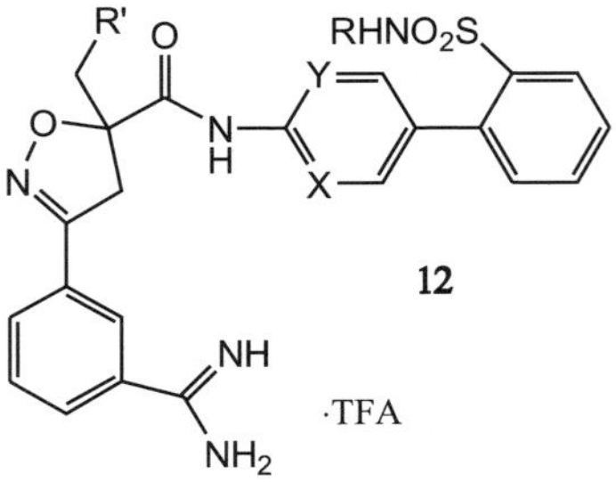

Scheme 3

that all the parts, where substituents have been entered, hydrophobic contacts have been made.

b) Biaryl-substituted Isoxazoline Derivatives

In biaryl-substituted isoxazoline derivatives (13–14) [75], a biaryl moiety was designed to interact with the S4 aryl-binding domain of the FXa active site. In this case, hydrophilicity was found to be the most important term (Eq. 10) [72].

$$\log 1/K_i = -\ 0.328(\pm 0.162)C \log P$$
$$+\ 0.689(\pm 0.371)MR\text{-}R_2 + 7.956(\pm 0.757)$$
$$n = 16,\ r = 0.932,\ r^2 = 0.868,\ q^2 = 0.806,$$
$$s = 0.308,\ F_{2,13} = 8.047,\ \alpha = 0.01 \tag{10}$$

MR-R_2 is the molar refractivity for the o-substituents. R_2 substituents are likely to contact a hydrophobic space.

Scheme 4

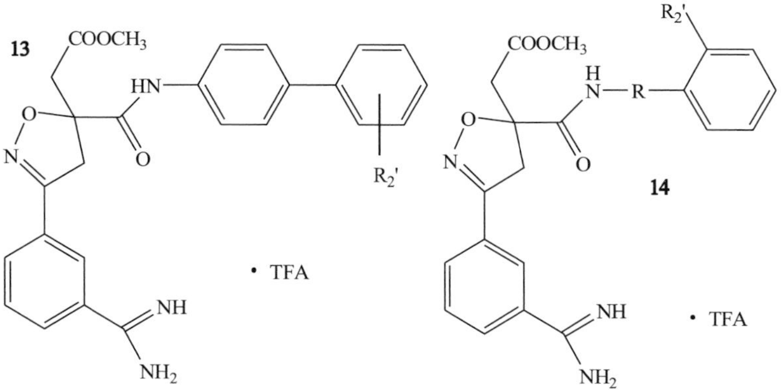

c) Pyrazolyl Analogues

The work of Pinto et al. and some others [76, 77] on the optimization of the heterocyclic core had led to the discovery of a novel pyrazole SN429. Further optimization of the pyrazole core substitution and the biphenyl P4 culminated in the discovery of a new series of FXa inhibitors (15–19) for which Eq. 11 was formulated [72]. The continuation of research on FXa inhibitors led to new variety of compounds. However, the lack of chirality of the isoxazole derivatives, such as SA862 [78] and its high affinity for FXa, made it an attractive template for further optimization. A significant study was performed for the use of other five-membered heterocyclic templates in which

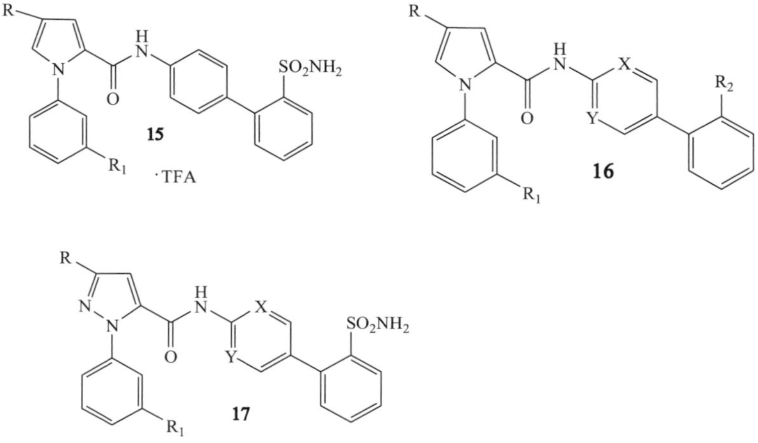

Scheme 5

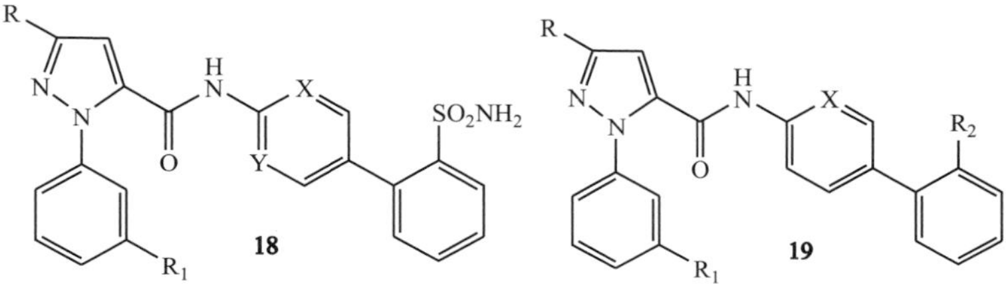

Scheme 6

the point of attachment to the P1 substituent was through a nitrogen atom in the heterocycle [79, 80].

$$\log 1/K_i = 0.985(\pm 0.363)\text{CMR} + 8.656(\pm 1.797)\text{MR-R}_1 +$$
$$0.783(\pm 0.378)\text{B}_{5\text{-R}} - 13.233(\pm 5.150)$$
$$n = 25, \; r = 0.929, \; r^2 = 0.863, \; q^2 = 0.784,$$
$$s = 0.365, \; F_{3,21} = 10.625, \; \alpha = 0.01 \tag{11}$$

The most significant term in Eq. 11 is the molar refractivity of the m-substituent at the phenyl ring. Since MR-R$_1$ is primarily a measure of bulk and of polarizability of the substituent, the positive coefficients with both terms MR-R$_1$ and CMR suggest that size of R$_1$-substituent as well as of the whole molecule will be conducive to the activity. The parameter B$_{5\text{-R}}$ is the Sterimol parameter for the maximum-width of the first atom of the group R at the N-heterocyclic ring. This points out a favorable role of the first atom through some steric interaction.

d) Benzamidine Derivatives Containing Heterocyclic Core

In an attempt to further optimize FXa inhibitors, some authors [81–84] prepared additional C-substituted core analogues and N-substituted core analogues (**20, 21**) focused on the idea that the nature of the heterocyclic core was
critical to binding potency. For these compounds, Eq. 12 was derived [72],
which seems strange because it contains no π or C log P term.

$$\text{MgVol} \log 1/K_i = 0.047(\pm 0.025)\text{MgVol} - 1.156(\pm 0.389)\text{I-I}_M +$$

$$1.579(\pm 0.489)\text{I-NCORE} - 12.686(\pm 12.072)$$

$$n = 12, \ r = 0.952, \ r^2 = 0.9907, \ q^2 = 0.795,$$

$$s = 0.272, \ F_{3,14} = 5.941, \ \alpha = 0.01 \tag{12}$$

refers to the molecular volume and its positive sign indicates that the larger is
a molecule, the more would be it Fxa inhibition potency. The indicator variable I-IM takes value of 1 for the compounds that contain an an imidazole
ring in the core. The indicator variable I-NCORE refers to the presence of an
N-substituted core. However, while the former indicates a negative effect of
imidazole ring, the latter indicates a positive effect of N-substituted core. No
parameterization has been done for the presence of another heterocyclic ring.

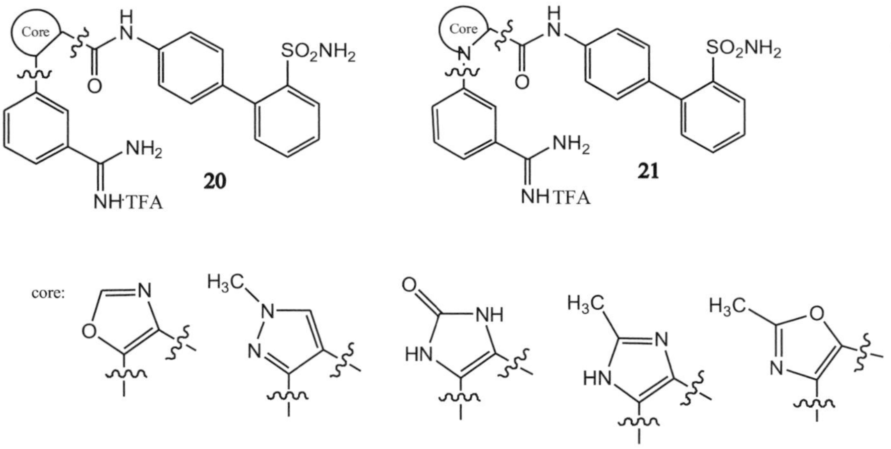

Scheme 7

Non-Amidine 1,2-Benzamidobenzene Derivatives

Masters and his group [82] reported inhibitors of FXa in which they used selected hydrophobic groups for occupying both the S1 and S4 binding domains
of the enzyme. They described the SAR of altering the linkage of the 1-(4-
pyridyl)piperidine to the central ring to optimize the placement of this group

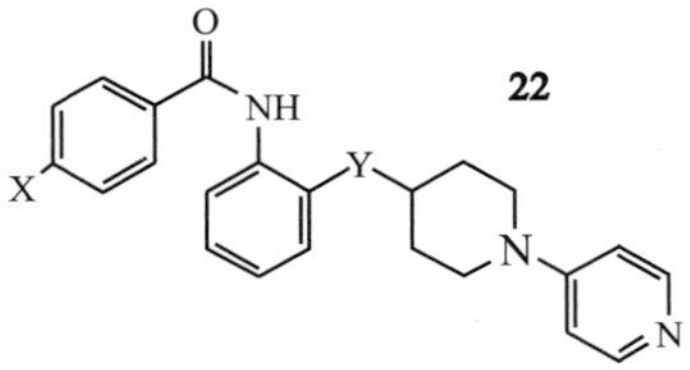

Scheme 8

in the S4 site of FXa and prepared various urethane and urea derivatives (22). For these compounds, Eq. 13 was derived [72].

$$\log K_{ass} = 12.516(\pm 7.377)C \log P - 1.804(\pm 0.972)(C \log P)^2 - 14.149(\pm 13.779)$$

$$n = 9, \; r = 0.936, \; r^2 = 0.875, \; q^2 = 0.679, \; s = 0.406, \; F_{2,6} = 21.085, \alpha = 0.01$$

$$C \log P_o = 3.469(\pm 0.309), \; (3.032 - 3.6500) \tag{13}$$

Tetrahydroisoquinoline

A long discussion was had concerning the binding sites of Factor Xa, the S1 and S4. In principle, these two binding modes were conceivable for the novel dibasic FXa inhibitors. Molecular modelling experiments based on the X-ray structures of uninhibited FXa and the DX-9065a/FXa complex have shown one binding mode: the tetrahydroisoquinoline fills the S1 pocket even better than the naphthalene moiety of DX-9065a as the iminomethyl piperidine residues occupy the S4 site [83]. Both pockets are consisted of hydrophobic walls and a negatively charged bottom. Hence, two binding modes are possible for the dibasic inhibitors. It has been also reported that the S1 pocket is accessible only from the top, whereas one wall of the S4 site is open to the solvent, and thus it can potentially accommodate bulkier groups which will not be able to enter the S1 site. Kurczierz et al. [84] made a significant study with assumption that a bioisosteric substitution of the 2,7-di-C,C-substituted naphthamidine moiety with the 2-carbamimidoyl-1,2,3,4-tetrahydro-isoquinolin-7-yloxy template would provide a novel P1 residue suitable for FXa inhibition (23, 24). Thus, (1,2,3,4-tetrahydroisoquinolin-7-

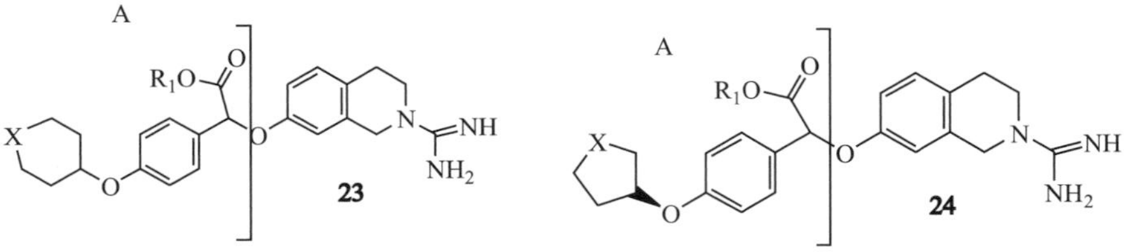

Scheme 9

yloxy)phenyl acetic acid derivatives ("TIPAC derivatives") were synthesized as a novel class of low molecular weight inhibitors of FXa.

$$\text{Log} 1/K_i = 0.203(\pm 0.115)\text{MR-A} + 1.112(\pm 0.297)\text{I-XCN} + 5.307(\pm 0.462)$$
$$N = 24, \; r = 0.902, \; r^2 = 0.813, \; q^2 = 0.772,$$
$$s = 0.399, \; F_{2,21} = 6.686, \; \alpha = 0.01 \tag{14}$$

For these compounds, Eq. 14 was derived, in which MR-A refers to the molar refractivity of the whole group A. No role for lipophilicity was found (no π or $C \log P$ term). It is shown that MR-A and $C \log P$ were correlated with $r = 0.597$. I-XCN, an indicator variable which takes the value of 1 for the existence of an amidino group $(C(= NH)NH_2)$ in the substituent X, was found to be significant. It shows an appreciable effect of an amidine group at X position. On the contrary, no role for the nature of the alicyclic ring (five- or six-membered) was shown.

Phenylalanine Derivatives

Using the 3-amidino-phenylalanine templates as a key building block, Stürzebecher et al. [54, 85] synthesized some novel inhibitors of FXa (25) and evaluated their biological activity. For these compounds, Eq. 15 was formulated [72], in which MR-R_1 is the molar refractivity of the substituent R_1. Indicator variable I-RAM indicates the existence of an amidine group, whereas I-R_2N is an indicator variable for the compounds where a bond between carbonyl group and N exists in substituent R_2. The negative coefficient of I-R_2N means that the presence of a bond between the carbonyl group and N is correlated with low inhibitory activity. Lipophilicity and electronic effects do not seem to play any role.

$$\log 1/K_i = 0.768(\pm 0.284)\text{I-RAM} + 0.249(\pm 0.080)\text{MR-}R_1 -$$
$$1.062(0.284)\text{I-}R_2\text{N} + 3.505(\pm 0.577)$$
$$n = 37, \; r = 0.900, \; r^2 = 0.810, \; q^2 = 0.694,$$
$$s = 0.263, \; F_{3,33} = 9.995, \; \alpha = 0.01 \tag{15}$$

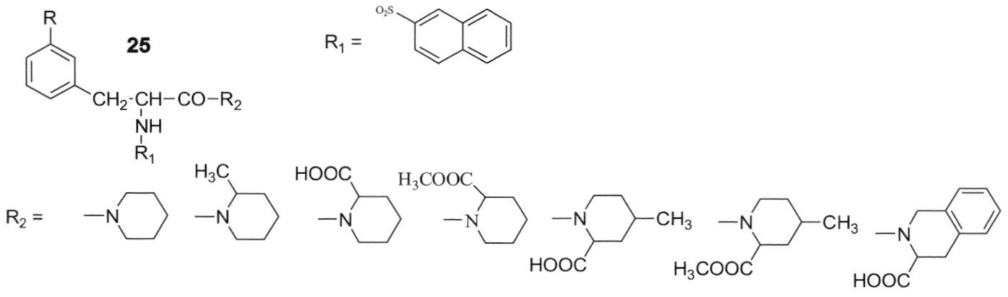

Scheme 10

Five- or Six-Membered Heteroaryl Derivatives

Aminopyridyl moieties were identified as important replacements for the *trans*- aminocyclohexyl group at P1 in the L-371,912 template. Feng et al. [86] reported a new class of thrombin inhibitors incorporating aminopyridyl moieties (26). These researchers anticipated that developing potent inhibitors with neutral P1 would require the use of these lipophilic residues at P3 to add more hydrophobic binding energy to compensate losses at P1. For these inhibitors, Eq. 16 was derived, where MR-R_1, MR-R_2, MR-R_3 are the molecular refractivities of R_1, R_2, and R_3 substituents [72]. The most significant term is MR-R_1, whose negative coefficient suggests a steric hindrance by R_1-substituents. $C \log P$ could not replace MR. It is thus clear that bulk of the substituents control the activity.

$$\log 1/K_i = 0.690\pi(\pm 0.316)\text{MR-}R_2 + 0.419(\pm 0.184)\text{MR-}R_3 -$$
$$0.768(\pm 0.504)\text{MR-}R_1 + 3.297(\pm 2.331) \quad (10)$$
$$n = 11, \ r = 0.930, \ r^2 = 0.865, \ q^2 = 0.676,$$
$$s = 0.301, \ F_{3,7} = 3.675, \alpha = 0.1 \quad\quad\quad\quad\quad\quad\quad (16)$$

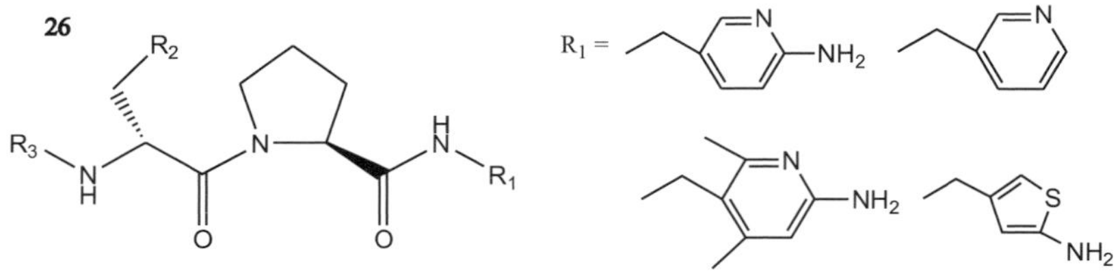

Scheme 11

Amido-(Propyl and Allyl)-Hydroxybenzamidine Derivatives

Pauls et al. [87] designed and synthesized amido-(propyl and allyl)-hydroxybenzamidine coagulation FXa inhibitors (27, 28), for which the correlation obtained was as shown by Eq. 17 [72], in which no hydrophobic and electronic parameters had surfaced. In this equation, however, the indicator variable Idpyr, which has been used for the existence of a pyridine ring in the molecular structure and IdH, which has been used for the compounds for which there is a possibility of hydrogen bonding formulation describe the effect of the pyridine ring and the hydrogen bonding.

$$\log 1/K_i = 0.447(0.389)\text{Idpyr} + 1.264(0.329)\text{IdH-H} + 6.949(\pm 0.221)$$
$$n = 19, \ r = 0.900, \ r^2 = 0.810, \ q^2 = 0.704,$$
$$s = 0.324, \ F_{2,16} = 2.967, \ \alpha = 0.1 \quad\quad\quad\quad\quad\quad\quad (17)$$

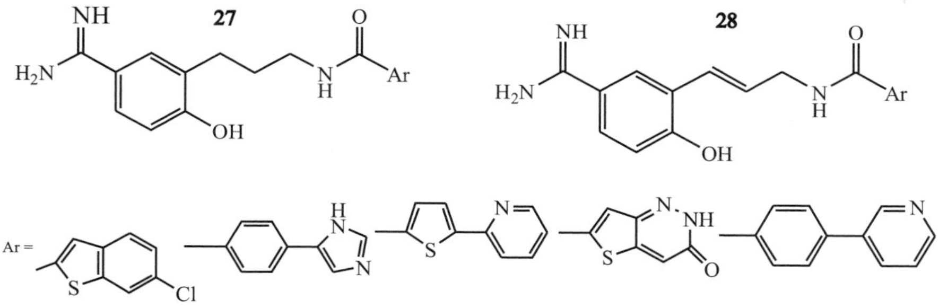

Scheme 12

Multi-Centered Short Hydrogen Bond Binding Mode Derivatives

Recently, a new and unusual high-affinity binding mode for the serine protease inhibitor bis (5-amidino-2-benzimidazolyl)methane (BABIM) was identified whereby the azine ion can be recruited to mediate binding between the inhibitor and the enzyme. The zinc ion is tetrahedrally coordinated between two chelating nitrogen atoms of the inhibitor and the active site residues. As a consequence, highly potent and selective zinc-dependent inhibitors for FXa have been developed in which the hydrogen bonds that are formed between Ser195, the inhibitor hydroxyl oxygen and a water molecule trapped in the oxyanion hole are less than 2.3 Å compared to ordinary hydrogen bonds (> 2.6 Å) [54]. For the inhibitors belonging to **29** and **30**, the equation obtained was as shown by Eq. 18. In this equation, I-INDOLE is an indicator variable for indolyl group in the molecule whereas the indicator variable I-2-OH is

$$\log 1/K_i = 1.062(\pm 0.421)\text{I-INDOLE} + 1.931(\pm 0.600)\text{I-2-OH} -$$
$$0.754(\pm 0.395)\text{ES}_{-3} + 3.114(\pm 0.620)$$
$$n = 31, \ r = 0.900, \ r^2 = 0.810, \ q^2 = 0.712,$$
$$s = 0.471, \ F_{3,27} = 5.326, \ \alpha = 0.01 \tag{18}$$

concerned with compounds where a 2-OH group is present. The Taft steric parameter ES_{-3} for substituent of 3-position is the most important parameter in the stepwise development (40%) and indicates an unfavorable steric effect for the 3-substituents on the phenyl ring.

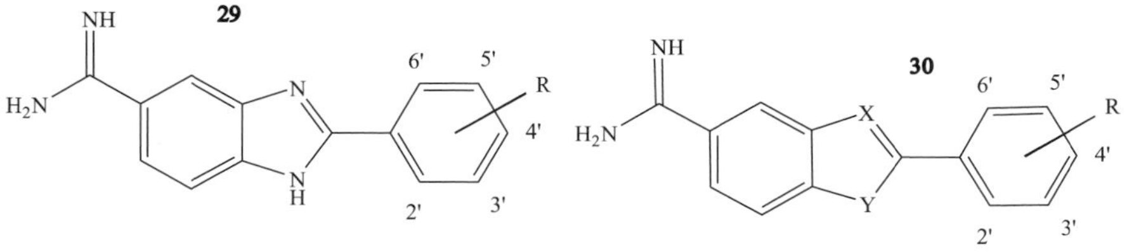

Scheme 13

For the same set of compounds, another correlation was obtained (Eq. 19), where ES_{-3} was replaced by MR-3 but an additional parameter σ_{-m} was to be included, which indicated a favorable effect of electron-withdrawing meta-substituents.

$$\log 1/K_i = 1.096(\pm 0.391)\text{I-INDOLE} + 1.874(\pm 0.581)\text{I-2-OH} +$$

$$0.208(\pm 0.182)\text{MR-3} + 1.933(\pm 0.910)\sigma_{-m} + 3.163(\pm 0.608) \quad (12a)$$

$$n = 31, \; r = 0.916, \; r^2 = 0.839, \; q^2 = 0.717,$$

$$s = 0.441, \; F_{4,26} = 66.725, \; \alpha = 0.01 \tag{19}$$

Benzoxazinones

Dudley et al. [88] through a combination of SAR studies and molecular modelling tried to synthesize a series of benzoxazinones analogues of 3-(4-[5(2R,6S)-2,6-dimethyltetrahydro-1(2H)-pyridinyl]pentyl]-3-oxo-3,4-dihydro-2H-1,4-benzoxazin-2-yl]-1-benzenecarboximidamide (31). X-ray crystallographic studies identified that the carbonyl group of the benzoxazinone accepted a hydrogen bond from GLy216NH and the aryl ring of the benzoxazinone stacked against Gly219 and made van der Waals contact with the Cys191-Cys220 disulfide bridge. A minimal conformational change of the flexible pentyl chain and modelling with FXa orientated the dimethylpiperidine in the "aryl-binding site" of FXa [83]. It was presumed therefore that the piperidine ring made a cationic-p interaction with the protein affording enhanced binding. From these data Eq. 20 was derived [72]. In this equation, B_{5-X3}, the Sterimol parameter for the largest width of the first atom of X-substituent at position 3 is the most significant term. The m-position on phenyl ring seems to be much more important. The indicator variable I-ZDMP takes a value of 1 for the compounds with the cis-2,6-diMe-piperidinyl at the Z substituent.

$$\log 1/IC_{50} = 0.228(\pm 0.226)C \log P + 1.025(\pm 0.831)\text{I-ZDMP} +$$

$$1.371(\pm 0.395)B_{5-X3} + 3.561(\pm 1.856) \quad (13)$$

$$n = 17, \; r = 0.941, \; r^2 = 0.886, \; q^2 = 0.791,$$

$$s = 0.531, \; F_{3,13} = 33.663, \alpha = 0.01 \tag{20}$$

31

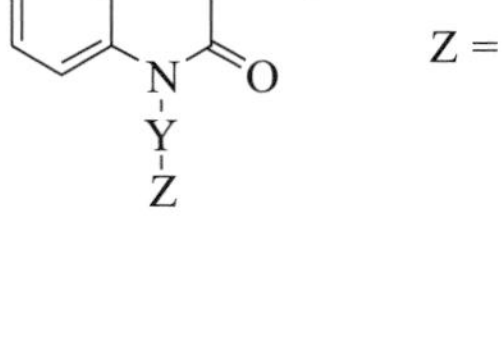

Z = cis-2,6-diMe-piperidinyl

Scheme 14

Amino Isoquinolines

Molecular modelling studies led to the design of conformationally con-strained diaryl ethers as well as benzopyrrolidinone derivatives [89]. This effort was extended to the synthesis of benzopyrrolidinone-based amino iso-quinolines (**32–35**). Analyzing the data of these compounds, Eq. 21 was de-rived [72].

$$\log 1/IC_{50} = 0.220(+0.131)C \log P + 0.339(\pm 0.245)I\text{-}A -$$
$$2.782(\pm 0.389)I\text{-}ISOQ - 0.439(\pm 0.338)I\text{-}R + 7.397(\pm 0.460) \quad (14)$$
$$n = 41, \ r = 0.926, \ r^2 = 0.857, \ q^2 = 0.818,$$
$$s = 0.334, \ F_{4,37} = 54.040, \ \alpha = 0.01 \quad\quad\quad\quad\quad\quad\quad (21)$$

The fact that $C \log P$ was used to model hydrophobicity implies that all the parts where substituents have been entered can make hydrophobic contacts. The existence of only a linear correlation between $\log 1/IC_{50}$ and $C \log P$ sug-gests that the $C \log P$ values were not great enough to establish the upper limit for the rate of binding. Three indicator variables are included in the deriva-tion of Eq. 21. I-ISOQ is an indicator variable (the most significant term) that takes value of 1 for the existence of an isoquinoline ring at the ring C. The variable I-R takes a value of 1 for the existence of a substituent (R) other than hydrogen on the amine group. I-A is an indicator variable for the compounds, where a $(C('OO)NH)$ is present in the molecular structure. The negative co-efficients of indicators I-ISOQ and I-R mean that the presence of the specific structure moieties are not conducive to the activity.

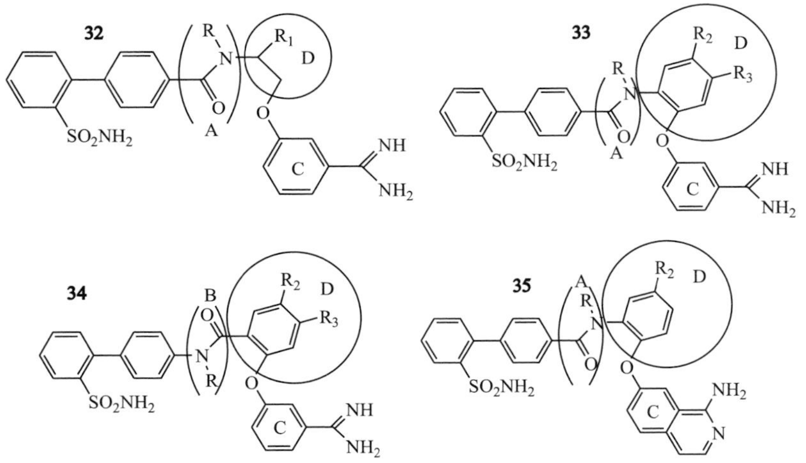

Scheme 15

Naphthalenosulphonamide Derivatives

Hirayama et al. [90] made the efforts to discover FXa inhibitors, containing carboxamide and sulfonamide linkers (**36–41**). From their data Eq. 22 was formulated [72], where MR_B is the most significant term followed by the indicator variable I-NAPH, used for the presence of a naphthyl group in the compound. Group B is indicated in the structures.

$\log 1/IC_{50} = 1.309(\pm 0.783)\text{I-NAPH} + 0.434(\pm 0.407)\text{MR-B} +$
$5.608(\pm 0.951)$ *(15)*

$n = 14,\ r = 0.823,\ r^2 = 0.677,\ q^2 = 0.469,$

$s = 0.591,\ F_{2,11} = 11.18,\ \alpha = 0.01$ \hfill (22)

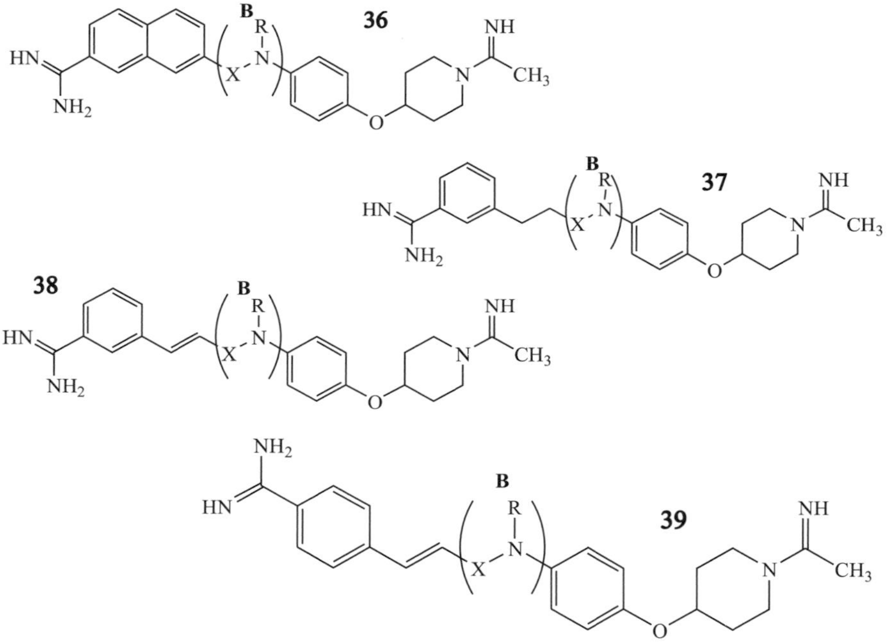

Scheme 16

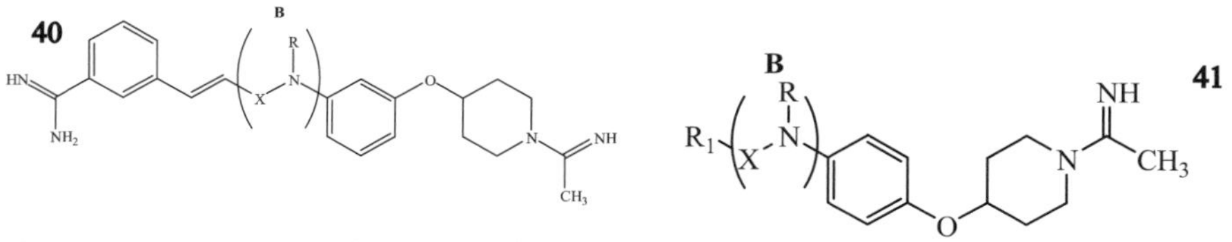

Scheme 17

Tetrazoles

A series of tetrazole FXa inhibitors containing benzamidine mimics as the P1 substrate was studied [91]. Pyrazoles, triazoles and tetrazoles with a *m*-benzamidine P1 moiety linked to a nitrogen in a five-membered heterocycle (**42–44**) have been shown to be potent FXa inhibitors [92]. For these inhibitors, Eq. 26 was derived which exhibited a dominant role of molar refractivity of P1 group.

$$\log 1/K_i = 6.701(\pm 3.758)\text{MR-P1} - 14.686(\pm 13.182)$$
$$n = 10, \ r = 0.824, \ r^2 = 0.679, \ q^2 = 0.467,$$
$$s = 0.485, \ F_{1,8} = 16.912, \ \alpha = 0.01 \tag{23}$$

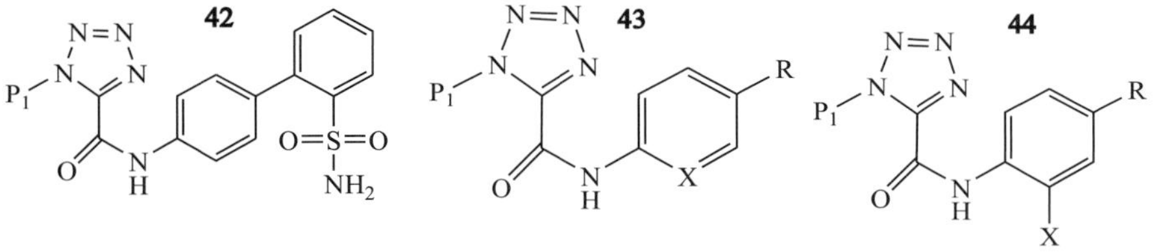

Scheme 18

Glucolic-Mandelic Derivatives

Su et al. [93] designed, synthesized, and studied in vitro a series of low molecular weight dibasic noncovalent FXa inhibitors (**45a, 45b**). The FXa inhibitory activity of these compounds was shown to be correlated as exhibited by Eq. 24. In this case, the molar refractivity of the group B (shown in the structures) was also shown to be important. The effect of varying the number (*n*) of methylene units between the O-ether and the CO group was not found to be important.

$$\log 1/\text{IC}_{50} = -0.915(\pm 0.483)C \log P + 1.305(\pm 0.540)\text{MR}_B + 5.544(\pm 2.265)$$
$$n = 18, \ r = 0.800, \ r^2 = 0.639, \ q^2 = 0.452,$$
$$s = 0.792, \ F_{2,15} = 8.134, \ \alpha = 0.01 \tag{24}$$

2.2.1
Statistics

With each QSAR, the following statistics are given: the 95% confidence limits for each term in parentheses; the correlation coefficient r between observed values of the dependent and the values predicted from the equation; r^2 the squared correlation coefficient; s the standard deviation; q^2 defines the cross-validated r^2 (indication of the quality of the fit); and the F-values are given for

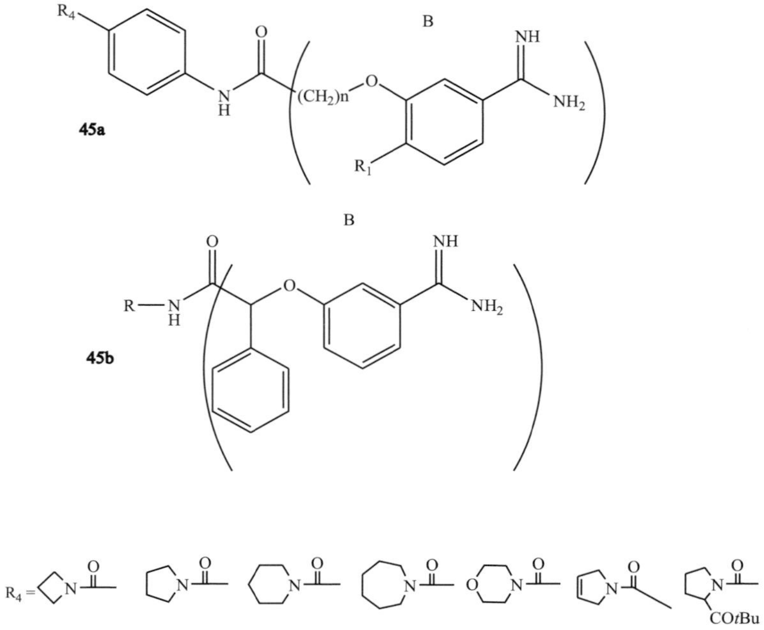

Scheme 19

the individual terms. *F* value is a measure of the level of statistical significance of the regression model. Log P_o represents the optimum hydrophobicity.

2.3
Discussion and Conclusion

The equations generated by QSAR analysis are an indication of the properties of the substituent groups that make a particular molecule a better or worse inhibitor. The overall essence of the above QSAR studies is the high dependence of the binding of substrates or inhibitors on the molar refractivity. Molar refractivity is related to volume and polarizability [94]. Polarizability is another kind of electronic effect that controls the activity. It is the conversion factor between an applied electronic field and the induced dipole moment. Since most molecules are asymmetric, it is a three-dimensional tensor. These results show that steric properties increase affinity. It is commonly assumed in QSAR studies that when CMR/MR appears with a positive sign, it indicates favorable polar interactions, whereas when it with the negative sign, unfavorable bulk effects are indicated. As pointed out previously, negative steric terms imply that the critical effects are on/in an active site on a macromolecule. The lack of importance of hydrophobic terms implies that the site is not hydrophobic nor the hydrophobicity is required for the drug entry to reach the active site, but this conclusion is born out only by the QSAR for in vitro studies and not from a whole animal.

In most of the cases CMR, MR, and $C \log P$, π were not found to be correlated. The high correlation with MR led us to assume that the binding pocket around the active site in FXa is not "typically hydrophobic". It has been presented [95] as evidence for another type of "hydrophobic bonding" in which groups surrounded by flickering clusters of water are held together in solution without desolvation playing the major role. Yoshimoto and Hansch [96] working on the serine proteinases, therefore, assumed that a high correlation with MR/CMR reflects this type of interaction.

The catalytic triad is formed by Asp102, His57, and Ser195, and is not typically hydrophobic, though these are mostly hydrophobic. The negative hydrophobic effect is striking. In an attempt to explain the result, the following should be considered. As compounds become more lipophilic, they also become bulkier, and the interaction with (or the entrance into) the gorge may get difficult. In addition, the hydrophobic groups undergo hydrophobic collapse onto the protein surface or they are unable to find a complementary patch of protein surface on which to undergo the hydrophobic collapse. The negative coefficient of $\log P$ term indicates that less hydrophobic compounds are favored and this might be related with the dynamic process that leads the ligands to the active site inside the gorge. It might be that more hydrophobic compounds become trapped at some peripheral binding site, instead of reaching the catalytic site. In two cases, where the relationship between $\log (1/K_i)$ or $\log (K_{ass})$ and $\log P$ was well approximated by parabola, the role of the lipophilic character of FXa inhibitors can be at least roughly separated from the electronic and steric requirements.

To describe the effects of some structural features that cannot be accounted for by any physicochemical parameters, few indicator variables had to be used. From these parameters we could find that saturated heterocyclic moiety (e.g., piperidine or piperazine) could be favorable for hydrophilic groups. This is in agreement with the fact that at this location piperazyl and piperazylsulfonyl derivatives possess enhanced binding affinity toward FXa compared to the piperidine derivatives. It could also be shown that by enclosing the amidino group in a ring with no opportunity for salt bridge formation decreased the activity. Also, substitution on the naphthyl or benzofuran ring by a methoxy or methyl group decreased the activity, implying that the amidino moiety was in a tight pocket forming a salt bridge.

The major electronic factor influencing the activity has been Hammett's constant σ. In some cases, the steric factors were also found to be important, e.g., B_5 the Sterimol parameter for the largest width of substituents and L the Sterimol parameter for the length of the substituents. The positive coefficients of these parameters give an important positive contribution of width and length of substituents to the inhibition. The values of Taft steric parameter E_s are all negative (except H = 0), and thus the negative sign of E_s-term indicates that the larger the substituent the greater the FXa inhibition.

3
Thrombin Inhibitors with Heterocyclic Moieties

3.1
3D-QSAR and Molecular Modeling Studies on Thrombin Inhibitors

The trypsin-like serine protease thrombin is a multifunctional key enzyme at the final step of the coagulation cascade and is involved in the regulation of haemostasis and thrombosis. Synthetic thrombin inhibitors have a long history; initial compounds were derived from electrophilic ketone and aldehyde-analogues of arginine. First, potent leads of non-covalent inhibitors were developed in the early eighties, which were further optimized in the nineties after the X-ray structure of thrombin became available. In the meantime, a huge number of highly active and selective inhibitors has been published. The success of these compounds as new anticoagulants will depend also from the future progress in the development of orally active Factor Xa inhibitors. In addition, they do not need cofactors, they are not influenced by platelet factors, and they are able to inhibit also clot-bound thrombin. Probably also the design of dual thrombin and FXa inhibitors might be a promising new strategy in the future development of potent anticoagulants. Ximelagatran is the first orally available thrombin inhibitor that has been approved in France for the prevention of venous thromboembolism.

Pooling the combined research in the thrombin area provides numerous libraries of inhibitors with diverse structures and a range of affinities for thrombin spanning 9 orders of magnitude (millimolar to picomolar). Thrombin, in contrast to the other coagulation proteinases shows only a small scale of allosteric changes [97, 98].

Bursi and Grootenhuis [99] have previously correlated theoretical and experimentally determined binding data for a series of thrombin inhibitors. Using molecular mechanics minimization of inhibitors in the receptor structure (in which they had been cocrystallized) a statistically significant correlation ($R = 0.74$) with a resolution of 2.5 Å. Comparative molecular field analysis (CoMFA) gave better result ($R = 0.95$) and cross-validated r^2 (q^2) = 0.46, when used with high-quality ab initio charges and minimized conformations of crystal structure of inhibitors (crystal structure alignment).

The work described by Deadman et al. [100] considered a subset of the above set of thrombin inhibitors. A training set of 16 homologous non-peptide inhibitors whose conformations had been generated in continuum solvent (MacroModel) and clustered into conformational families (XCluster) was regressed against this pharmacophore so as to obtain a 3D-QSAR mode.

The structure-based pharmacophore was built using the crystallographic co-ordinates of **argatroban** complexed with thrombin [101] and the functional group definitions contained within the CATALYST software to predict the effects on the K_i values of structural modification of a set of homologous 4-aminopyridine (4-AP) thrombin inhibitors.

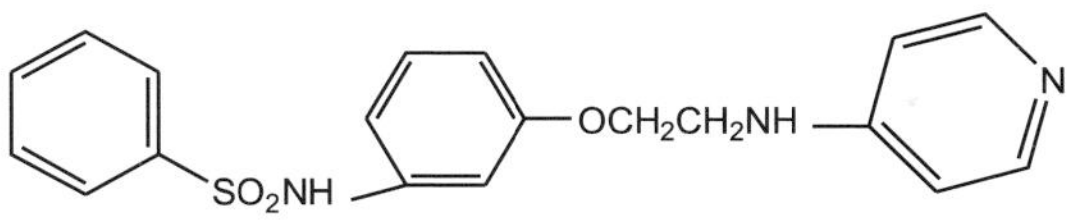

argatroban

Scheme 20

CATALYST attempted to produce a 3D-QSAR model that correlated the es-timated activities with the measured activities. To test the robustness of the resulting QSAR model, the synthesis of a series of non-peptide thrombin in-hibitors based on arylsuphonyl derivatives of an aminophenol ring linked to a pyridyl-based S1 binding group was undertaken. These compounds served as a test set. The crystal structure for the novel symmetrical disulfonyl com-pound in complex with thrombin had been solved. Its calculated binding mode was in general agreement with the crystallographically observed one, and the predicted K_i value was in close accord with the experimental value.

CoMFA, advance CoMFA, and CoMSIA have been carried out on a series of pyrroloquinazolines [102] for their thrombin receptor antagonistic activity. The predicted activities by highly significant CoMFA ($q^2 = 0.66$) and CoMSIA ($q^2 = 0.67$) models were in good agreement with observed activities and the models might be useful for optimization of thrombin receptor antagonistic activity. Molecular modelling techniques CoMFA, advance CoMFA and CoM-SIA were adopted on a Silicon Graphics Octane R12000 workstation using SYBYL6.9 molecular modelling software. CATALYST running on a SGI O2 was used for common feature hypothesis generation.

In Sect. 2.1, we have presented the 3D-QSAR CoMFA studies on a se-ries of indole/benzoimidazole-5-carboxamidines as FXa inhibitors. Herein we present the 3D-QSAR/CoMFA models developed for the same series of indole/benzoimidazole-5-carboxamidines [69] as thrombin inhibitors, using 29 training set molecules and validated with seven test set molecules (Fig. 5).

The CoMFA model obtained with only the steric field showed r^2 (cross-validated) = 0.520 from nine components, r^2 (non cross-validated) = 0.969, F value = 55.811, $r^2_{bs} = 0.985$, while the model with only the electrostatic field showed r^2 (cross-validated) = 0.425 from five components, r^2 (non cross-validated) = 0.992, F-value = 51.296, $r^2_{bs} = 0.958$. The CoMFA model generated from both steric and electrostatic fields exhibited r^2 (cross-validated) = 0.504 with three components, r^2 (non-cross-validated) = 0.845, F value = 39.962,

$r_{bs}^2 = 0.896$, r^2 (predicted) $= 0.860$ with 54.5% steric and 45.5% electrostatic field contributions. All the molecular modelling and 3D-QSAR studies were performed with the standard protocol using SYBYL 6.7 software.

Riester et al. [103] presented a series of thrombin inhibitors that were generated by using powerful computer-assisted multiparameter optimization process, which was organized in design cycles starting with a set of 170 ran-

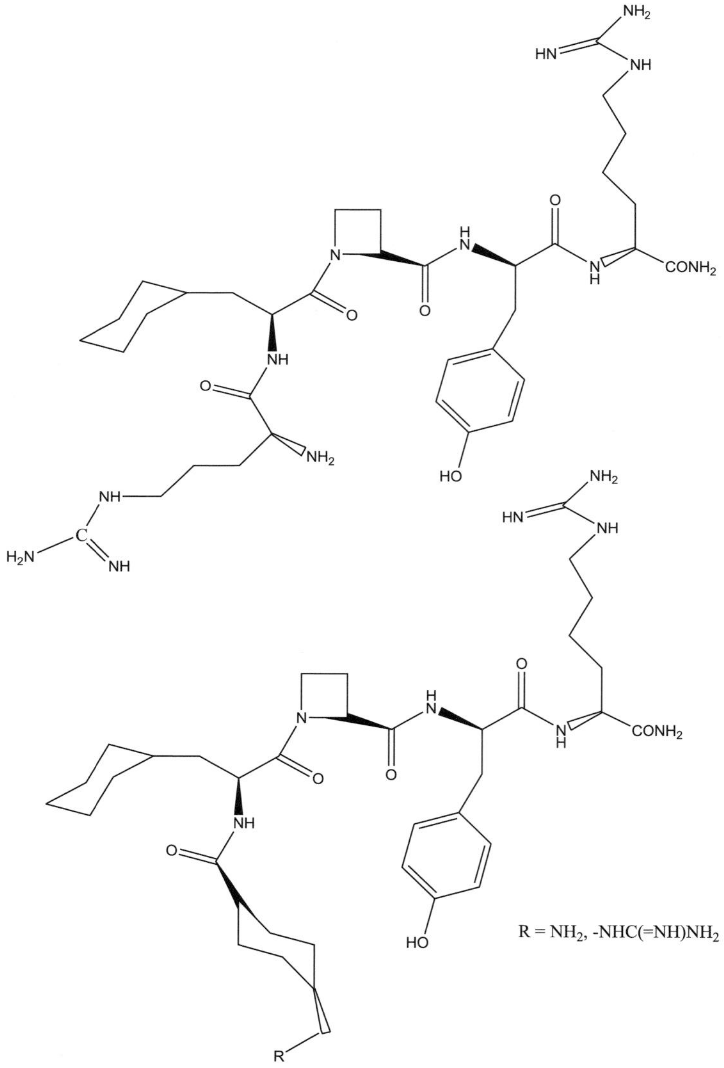

Fig. 7 Structures of thrombin inhibitors identified in design cycles seven and eight of the computer-assisted optimization procedure

domly chosen molecules (peptides with sequence lengths between three and ten amino acids). Each cycle combined combinatorial synthesis, multiparameter characterization of compounds in a variety of bioassays, and algorithmic processing of the data to devise a set of compounds to be synthesized in the next cycle. After algorithmic processing of the data, a set of 96 compounds was determined by the algorithm and synthesized subsequently. The identified lead compounds were by far the most selective synthetic inhibitors (Fig. 7).

3.2
2D-QSAR on Thrombin Inhibitors

It is well known that synthetic arginine esters, such as Nα-tosyl-L-arginine methyl ester (TAME), are hydrolyzed by thrombin and inhibit the clotting activity of thrombin. Since the binding specificities of arginine derivatives would be determined by the structure of both sides of arginine, i.e., amino as well as carboxylic sides, a series of studies were undertaken by Okamoto et al. [104–106] to obtain potent and specific inhibitors of thrombin by modifications of the Nα-substituent and methyl ester positions of TAME.

Gupta et al. [107, 108] performed a significant QSAR study showing that hydrophobicity could be an important factor in thrombin inhibition for TAME derivatives (46–51) (Eqs. 25–29). In this analysis, log P values were calculated according to Hansch and Leo method [109].

$$pI_{50} = 5.960 + 1.053(\pm 0.423) \log P$$
$$n = 9, \ r = 0.803, \ s = 0.486 \tag{25}$$

$$pI_{50} = 5.789 + 0.797(\pm 0.105) \log P$$
$$n = 7, \ r = 0.959, \ s = 0.221 \tag{26}$$

$$pI_{50} = 5.723 + 0.531(0.082) \log P$$
$$n = 12, \ r = 0.899, \ s = 0.384 \tag{27}$$

$$pI_{50} = 6.116 + 0.532(\pm 0.111) \log P \quad (21)$$
$$n = 11, \ r = 0.847, \ s = 0.528 \tag{28}$$

$$pI_{50} = 6.608 + 0.592(\pm 0.133) \log P - 0.456(\pm 0.154)(\log P)^2 \quad (22)$$
$$n = 18, \ r = 0.816, \ s = 0.408 \tag{29}$$

A QSAR study based on a series of tripeptidyl thrombin inhibitors (52) [110] was performed. Employing partial least squares (PLS) analysis with Sybyl/QSAR, the researchers developed Eqs. 30–33 for the inhibitory activities using the interaction energies E_{total} (total interaction energy), E_{steric} (steric interaction energy), and E_{elec} (electrostatic interaction energy) as descriptors. The

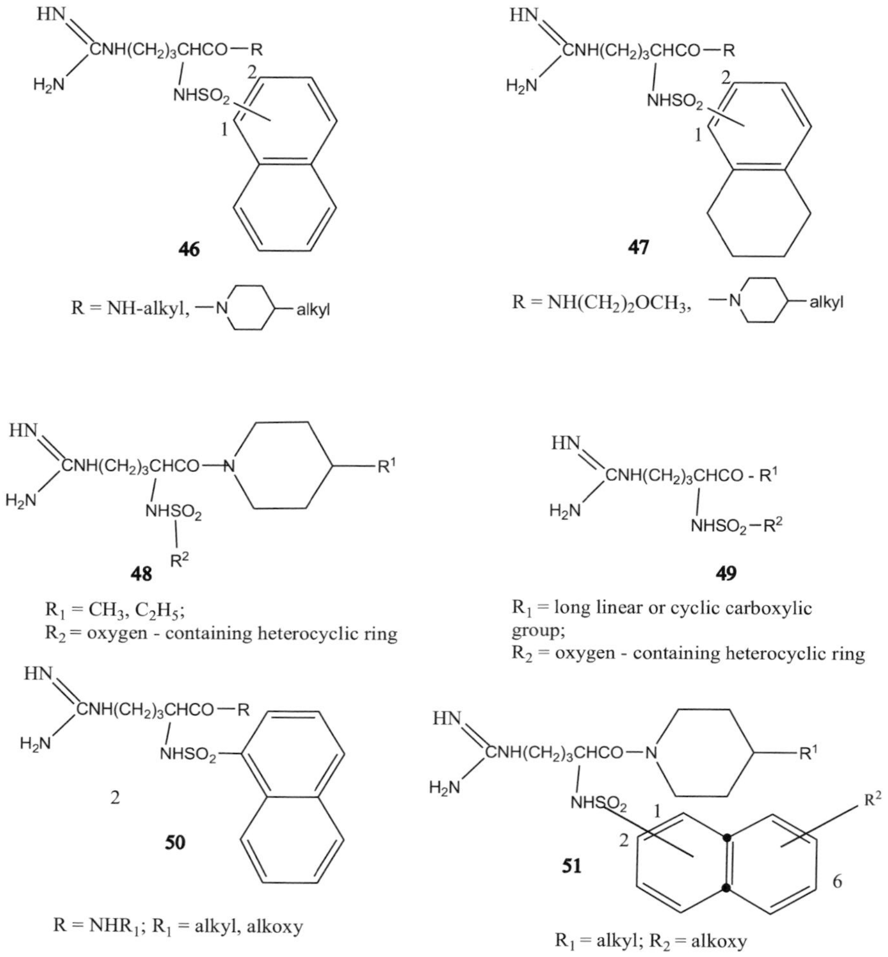

46

R = NH-alkyl, —N⟨ ⟩—alkyl

47

R = NH(CH₂)₂OCH₃, —N⟨ ⟩—alkyl

48

$R_1 = CH_3, C_2H_5$;
R_2 = oxygen - containing heterocyclic ring

49

R_1 = long linear or cyclic carboxylic group;
R_2 = oxygen - containing heterocyclic ring

50

R = NHR₁; R₁ = alkyl, alkoxy

51

R_1 = alkyl; R_2 = alkoxy

Scheme 21

best correlation was found with the total interaction energy (Eq. 30), in which the steric component seems to be less important (Eq. 31) than the electronic component (Eq. 32). Thus only electronic effect was shown to control the activity [111].

$$pI_{50} = 2.882 - 0.025 E_{total}$$
$$r^2 = 0.598, \ F = 22.297, \ s = 0.248 \tag{30}$$

$$pI_{50} = 4.836 - 0.021 E_{steric}$$
$$r^2 = 0.248, \ F = 4.166, \ s = 0.346 \tag{31}$$

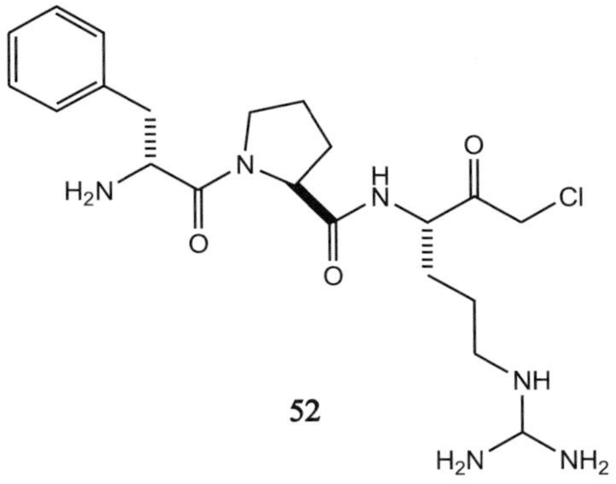

52

PPACK - D-Phe -Pro-Arg

Scheme 22

$$pI_{50} = 5.870 - 0.027E_{elec}$$
$$r^2 = 0.598, \ F = 8.492, \ s = 0.309 \tag{32}$$

$$pI_{50} = 3.220 - 0.021E_{steric} - 0.028E_{elec}$$
$$r^2 = 0.593, \ F = 9.487, \ s = 0.259 \tag{33}$$

$N\alpha$-(Arylsulfonyl)-L-Arginine Amide Derivatives

Equation 34 was derived [112] for a series of $N\alpha$-(arylsulfonyl)-L-arginine amide derivatives (**53**) [105]. The inhibitory activity was measured in the assay system of clotting activity.

$$\log(1/IC_{50}) = 0.836(\pm0.359)C \log P + 4.016(\pm0.741)$$
$$n = 6, \ r = 0.955, \ r^2 = 0.912, \ q^2 = 0.709,$$
$$s = 0.237, \ F_{1,4} = 41.84, \alpha = 0.01 \tag{34}$$

$$\log(1/IC_{50}) = 6.362(\pm4.023)(CMR) - 0.243(\pm0.155)(CMR)^2 +$$
$$1.302(\pm0.329)IdN - 37.094(\pm26.085)$$
$$n = 56, \ r = 0.770, \ r^2 = 0.593, \ q^2 = 0.525, \ s = 0.596, \ F_{3,52} = 10.05,$$
$$\alpha = 0.01 \ (CMR)_o = 13.068(\pm0.452), \ 12.659 - 13.563 \tag{35}$$

For in vitro data, Eq. 35 was derived. Parabolic dependence on CMR (molar refractivity) provided an optimum value for molar refractivity. MR represents the molar refractivity of the molecules. IDN is an indicator variable, which takes the value of 1 for the compounds that contain an heterocyclic ring with nitrogen. It seems that this type of ring is the best substituent compared to the

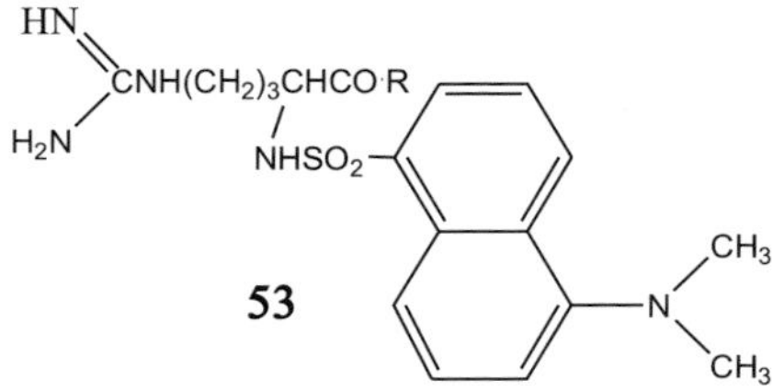

53

Scheme 23

other substituents for in vitro action. However, Eq. 35 [112] is not very significant, but a better correlation was obtained when $C \log P$ was used (Eq. 36). In Eq. 36,

$$\log(1/IC_{50}) = 0.776(\pm0.305)C \log P + 0.914(\pm0.398)\text{I-2}$$
$$- 2.105(\pm1.165)\text{MR-RR}_1 + 10.777(\pm3.812)$$
$$n = 27, \ r = 0.836, \ r^2 = 0.699, \ q^2 = 0.483,$$
$$s = 0.498, \ F_{3,23} = 13.97, \ \alpha = 0.01 \tag{36}$$

I-2 is an indicator variable used for compounds where $NHSO_2$ is attached to the 2-position of the naphthyl ring and MR-RR1 expresses the molar refractivity of the heterocyclic ring and its R_1 substituent. While the negative MR term suggests that a very bulky molecule would not be advantageous, the positive coefficient I-2 suggests that an $NHSO_2$ group attached to position will be conducive to the activity.

For another series of compounds belonging to **54**, Eq. 37 was obtained which exhibited the dependence of activity on $\log P$ only.

$$\log(1/IC_{50}) = 1.005(\pm0.349)C \log P + 3.368(\pm0.856)$$
$$n = 11, \ r = 0.908, \ r^2 = 0.824, \ q^2 = 0.744,$$
$$s = 0.375, \ F_{1,9} = 42.33, \ \alpha = 0.01 \tag{37}$$

For a combine of all the compounds for which Eqs. 34–37 were obtained, Eq. 38 was obtained [112]. In Eq. 38, I-1 is an indicator variable for the compounds in which $NHSO_2$ group is attached to 1-position of the naphthyl ring and IdN is an indicator variable for compounds containing heterocyclic ring with nitrogen. The negative coefficient of I-1 exhibited that if $NHSO_2$ group is attached to 1-position of naphthyl ring it would be detrimental to the activity.

$$\log(1/IC_{50}) = 0.255(\pm0.165)C \log P - 1.005(\pm0.308)\text{I-1} +$$
$$1.065(\pm0.273)\text{IdN} + 0.455(\pm0.257)\text{MR-X} + 2.468(\pm1.214)$$
$$n = 105, \ r = 0.746, \ r^2 = 0.557, \ q^2 = 0.511,$$
$$s = 0.669, \ F_{4,100} = 9.30, \ \alpha = 0.01 \tag{38}$$

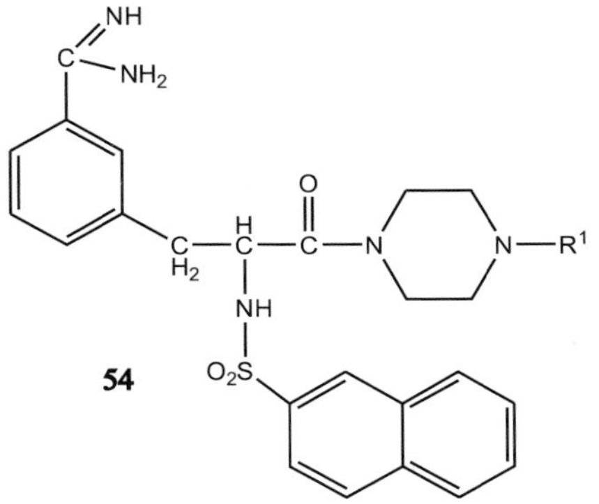

Scheme 24

Carboxyl-Containing Amide Derivatives of $N\alpha$-Substituted L-Arginine Derivatives

For another group of $N\alpha$-substituted L-arginine derivatives (**47**), Eq. 9 was formulated using the molar refractivity of the whole molecule [112], the CMR was found to have high correlation with $C \log P$ ($r = 0.837$). Hence, another equation was derived (Eq. 40), where

$$\log(1/IC_{50}) = 0.802(\pm 0.212)CMR - 4.455(\pm 2.883)$$
$$n = 10, \ r = 0.951, \ r^2 = 0.905, \ q^2 = 0.719,$$
$$s = 0.236, \ F_{1,8} = 75.98, \ \alpha = 0.01 \tag{39}$$

$$\log(1/IC_{50}) = -2.034(\pm 0.243)Id\text{-}O2$$
$$+ 0.904(\pm 0.365)MR\text{-}R_1 + 3.586(\pm 1.234)$$
$$n = 10, \ r = 0.994, \ r^2 = 0.989, \ q^2 = 0.972,$$
$$s = 0.119, \ F_{2,7} = 34.3, \ \alpha = 0.01 \tag{40}$$

$MR\text{-}R_1$ exhibited the molar refractivity effect of only R_1-substituent. In Eq. 40, $Id\text{-}O2$ is an indicator variable that was assigned a value of 1 for a molecule containing a heterocycling ring with two oxygen atoms. Obviously such a heterocyclic ring did not seem to be of advantageous.

Aminoiminomethyl Piperidines

An extensive search for new thrombin inhibitors led to a variety of low molecular weight thrombin inhibitors. Screening of small basic molecule for binding in the recognition pocket of thrombin led to selective thrombin inhibitors – derivatives of D-phenylalanyl-3-[(aminomethyl)amidino]piperidine (**55–58**) [113]. For this series of compounds, the correlation obtained

55

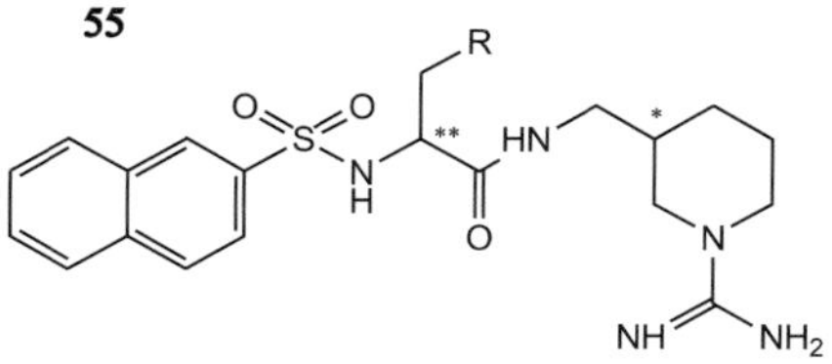

56

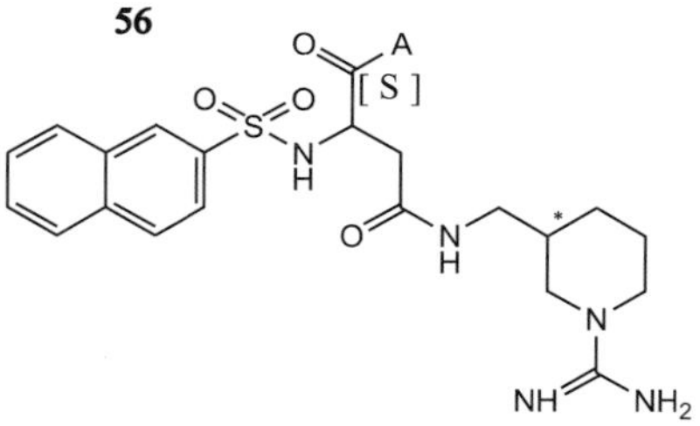

57

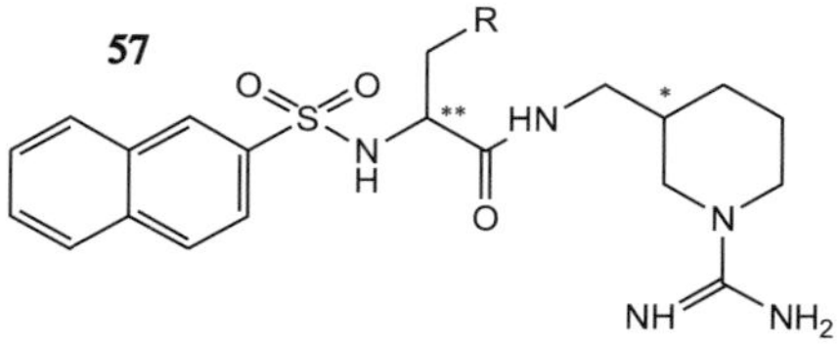

58

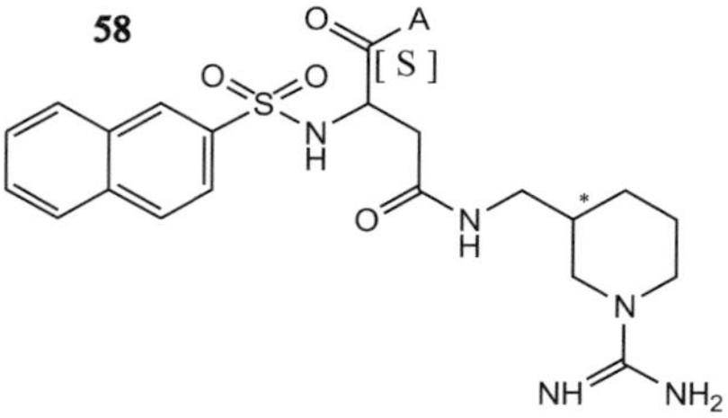

Scheme 25

was as shown by Eq. 41. In this equation, the indicator variable I^*S

$$\log(1/K_i) = 3.222(\pm1.074)(C \log P) - 0.901(\pm0.474)(C \log P)^2$$
$$+ 0.771(\pm0.568)I^*S + 7.436(\pm0.665)$$
$$n = 15, \ r = 0.950, \ r^2 = 0.902, \ q^2 = 0.727, \ s = 0.471, \ F_{3,11} = 2.972,$$
$$\alpha = 0.1 \ C \log P_o = 1.787(\pm0.582), 1.483 \text{ to } 2.467 \tag{41}$$

is assigned to take a value of 1 when the configuration is S at the carbon atom marked with an asterisk [*]. It seems that the S configuration is the best in relation to the in vitro activity. For a relatively larger group, the correlation obtained was as [112]:

$$\log(1/K_i) = 3.020(\pm0.944)(C \log P) - 0.765(\pm0.287)(C \log P)^2 +$$
$$0.878(\pm0.622)I^*S + 7.346(\pm0.848)$$
$$n = 24, \ r = 0.874, \ r^2 = 0.765, \ q^2 = 0.688, \ s = 0.657, \ F_{3,20} = 2.89, \ \alpha = 0.1$$
$$\text{For } C \log P_o = 1.973(\pm0.329), 1.719 \text{ to } 2.377 \tag{42}$$

A parabolic approach correlating the biological response with the overall lipophilicity is again the best correlation. The indicator I^*S continues to be important.

Dibasic Benzo[b]Thiophene Derivatives

A number of recent approaches to the discovery of orally active thrombin inhibitors have their origins in the tripeptide sequence D – Phe – Pro – Arg.

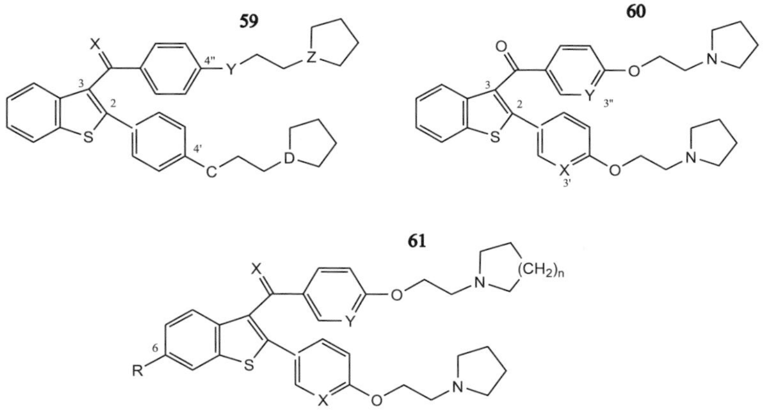

Scheme 26

Incorporation of a C-terminal electrophile in the form of an aldehyde was first accomplished by Bajusz [114] leading to the transition state analogue D – MePhe – Pro – Arg – H with the aid of X-ray crystallography and molecular modelling. A series of 2,3-disubstituted benzothiophene derivatives as more potent and selective inhibitors was synthesized and screened (59–61) [114], for which Eq. 43 was derived [114]. In this equation, indicator parameter IdXO was assigned a value of 1 for the presence of a carbonyl group at the position X of the compounds and MR-R is the molar refractivity for R group. The fact that IdXO has a negative sign means that conversion of C-3 ketone to the corresponding olefin results in higher activity. Equation 36 indicated that hydrophobicity is critical for the activity where as the negative sign with MR-R suggested steric hindrance. Possibly steric factors or conformational effect due to the hybridization state at the carbon atom are important for binding.

$$\log K_{ass} = -0.872(\pm 0.411) C \log P - 1.216(\pm 0.503) IdXO -$$
$$1.649(\pm 0.825) MR\text{-}R + 14.123(\pm 3.586)$$
$$n = 15, \ r = 0.912, \ r^2 = 0.831, \ q^2 = 0.673, \ r^2 = 0.831, \ s = 0.415,$$
$$F_{3,11} = 7.035, \ \alpha = 0.01 \tag{43}$$

Pyridinium-Sulfanilylguanidine Moieties

Recently some non-basic S1 anchoring groups have been incorporated in the molecules of some thrombin inhibitors [54, 115, 116]. The presence of guanidine-/benzamidino moieties in such compounds was critical. Three series of derivatives using benzamidine and sulfaguanidine as lead molecules have been synthesized and tested (62–64) against human thrombin. Forty

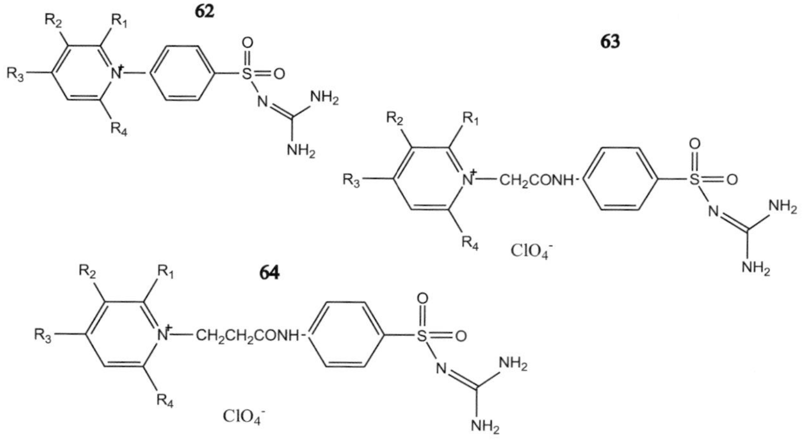

Scheme 27

eight different compounds were categorized in three groups, depending on their general structure. Equation 44 was obtained [112] for **62**, which showed that L-R$_3$, indicating the length of the substituent R$_3$, is an important factor after the steric parameter E_s for substituent R$_4$.

$$\log(1/K_i) = 0.080(\pm0.016)\text{L-R3} + 0.313(\pm0.109)E_{s\text{-R4}} + 7.296(\pm0.166) \ (37)$$
$$n = 14, \ r = 0.971, \ r^2 = 0.942, \ q^2 = 0.885,$$
$$s = 0.046, \ F_{2,11} = 39.42, \ \alpha = 0.01 \tag{44}$$

The $E_{s\text{-R4}}$ term indicates an unfavorable steric effect for R$_4$ substituents in the 6-position of the pyridyl ring pointing out the detrimental effect of the o-group. Electronic factors were not found to play any definite role.

For the second group with the general structure **63**, Eq. 45 was derived [112]:

$$\log(1/K_i) = -0.143(\pm0.113)\text{MR-R}_1 + 0.117(\pm0.070)\text{MR-R}_3 +$$
$$0.094(\pm0.055)\text{L-R}_4 + 6.943(\pm0.196)$$
$$n = 14, \ r = 0.933, \ r^2 = 0.871, \ q^2 = 0.701,$$
$$s = 0.095, \ F_{3,10} = 7.91, \ \alpha = 0.01 \tag{45}$$

Equation 45 contains only molar refractivity terms for R$_1$ and R$_3$ substituents and the Verloop Sterimol parameters L for the length of substituent R$_4$. The negative sign with MR-R$_1$ suggests a steric hindrance directly or through a conformational change. MR-R$_2$ is the most significant parameter.

For the third group **64**, Eq. 46 was derived [112].

$\log(1/K_i) = -0.149(\pm0.064)\text{L-R1} + 0.102(\pm0.051)\text{L-R3} +$
$0.106(\pm0.057)\text{L-R4} + 7.002(\pm0.260)$ *(39)*
$n = 14,\ r = 0.941,\ r^2 = 0.885,\ q^2 = 0.763,$
$s = 0.092,\ F_{3,10} = 19.88,\ \alpha = 0.01$ (46)

This equation indicates that the activity increases with the decrease in the length of the R_1 group and with the increase in the length of R_3 and R_4 substituents. A major concern with Eq. 46 is, however, that the L-R_1, L-R_3, L-R_4 parameters are highly collinear with $C \log P$ ($C \log P$/L-R_1, $r = 0.778$; $C \log P$/L-R_3, $r = 0.783$; $C \log P$/L-R_4, $r = 0.901$), indicating that at the site of action compounds might be making hydrophobic contacts.

For a combine of all these 3 sets, Eq. 47 was formulated [112].

$\log(1/K_i) = 0.116(\pm0.018)\text{CMR} - 0.162(\pm0.036)\text{L}_{-\text{R1}} +$
$0.208(\pm0.124)E_{\text{s-R3}} + 6.748(\pm0.271)$
$n = 45,\ r = 0.900,\ r^2 = 0.810,\ q^2 = 0.774,$
$s = 0.099,\ F_{3,41} = 11.22,\ \alpha = 0.01$ (47)

From this linear correlation, some important conclusions could be obtained. The biological activity of these compounds is related to the molar refractivity of the whole molecule, the Sterimol parameter L for the group R_1 and the steric parameter E_s for the group R_3. CMR is the most important parameter accounting for the volume and the steric effect. The Sterimol parameter L-R_1 brings out a negative effect of the R_1 group. Recalling that E_s values are negative, the positive term with $E_{\text{s-R3}}$ implies a negative steric effect.

Aminopyridyl Moieties

Feng et al. [86] reported novel aminoaryl P1 substituents (**26**) varying in size and basicity. They anticipated that developing potent inhibitors with a neutral P1 would require the use of lipophilic aminoacids and benzenesulfonyl group on the *N*-terminus as lipophilic residues at P3 to add more hydrophobic binding energy to compensate losses at P1. The correlation obtained was Eq. 48 [112].

$\log(1/K_i) = -1.573(\pm0.661)\text{ISN} + 0.813(\pm0.322)\text{MR-R2} + 4.127(\pm1.442)$
$n = 20,\ r = 0.847,\ r^2 = 0.717,\ q^2 = 0.571,$
$s = 0.588,\ F_{2,17} = 14.20,\ \alpha = 0.01$ (48)

Equation 48 is not very significant. However, it indicates a positive effect of molar refractivity of R_2 substituent and a negative effect of a thiazolyl ring in the molecule through the indicator variable ISN, which has a value of 1 for the presence of a thiazolyl ring. The R_2 substituent may be involved in some dispersion interaction.

D-Amino Acid Series

Amidinophenyl alanine in which the benzamidine moiety imitates the guanidinoalkyl side chain of arginine was used as a key structural element for the development of thrombin inhibitors. Na-tosylated 3-amidinophenylalanyl piperidide (3-TAPA) [117] is another promising structure. The thrombin 3-TAPA complex indicated more space available for substituents at both the toluene ring and the piperidine moiety. Stürzebecher et al. [54, 89] exchanged the piperazine moiety by a piperazine ring to facilitate the incorporation of different substituents on the second nitrogen of the piperazine moiety [52, 118]. These new derivatives (65, 66) were used for the derivation of Eq. 49 [112].

$$\log(1/K_i) = -1.090(\pm 0.649)\text{ICO} + 0.535(\pm 0.253)\text{MR-R2} -$$
$$0.461(\pm 0.285)\text{B5-R2} + 5.956(\pm 1.421) \quad (42)$$
$$n = 24, \ r = 0.851, \ r^2 = 0.724, \ q^2 = 0.584,$$
$$s = 0.674, \ F_{3,20} = 4.08, \ \alpha = 0.05 \tag{49}$$

In Eq. 49, the B5-R_2 term exhibits a negative steric effect for R_2 substituents. The indicator variable ICO used for compounds where a carbonyl group is present in the R_1 substituent seems to decrease the inhibition constant, whereas the bulk of the substituent R_2, described by the molar refractivity, is shown to be conducive. The R_2 substituents might contact some polar space.

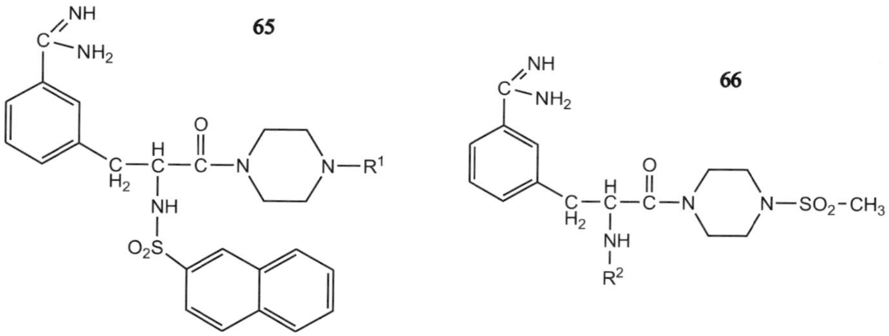

Scheme 28

1-Benzyl-3-(5-hydroxymethyl-2-furyl) Indazole

For a series of 1-Benzyl-3-(5-hydroxymethyl-2-furyl)indazole analogues (67–69) [54] acting as antiplatelet agents, the following equation was derived [112].

$$\log(1/IC_{50}) = 0.203(\pm 0.121)C \log P$$
$$+ 0.197(\pm 0.183)\text{MR-R} + 2.850(\pm 0.539)$$
$$n = 11, \; r = 0.908, \; r^2 = 0.825, \; q^2 = 0.211,$$
$$s = 0.110, \; F_{2,8} = 3,125, \; \alpha = 0.1 \tag{50}$$

This equation shows the effect of the hydrophobicity. Molecules may be involved in hydrophobic interactions. MR-R is the molar refractivity of the substituent R. The positive coefficient of this term indicates that R substituents might contact the polar space.

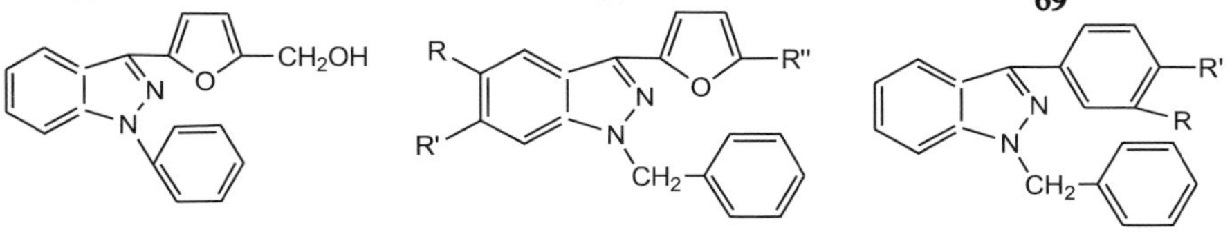

Scheme 29

3.3
Conclusion

It is known from X-ray crystallographic studies that Tyr-Pro-Trp loop, unique to thrombin, helps to form two hydrophobic moieties in which two hydrophobic moieties of the inhibitors are bound. The hydrophobic P-pocket, proximal to the catalytic center, corresponds to the S2 subsite and binds the inhibitor through the basic group (piperidide, piperizide, etc.) The P-pocket distal to the catalytic center is also hydrophobic and binds the aryl moieties of the inhibitors. The basic amino acid chain is placed in the P1 position with its carbonyl moiety binding in the oxyanion hole.

As in Sect. 2.2, the overall essence of the QSAR studies on thrombin inhibitors is the high dependence of the binding of substrates or inhibitors on the molar refractivity. Generally, there was high collinearity between MR and/or CMR and $C \log P$. This led the researchers to assume that the binding pocket around the active site in thrombin is not typically hydrophobic.

The hydrophobicity plays only a partial role in the QSAR of thrombin inhibitors. It appears that some of the molecules must be interacting with a hydrophobic portion of the enzyme. The negative hydrophobic effect is striking. An attempt to explain this result was made in Sect. 3.3. The active site gorge is a deep hydrophobic cavity at the bottom of which the catalytic site is hosted. It might be that the hydrophobic contacts of the inhibitor molecules with the walls of the cavity prevent or slow down the access of the cationic moiety to its binding site. If this occurs and the ionic bond is not established, the overall interaction might be weaker, resulting into lower inhibitory activity.

Probably substituents on the aryl part do not have contact with the enzyme surface but remain outside the gorge exposed to the solvent. As a consequence, given that changes in the log P values are mostly due to the different aryl substituents, if they are all or in part exposed to the solvent, their desolvation is not required and lower log P can be more favourable.In the cases where no role for lipophilicity is found, there is a considerable collinearity problem, an overlap between substituent bulk and hydrophobicity (a most common problem through out drug development). This overlap makes it difficult to separate steric and hydrophobic effects.

In most of the equations the CMR and MR terms are present.

For some structural features indicator variables have been used as a device to account for the effect of a specific feature that cannot be accounted for by any specific physicochemical parameters. Electronic parameters indicative of dipole–dipole or charge–dipole interactions, charge transfer phenomena, or hydrogen bond formation, are not found to govern the thrombin inhibition. No significant correlations were obtained with any electronic term although the SO_2 group might contribute to an electronic effect with the neighboring groups. However, there was no sufficient variation (alkyl groups only) in substituents and thus no role for this property could be seen.

The QSARs do bring out flexibility of the ligand–enzyme interaction. Steric effects are modelled by the variables MR, E_s, etc., which show that contact between enzyme and ligand does not destroy activity completely but simply reduces it in proportion to the substituent size. Probably both ligand and enzyme are gradually moved from their preferred positions.

4
General Conclusions

The success of compounds as new anticoagulants will depend on future progress in the development of dual thrombin and Factor Xa inhibitors, orally active with less side effects. The examination of the quantitative models for the structure activity relationships of several series of FXa and thrombin inhibitors with heterocyclic moieties allowed us to point the physicochemical properties involved in the inhibitory activity. The conclusions can be summarized as follows:

a) Hydrophobicity plays a critical role. Hydrophobic interactions might favor the binding of inhibitors to the cavity but in some cases they can also be an obstacle to reach the deep anchoring site.
b) Steric factors are also significant but collinearity with lipophilicity does not allow one to draw a clear picture.

Taking a closer look at the equations, common features within each class could be revealed and at least partially rationalized.

References

1. Babine RE, Bender SL (1997) Chem Rev 97:1359
2. Ripka AS, Rich DH (1998) Curr Opin Chem Biol 2:441
3. Shaw E (1990) Adv Enzymol Relat Areas Mol Biol 63:271
4. Leung D, Abbenante G, Fairlie DP (2000) J Med Chem 43:305
5. Schechter I, Berger A (1967) Biochem Biophys Res Commun 27:157
6. Hedstrom L (2002) Chem Rev 102:4501
7. Stubbs MT, Bode W (1994) Curr Opin Struct Biol 4:823
8. Claeson G (1994) Blood Coag Fibrinolysis 5:411
9. Prager NA, Abenschein DR, McKenzie CR, Einsanberg PR (1995) Circulation 92:962
10. Furie B, Furie BC (1988) Cell 53:505
11. Hammer RH (1995) Anticoagulants, Coagulants and Plasma Extenders. In: Foye WO, Lemke TL, Williams DA (eds) Principles of Medicinal Chemistry. Waverly BI PVt. Ltd., New Delhi, p 388
12. Davie EW, Fujikawa K, Kisiel W (1991) Biochemistry 30:10363
13. Vaz RJ, McLean LR, Pelton JT (1998) J Comp Aid Mol Des 12:99
14. Leadley RJ Jr (2001) Curr Top Med Chem 1:151
15. Turk D, Stürzebecher J, Bode W (1991) FEBS 287:133
16. Narasimhan LS, Rubin JR, Holland DR, Plummer JS, Rapundalo ST, Edmunds JE, St-Denis Y, Siddiqui MA, Humblet C (2000) J Med Chem 43:361
17. Linusson A, Gottfries J, Olsson T, Örnskov E, Folestad S, Norden B, Wold S (2001) J Med Chem 44:3424
18. Bode W, Mayr I, Baumann U, Huber R, Stone SR, Hofskeenge J (1989) EMBO J 8:3467
19. Padmanahan K, Padmanahan KP, Tulinsky A, Park CH, Bode W, Huber R, Blakenship DT, Cardin AD, Kisiel W (1993) J Mol Biol 232:947
20. Brandstetter H, Kuhne A, Bode W, Huber R, Von der Saal, Wirthensohn K, Engh RA (1996) J Biol Chem 271:29988
21. Maignan S, Guilloteau JP, Pouzieux S, Choi-Sledeski YM, Becker MR, Klein SI, Ewing WR, Pauls HW, Spada AP, Mikol V (2000) J Med Chem 43:3226
22. Dougherty DA (1996) Science 271:163
23. Gallivan JP, Dougherty DA (1999) Proc Natl Acad Sci USA 96:9459
24. Matter H, Defossa E, Heinelt U, Blohm PM, Schneider D, Müller A, Herok S, Schreuder H, Liesum A, Brachvogel V, Lonze P, Walser A, Al-Obeidi F, Wildgoose P (2002) J Med Chem 45:2749
25. Yee YK, Tebbe AI, Linebarger JH, Beight DW, Craft TJ, Gifford-Moore D, Goodson T, Herron DK, KlimKowski VJ, Kyle JA, Sawyer JS, Smith GF, Tinsley JM, Towner RD, Weir L, Wiley MR (2000) J Med Chem 43:873

26. Buckman BO, Mohan R, Koovakkat S, Liang A, Trinh L, Morrissey MM (1998) Bioorg Med Chem Lett 8:2235
27. Maignan S, Mikol V (2001) Curr Top Med Chem 1:161
28. Ewing WR, Pauls HW, Spada AP (1999) Drugs Future 24:771
29. Choi-Sledeski YM, MCGarry DG, Green DM, Mason HJ, Rewcker MR, Davis RS (1999) J Med Chem 42:3572
30. Shaw KJ, Guilford WJ, Dallas JL, Koovakkat SK, Mc Carrick MA, Liang A, Light DR, Morrisey MM (1998) J Med Chem 41:3551
31. Daura X, Haaksma E, Van Gunsteren WF (2000) J Comput Aided Mol Des 14:507
32. Koshland DE (1958) Proc Natl Acad Sci USA 44:98
33. Jorgensen WL (1991) Science 254:951
34. Morton A, Matthews BW (1995) Biochemistry 34:8576
35. Andrews PR, Craik DJ, Martin JL (1984) J Med Chem 27:1648
36. Cohen NC (1996) Guidebook on Molecular Modeling in Drug Design. Academic, New York
37. Cooper DN, Millar DS, Wacey A, Pemberto S, Tuddenham EGD (1997) Thromb Haemost 78:161
38. Lapatto R, Krengel U, Schreuder HA, Arkema A, De Boer B, Kalk KH, Hol WG, Grootenhuis PDJ, Mulders JW, Dijkema R, Theunissen HJ, Dijkstra BW (1997) EMBO J 16:5151
39. Al-Obeidi F, Ostrem JA (1999) Expert Opin Ther Pat 74:635
40. Morishima Y, Tanabe K, Terada Y, Hara T, Kunitada S (1997) Thromb Haemost 78:1366
41. Wong PC, Crain EJ, Nguan O, Watson CA, Racanelli A (1996) Thromb Res 83:117
42. Lamba D, Bauer M, Huber R, Fischer S, Rudolph R, Kohert U, Bode W (1996) J Mol Biol 258:117
43. Spraggon G, Phillips C, Nowak UK, Ponting CP, Saunders D, Dobson CM, Stuart DI, Jones EY (1995) Structure 3:681
44. Katz BA, Mackman R, Luong C, Radika K, Martelli A, Sprengeler PA, Wang J, Chan H, Wong L (2000) Chem Biol 7:299
45. Rai R, Sprengeler PA, Elrod KC, Young WB (2000) Curr Med Chem 1:35
46. Katakura S, Nagahara T, Hara T, Iwamoto MA (1993) Biochem Biophys Res Commun 197:965
47. Katakura S, Nagahara T, Hara T, Kumitada S, Iwamoto MA (1995) Eur J Med Chem 30:387
48. Lin Z, Johnson ME (1995) FEBS Lett 370:1
49. Stubbs MT, Huber R, Bode W (1995) FEBS Lett 375:103
50. Böhm M, Stürzebecher J, Klebe G (1999) J Med Chem 42:458
51. Stürzebecher J, Vieweg H, Wilkström P (1991) WO 92/08709
52. Stürzebecher J, Vieweg H, Wilkström P (1993) WO 94/18185
53. Stürzebecher J, Prasa D, Vieweg H, Wilkström P (1995) J Enzyme Inhib 9:87
54. Stürzebecher J, Prasa D, Hauptmann J, Vieweg H, Wilkström P (1997) J Med Chem 40:3091
55. Labute P (2000) J Mol Graph Model 18:464
56. Murcia M, Ortiz AR (2004) J Med Chem 47:805
57. Matter H, Will DW, Nazare M, Schreuder H, Lax V, Wehner V (2005) J Med Chem 48:3290
58. Maignan S, Guilloteau JP, Choi-Sledeski YM, Becker MR, Ewing WR, Pauls HW, Spada AP, Mikol V (2003) J Med Chem 46:685

59. Adler M, Kochanny MJ, Ye B, Rumnenik G, Light DR, Biancalana S, Whitlow M (2002) Biochemistry 41:15514
60. Kamata K, Kawamoto H, Honma T, Iwama T, Kim SH (1998) Proc Natl Acad Sci USA 95:6630
61. Nar H, Bauer M, Schmid A, Stassen JM, Wiennen W, Priepke HWM, Kaufmann IK, Rise UJ, Hauel NH (2001) Structure 9:29
62. Adler M, Davey DD, Philips GB, Kim SH, Jancarik J, Rumennik G, Light DR, Whitlow R (2000) Biochemistry 39:12534
63. Cramer RD, Patterson DE, Bunce JE (1988) J Am Chem Soc 110:5959
64. Clark M, Cramer RD, Jones DM, Patterson DE, Simeroth PE (1990) Tetrahedron Comput Methodol 3:47
65. Kubinyi H (ed) (1993) 3D-QSAR in Drug Design: Theory, Methods and Applications. ESCOM, Leiden, The Netherlands
66. Klebe G, Abraham U, Mietzner T (1994) J Med Chem 37:4130
67. Roy K, De AU, Sengupta C (2002) Drug Des Disc 18:23
68. Roy K, De AU, Sengupta C (2002) Drug Des Disc 18:33
69. Bhongade BA, Gouripur VV, Gadad AK (2005) Bioorg Med Chem 13:2773
70. Taha MO, Qandil AA, Zaki DD, AlDamen MA (2005) Eur J Med Chem 40:701
71. Phillips G, Guilford WJ, Buckman BO, Davey DD, Eagen KA, Koovakkat S, Liang A, McCarrick M, Mohan R, Ng HP, Pinkerton M, Subramanyam B, Ho E, Trinh L, Whitlow M, Wu S, Xu W, Morrisey MM (2002) J Med Chem 45:2484
72. Kontogiorgis CA, Hadjipavlou-Litina D (2004) Med Res Rev 24:687
73. Wu S, Guilford WJ, Chou Y-L, Griedel BD, Liang A, Sakata S, Shaw KJ, Trinh l, Xu W, Zhao Z, Morrisey MM (2002) Bioorg Med Chem Lett 12:1307
74. Quan ML, LiauwAY, Ellis CD, Pruit JR, Carini DJ, Bostrom LL, Huang P, Harrison K, Knabb RM, Thoolen MJ,Wong PC, Wexler RR (1999) J Med Chem 42:2752
75. Quan ML, Ellis CD, Liauw AY, Alexander RS, Knabb RM, Lam GN, Wright MM, Wong PC, Wexler RR (1999) J Med Chem 42:2760
76. Pinto DJP, Orwat MJ, Wang S, Fevig JM, Quan ML, Amparo E, Cacciola J, Rossi KA, Alexander RS, Smallwodd AM, Luettgen JM, Liang L, Aungst BJ, Wright BR, Knabb RM, Wong PC, Wexler PP, Lam PYS (2001) J Med Chem 44:566
77. Smith GF, Shuman RT, Craft TJ, Gifford DS, Kurz KD, Jones ND, Chirgadze N, Hermann RB, Coffman WJ, Sandusky GE, Roberts E, Jackson CVA(1996) Semin Thromb Hemostasis 22:173
78. Hirsh J (1991) New Engl J Med 324:1865
79. Pruitt JR, Pinto DJ, Quan ML, Estrella MJ, Bostrom LL, Knabb RM, Wong PC, Wexler RR (2000) Bioorg Med Chem Lett 10:685
80. Stein PD, Grandison D, Hua TA (1994) Postgrad Med J 70 Suppl 1:S72
81. Fevig JM, Pinto DJ, Han Qi, Quan ML, Pruitt JR, Jacobson IC, Galemmo RA Jr, Wang S, Orwat MJ, Bostrom LL, Knabb RM, Wong PC, Lam PYS, Wexler RR (2001) Bioorg Med Chem Lett 11:641
82. Masters JJ, Franciskovich JB, Tinsley JM, Cambell JB, Craft TJ, Froelich LL, Gifford-Moore DS, Herron DK, Klimkowski VJ, Kurz KD, Metz JT, Ratz AM, Shuman RT, Smith GF, Towner RD, Wiley MR, Wilson A, Yee YK (2000) J Med Chem 43:2087
83. Hirsh J, Poller L (1994) Arch Int Med 154:282
84. Kurczierz R, Grams F, Leinert H, Marzenell K, Engh RA, Von der Saal RA (1998) J Med Chem 41:4983
85. Brandstetter H, Kühne A, Bode W, Huber R, Von der Saal W, Wirthensohn K, Engh RA (1996) J Biol Chem 271:29988.

86. Feng DM, Gardell SJ, Lewis SD, Bock MG, Chen Z, Freidinger RM, Naylor-Olsen AM, Ramjit HG, Woltmann R, Baskin EP, Lynch JL, Lucas R, Shafer JA, Dancheck BK, Chen W, Mao S, Krueger JL, Hare TR, Mulichak AM, Vacca JP (1997) J Med Chem 40:3726

87. Gong Y, Pauls HW, Spada AP, Chekaj M, Liang G, Chu V, Colussi DJ, Brown KD, Gao J (2000) Bioorg Med Chem Lett 10:217

88. Dudley DA, Bucker AM, Chi L, Cody WL, Holland DR, Ingasiak DP, Janiczek-Dolphin N, McClanahan TB, Mertz TE, Narasimhan LS, Rapundalo ST, Trautschold JA, Van Huis CA, Edmunds JJ (2000) J Med Chem 43:4063

89. Zhang P, Zuckett JF, Woolfrey J, Tran K, Huang B, Wong P, Sinha U, Park G, Reed A, Malinowski J, Hollenbach S, Scarborough RM, Zhu B-Y (2002) Bioorg Med Chem Lett 12:1657.

90. Hirayama F, Koshio H, Ishihara T, Watanuki S, Hachiya S, Kaizawa H, Kuramochi T, Katayama N, Kurihar H, Taniuchi Y, Sato K, Sakai-Moritani Y, Kaku S, Kawasaki T, Matsumoto Y, Sakumoto S, Tsukamoto S (2002) Bioorg Med Chem 10:2597

91. Quan ML, Ellis CD, He MY, Liauw AY, Woerner FJ, Alexander RS, Knabb RM, Lam PYS, Luettgen JM, Wong PC, Wright MR, Wexler RR (2003) Bioorg Med Chem Lett 13:369

92. Kurihara H, Taniuchi Y, Sato K, Sakai-Moritani Y, Kaku S, Kawasaki T, Matsumoto Y, Sakamoto S, Tsukamoto S (2002) Bioorg Med Chem 10:2597

93. Su T, WuY, Doughan B, Kane-Maguire K, Morlowe CK, Kanter JP, Woolfrey J, Huang B, Wong P, Sinha U, Park G, Malinowski J, Hollenbach S, Scarborough RM, Zhu B-Y (2001) Bioorg Med Lett 11:2279

94. Hansch C, Leo AJ (1995) Exploring QSAR: Fundamentals and applications in chemistry and biology. American Chemical Society, Washington, DC

95. Hansch C, Caldwell J (1991) J Comp Aid Mol Des 5:441

96. Yoshimoto M, Hansch C (1976) J Org Chem 41:2269

97. Guinto ER, Vindigni A, Ayala YM, Dang QD, Di Cera E (1995) Proc Natl Acad Sci USA 92:1185

98. Goodwin CA, Deadman JJ, Le Bonniec BF, Elgendy S, Kakkar VV, Scully MF (1996) Biochem J 315:77

99. Bursi R, Grootenhuis PD (1999) J Comput Aided Mol Design 13:221

100. Greenidge AP, Merette SAM, Beck R, Dodson G, Goodwin CA, Scully M, Spencer J, Weiser J, Deadman J (2003) J Med Chem 46:1293

101. Merops database, http://www.MEROPS.Sanger.Ac.Uk.code S01.217

102. Dixit A, Kashaw SK, Gaur S, Saxena AK (2004) Bioorg Med Chem 12:3591

103. Riester D, Wirsching F, Salinas G, Keller M, Gebinoga M, Kamphausen S, Merwirth C, Goetz R, Wiesenfeldt M, Stürzebecher J, Bode W, Friedrich R, Thürk M, Schwienhorst A (2005) Proc Natl Acad Sci (USA) 12:8597

104. Sanderson PEJ, Naylor-Olsen AM (19980 Curr Med Chem 5:289

105. Kikumoto R, Tamao Y, Ohkubo K, Tezuka T, Tonomura S (1980) J Med Chem 23:830

106. Kikumoto R, Tamao Y, Ohkubo K, Tezuka T, Tonomura S (1980) J Med Chem 23:1293

107. Gupta SP (1987) Chem Rev 87:1183

108. Gupta SP, Prahhankar YS, Handa A (1984) In: QSAR in design of Bioactive compounds. Prous, Barcelona, p 175

109. Fujita T (1973) J Med Chem 16:923

110. Shuman TR, Rothenberger RB, Cambell CS, Smith GF, Gillford-Moore DS, Gesellchen PD (1993) J Med Chem 36:314

111. Vacca JP (1998) Annual Report in Medicinal Chemistry 33:81

112. Kontogiorgis CA, Hadjipavlou-Litina D (2003) Curr Med Chem 10:1241
113. Hilpert K, Ackermann J, Banner DW, Gaxt A, Gubernator K, Hadvary P, Labler L, Müller K, Schmid G, Tschopp TP, Van der Waterbeemd (1994) J Med Chem 37:3889
114. Bajusz S, Szell E, Bagdy D, Bardas E, Howarth G, Dioszegi M, Fittler Z, Szabo G, Juhasz A, Tomori E (1998) Biochem 37:12094
115. Sall DJ, Bastian JA, Brigga SI, Buben JA, Chirgadze NY, Clawson DK, Denney ML (1997) Structure-Activity Relationships and Binding Orientation 40:3489
116. Salemme FR, Spurlino J, Bone R (1997) 5:319
117. Lewis SD, Ng AS, Lyle EA, Mellott MJ, Appledy SD, Brady SF, Stauffer KJ, Sisko JT, Mao SS, Veber DF (1995) Thromb Haemostasis 74:1107
118. Markwardt F, Wagner G, Stürzebecher J, Walshmann P (1980) Thromb Res 17:425

Top Heterocycl Chem (2006) 4: 55–84
DOI 10.1007/7081_030
© Springer-Verlag Berlin Heidelberg 2006
Published online: 23 March 2006

Structural Information and Drug–Enzyme Interaction of the Non-Nucleoside Reverse Transcriptase Inhibitors Based on Computational Chemistry Approaches

Supa Hannongbua

Department of Chemistry, Faculty of Science, Kasetsart University, 10900 Bangkok, Thailand
fscisph@ku.ac.th

Abstract Non-nucleoside reverse transcriptase inhibitors, such as nevirapine, TIBO, HEPT, and efavirenz, are very specific to HIV-1 reverse transcriptase and have few side effects compared to NRTIs. However, mutation of the HIV-1 virus has caused drug resistance to develop and reduce the efficiency of these inhibitors for drug therapy. As the association of NNRTIs with the binding pocket of the enzyme is essential for the inhibition process, this interaction is of high interest. Potential energy surface is used for conformational analysis of these flexible NNRTIs. The interactions between the inhibitor molecules and the surrounding amino acids are the key to determining the binding affinity. Much work has been done by using the 3D-QSAR method, with detailed molecular structural analysis of HIV-1 inhibitors by theoretical calculations, including enzyme–inhibitor interactions. Accurate calculations of detailed interactions are demonstrated by multilayered integration or ONIOM method to give some insight into the particular interaction of the NNRTIs with the residues in the binding site. In addition, molecular dynamics simulations, Monte-Carlo simulations and protein-based inhibitor design methods have been applied extensively on these inhibitors. This review is an attempt to combine various QSAR and CAMD methods on NNRTIs into a common prediction model to support the design of new, more potent inhibitors, in particular, active against mutant enzyme prior to synthesis.

Keywords Drug-enzyme interaction · Molecular simulations ·
Non-nucleoside reverse transcriptase inhibitor · ONIOM ·
Quantum chemical calculations

Abbreviations

α-APA	(+ / –)-2,6-Dichloro-α-[(2-acetyl-5-methylphenyl)amino]benzamide
2D, 3D	Two dimensional, three dimensional
2D-QSAR	Two-dimensional quantitative structure–activity relationships
3D-QSAR	Three-dimensional quantitative structure–activity relationships
8-Cl TIBO	8-Chlorotetrahydroimidazo(4,5,1-jk)(1,4)-benzodiazepin-2(1H)-thione
9-Cl TIBO	9-Chloro-4,5,6,7-tetrahydro-5-methyl-6-(3-methyl-2-butenyl)imidazo(4,5,1-jk)(1,4)benzodiazepine-2-(1H)-thione
AIDS	Acquired immunodeficiency syndrome
AMBER	AMBER force field
ANN	Artificial neural network
Arg	Arginine
Asn	Asparagine
Asp	Aspartic acid
B3LYP	Beck's three-parameter exchange functional with Lee–Yang–Parr correlation
CAMD	Computer-aided molecular design
CC	Computational chemistry
cc-PVDZ	Correlation-consistent polarized valence double zeta
cc-PVTZ	Correlation-consistent polarized valence triple zeta
CoMFA	Comparative molecular field analysis
CoMSIA	Comparative molecular similarity indexes analysis
Cys	Cysteine
DFT	Density functional theory
DMA	N6-dimethylallyl
DNA	Deoxyribonucleic acid
dNTP	Deoxynucleotide triphosphate
EC_{50}	Median effective concentration
FEP	Free energy perturbation

GIAO	Gauge-independent atomic orbital
Glu	Glutamic acid
Gly	Glycine
HEPT	1-[(2-Hydroxyethoxy)methyl]-6-(phenylthio)thymine
HF	Hartree–Fock theory
His	Histidine
HIV-1	Human immunodeficiency virus type 1
HQSAR	Hologram quantitative structure–activity relationships
IC_{50}	Median inhibitory concentration
IMOMO	Integrated molecular orbital molecular mechanics
Ile	Isoleucine
Leu	Leucine
Lys	Lysine
MC	Monte Carlo
MD	Molecular dynamics
MM3	Molecular mechanics, Allinger force field version 3
MP2	Second-order Møller–Plesset perturbation theory
MT-4	Human membrane type-4
NMR	Nuclear magnetic resonance
NNRTIs	Non-nucleoside reverse transcriptase inhibitors
NRTIs	Nucleoside reverse transcriptase inhibitors
ONIOM	Our own N-layered integrated molecular orbital and molecular mechanics method
p51	Peptide having molecular weight 51 kDA
p66	Peptide having molecular weight 66 kDA
PASs	Pyrrolyl aryl sulfones
PDBS	Pharmacophore-based database searching
PES	Potential energy surface
PETT	Phenylethylthiazolylthiourea
Phe	Phenylalanine
PM3	Modified neglect of diatomic overlap, parametric method number 3
Pro	Proline
QC	Quantum chemistry
QM/MM	Quantum mechanics and molecular mechanics
QM/QM	Quantum mechanics and quantum mechanics
QSAR	Quantitative structure–activity relationships
RCSB PDB	Research Collaboratory for Structural Bioinformatics (RCSB), the non-profit consortium that manages the Protein Data Bank (PDB)
RNA	Ribonucleic acid
RNaseH	Ribonuclease H
RT	Reverse transcriptase
SEAL	Steric and electrostatic alignment
Ser	Serine
SI	Selectivity index
SMD	Steered molecular dynamics
SMF	Substructural molecular fragments
SOM	Self-organizing map
TCs	Thiocarbamates
Thr	Threonine

TIBO   *S*-(+)-4,5,6,7-Tetrahydro-5-methyl-6-(3-methyl-2-butenyl)-imidazo[4,5,1-jk]
            [1,4]-benzodiazepin-2-(1H)-thione
Trp   Tryptophan
Tyr   Tyrosine
UFF   Universal force field
Val   Valine
WT   Wild type

1
Introduction

1.1
Biochemical and Pharmacological Description of the Problem

The human immunodeficiency virus type 1 (HIV-1) has been identified as the causative agent of acquired immunodeficiency syndrome (AIDS). The reverse transcriptase (RT) converts the single-stranded viral RNA into double-stranded DNA. The multifunctional enzyme has the following activities: RNA-dependent DNA polymerase, RNaseH, and DNA-dependent DNA polymerase. As this enzyme is essential for the replication of HIV-1, it is an important target for drug design efforts [1]. More than half of the currently approved anti-AIDS drugs target the RT enzyme [2–4]. Two main classes of inhibitors have been found to be active against HIV-1 RT. First are the non-nucleoside inhibitors (NNRTIs) which act allosterically, are highly specific, and show lower cellular toxicity compared to the second main class. These are the nucleoside inhibitors (NRTIs), which are competitive inhibitors of the dNTP substrate. The structures of some selected NNRTIs are given in Fig. 1, including *S*-(+)-4,5,6,7-tetrahydro-5-methyl-6-(3-methyl-2-butenyl)-imidazo [4,5,1-jk][1,4]-benzodiazepin-2(1H)-thione (tivirapine), 11-cyclopropyl-4-methyl-5,11-dihydro-6H-dipyrido[3,2-b:2',3'-e][1,4]diazepin-6-one (Nevirapine), 1-[3-[(1-methylethyl)amino]-2-pyridin-yl]-4-[[5-[(methylsulfonyl)amino]-1H-indol-2-yl]carbonyl]-piperazine (delavirdine), 1-[(2-hydroxyethoxy)methyl]-6-(phenylthio)thymine (HEPT), and most recently, (4S)-6-chloro-4-cyclopropylethynyl-4-trifluoromethyl-1,4-dihydro-benzo[d][1,3]oxazin-2-one (efavirenz). Although NNRTIs are a surprisingly diverse set of chemical entities [5], they all bind to a common site on the enzyme near, but distinct from, the polymerase active site. They inhibit the chemical step of polymerization, whereas NRTIs bind directly to the active site of RT and block HIV-1 replication by terminating DNA chain elongation [1]. Structural information about HIV-1 RT can be obtained from X-ray crystallography and NMR spectroscopy [6,7], and from these a structural element common to all NNRTIs, namely a butterfly-like shape, has been postulated [8]. The binding event alters the conformation of critical

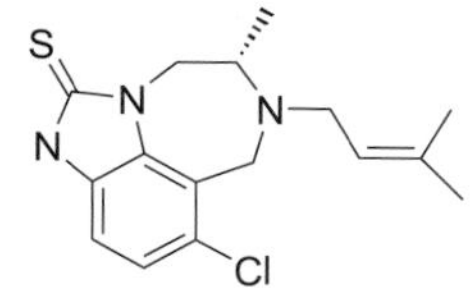

8-Cl *TIBO* : tivirapine 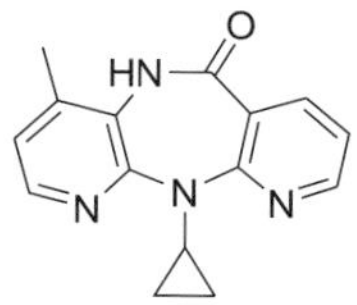   nevirapine 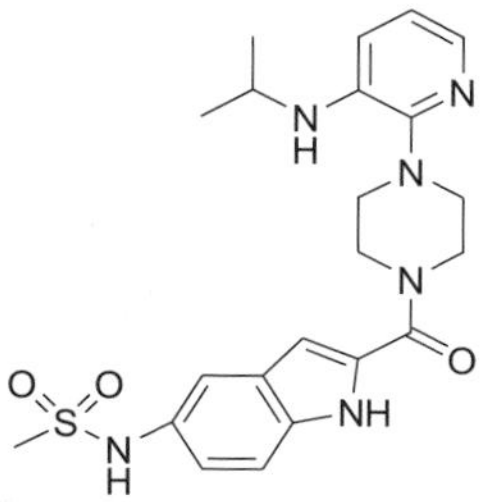   delavirdine

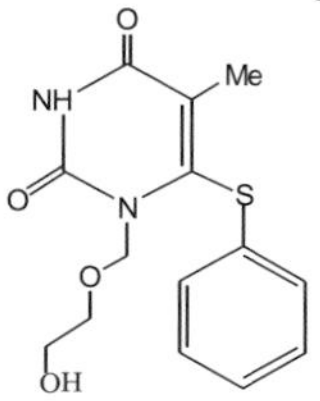

HEPT 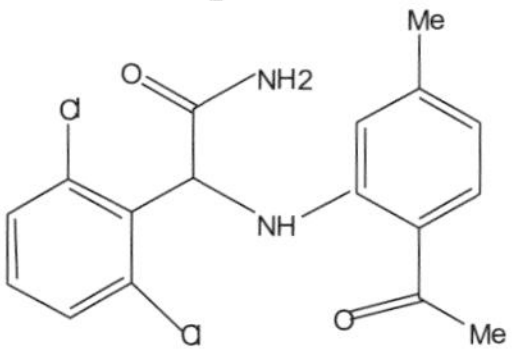   loviride

First generation NNRTIs

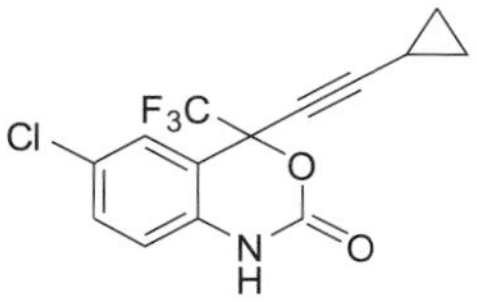

efavirenz 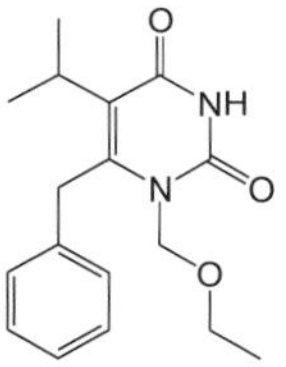   emivirine

Second generation NNRTIs

Fig. 1 Structures of selected non-nucleoside reverse transcriptase inhibitors (NNRTIs)

residues and thereby inhibits the ability of the enzyme to perform normal RT functions.

A common drawback of all currently known NNRTIs is the fast emergence of drug-resistant mutant HIV-1 [9, 10]. Nevirapine is a first generation inhibitor. Unfortunately it is highly sensitive to single point mutations of HIV-RT. The more recent compounds (second generation inhibitors) have a better resistance profile against mutation [5].

A great number of structures of HIV-1 RT either in free form or in complexes have been resolved by X-ray crystallography and their coordinates have been deposited in the Protein Data Bank (RCSB PDB) [11]. These structures include:

1. HIV-1 RT with no ligand bound
2. HIV-1 RT bound to double-stranded oligonucleotide template-primers both in the presence and in the absence of a deoxynucleotide triphosphate (dNTP) substrate
3. HIV-1 RT complexed with different NNRTIs
4. Structures of various HIV-1 RT mutants important for drug resistance, either alone or with different inhibitors

An overview of the available X-ray crystal structures has been given by Lawtrakul et al. [6, 7].

HIV-1 RT is a heterodimer containing two separate chains with identical amino acid sequences but of different lengths, with molecular weights of 66 kDa (p66) and 51 kDa (p51) [12]. Unit p66 contains the N-terminal polymerase part (pol, about 440 residues) and the RNaseH domain at the C-terminus (about 120 residues). The pol subunit is folded into four domains

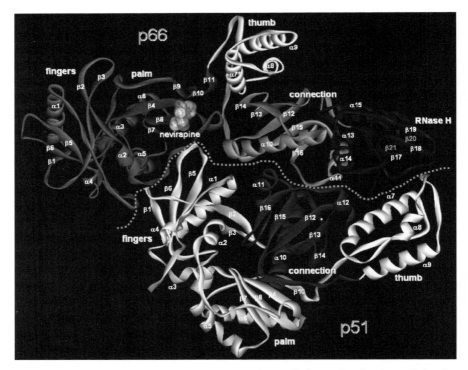

Fig. 2 Heterodimer of HIV-1 RT: p66 (*gray*) and p51 (*light gray*) subunits and the five domains: finger, palm, thumb, connection, and RNase H, complexed with nevirapine (spherical surface representation)

called "fingers", "thumb", "palm", and "connection". Overall p66 resembles a human right hand that holds on to the primer/template DNA duplex. The connection domain is located between the hand of the pol subunit and the RNaseH domain. Structural details of p66 including domain boundaries [13] and secondary structure [8] are outlined in Fig. 2.

The palm and connection domains can be described as five-stranded beta sheets with two alpha-helices on one side. The thumb domain consists of four helices. p51 lacks the RNaseH domain but still contains the connection domain. It is processed by proteolytic cleavage of p66.

For the interaction between the enzyme and NNRTIs, the contacts between the inhibitor molecules and the surrounding amino acids are the key to determining the binding affinity. This can be studied in detail because of the increasing number of available crystal structures of inhibitor enzyme complexes [6, 7], and with the help of computational chemistry which will be described later. This is beneficial in obtaining more understanding and knowledge of inhibitor–enzyme interactions of NNRTIs in order to design new potent drugs in particular ones active against mutant enzyme.

1.2
Binding of NNRTIs to HIV-1 RT

All NNRTIs, where the structure of the RT complex is currently known, bind to a region of the enzyme which is approximately 10 Å away from the catalytic site [6, 14]. This binding pocket is located between the sheet $\beta4$-$\beta7$-$\beta8$ and the sheet $\beta9$-$\beta10$-$\beta11$ of the p66 palm domain. It includes the $\beta5$-$\beta6$ loop (Pro97, Leu100, Lys101, Lys103), $\beta6$ (Ser105, Val106, Val108), the $\beta9$-$\beta10$ hairpin (Val179, Tyr181, Tyr188, Gly190, Asp192), the $\beta12$-$\beta13$ hairpin (Glu224, Phe227, Trp229, Leu234, Pro236) and part of $\beta15$ (Tyr318, Tyr319) (Rodgers 1995) (see also Fig. 2 for details). The internal surface of the pocket is predominantly hydrophobic with substantial aromatic character (Tyr181, Tyr188, Phe227, Trp229, Tyr232). It also includes a few hydrophilic residues (Lys101, Lys103, Ser105, Asp192, Glu224) and backbone atoms which are able to form hydrogen bonds with the inhibitor. A small part of the pocket is formed by residues from the p51 subunit (Thr135, Glu138). There is no NNRTI binding pocket in p51 itself. A solvent-accessible entrance to the cavity is formed by residues Leu100, Lys103, Val179, Ser191, and Glu138 from p51. The function of this inhibition pocket in regard to the activity of the enzyme is still unknown, but some mutations of amino acids of the pocket lead to changes in the activity.

NNRTIs are particularly attractive drug candidates because their binding site is unique to reverse transcriptase of HIV-1 and thus they are less likely to cause adverse side-effects by disrupting normal DNA polymerase activity [6].

However, a serious problem concerning all NNRTIs is the emergence of viral strains, which have point mutations in the region of the binding pocket, preventing these drugs from inhibiting RT activity. Some crystal structures of mutant type RT with inhibitors have been published [15, 16]. For a review of resistant mutations of RT, and a selection of mutations clustered around the inhibition pocket, which might be the reason for the decreased inhibitor activity, see Shafer et al. [17], and for a summary see Lawtrakul et al. [6, 7].

1.3
Applications of Computational Chemistry in Drug Design

Computational chemistry (CC) or quantum chemical (QC) calculations are nowadays key elements in computer-aided drug design research. The use of CC calculations provides a description of how molecules interact and form their 3D shape, which in turn determines molecular function [18]. Moreover, this approach can show the connecting link between experimentally determined structures and biological function [19]. In particular, CC calculations can be used to understand enzyme reactions and mechanisms [20–22], hydrogen bonding, polarization effects, spectra [23, 24], ligand binding and other fundamental processes both in normal and aberrant biological contexts [25–27]. With the advancement of parallel computing and progress in computer algorithm design, more realistic models of drug–enzyme interaction and even biological macromolecules are possible. Several excellent text books and reviews [18, 19, 28] are available as introductions to the basic theory.

In this chapter, the comprehensive utility of computational chemistry is exemplified based on NNRTIs. In Sect. 2, QSAR and 3D-QSAR studies of NNRTIs are reviewed. The 3D molecular geometry of the inhibitor is probably the key structural information on drug–enzyme interaction. In this context, Sect. 3 demonstrates the influence of conformational analysis, and the application of potential energy surfaces of the flexible inhibitors is highlighted. Another important area of CC is the hybrid quantum mechanics and molecular mechanics (QMMM), which can be applied on larg systems such as enzymes. Section 4 introduces a recent new hybrid QM/QM such as the ONIOM method for investigation of particular interactions between the binding site of HIV-1 RT and the NNRTIs. Section 5 reviews molecular simulations such as molecular dynamics (MD) and Monte Carlo simulations on NNRTIs, which can demonstrate structural information about the inhibitors in a more realistic system. Due to the increasing availability of crystal structures and molecular databases, protein-based drug design with the applicable molecular docking has been widely used to search for new NNRTIs. Therefore, in Sect. 6 a number of investigation on NNRTIs are reviewed. The chapter concludes with an outlook.

2
QSAR and 3D-QSAR of NNRTIs

Quantitative structure–activity relationships (QSAR) [29] and three-dimensional QSAR (3D-QSAR) are highly active areas of research in drug design. Scope and limitations, as well as comparison, of classical and 3D-QSAR are reported [30]. A number of QSAR studies were reported to identify important structural features responsible for the inhibition by NNRTIs [31–38]. 2D-QSAR of HEPT and TIBO were studied using various molecular properties of the inhibitors obtained from quantum chemical calculations by means of linear regression analysis [39, 40]. In contrast to dipyridodiazepinone analogs, the use of similar structural descriptors seemed to be insufficient to set up a proper QSAR model. Therefore, steric descriptors such as connectivity [41] and topological indices [42, 43] were used to improve the QSAR models. These new models showed improved predictive ability. An artificial neural network (ANN) was applied to develop nonlinear QSAR models of dipyridodiazepinone analogs [44]. The results reveal satisfactory models for the activity against both WT and Tyr181Cys HIV-1 RT. An additional 3D-QSAR study, based on comparative molecular field analysis (CoMFA), was also performed on these data sets [45, 46]. Common structural requirements of three different classes of HIV-1 RT inhibitors, TIBO, HEPT, and nevirapine, were obtained based on an extensive comparison of several QSAR approaches. A novel tool called hologram QSAR (HQSAR) was introduced on NNRTIs by Pungpo et al. [47].

2.1
Molecular Alignment in 3D-QSAR

The alignment of molecules and "active" conformation selection are the key to a successful 3D-QSAR model by CoMFA. Therefore, in this regard, several attempts have been made to find the most appropriate molecular alignment in order to obtain more reliable predictive ability from the model. Medina-Franco et al. [48] demonstrated the application of docking as molecular alignment to determine active conformation of pyridinone derivatives by using CoMFA and CoMSIA (comparative molecular similarity indexes analysis) [49] methods. Similar approaches were applied to other NNRTIs such as phthalimide derivatives [50] and TIBO derivatives [51, 52]. This indicates that a combination of ligand-based and receptor-based modeling is a powerful approach for building 3D-QSAR models. Another approach has been tackled by Ishiki et al. [53]. This required selecting the α parameter value which was used in the SEAL program for the alignment procedures. For validation, a set of 84 nevirapine derivatives was studied and submitted to a CoMFA analysis. Alignments were done using two approaches: rigid alignment, in which the compounds were fitted atom-by-atom onto a template considering only their dipyridodiazepinone moiety; and field based, in which the steric, electrostatic and hydrophobic fields, generated by the SEAL program were considered in the alignment. Although the studied compounds bear a fairly rigid moiety, in general, better CoMFA models were observed using the SEAL alignment procedure than those derived from rigid alignments.

2.2
Influence of Conformational Mobility in 3D-QSAR

Due to the flexibility of the inhibitors care has to be taken using 3D-QSAR as the geometry obtained from X-ray crystallography might be different from the active bound conformation (see Sect. 3 for more details). Kireev et al. [54] has shown that descriptors related to the conformational changes were found to be an important factor, underlying RT inhibitory activity in the HEPT series for example. Indeed, the QSAR model provides evidence concerning the conformational transformations the molecules may undergo during the inhibition process. The established relationships are supplementary to the experimental study on the binding of HEPT type inhibitors to RT as shown by Hopkins et al. [55]. The results suggest a quantitative interpretation of the structure–activity relationships, which otherwise cannot be explained within the framework of the crystal inhibitor–protein model.

3
Conformational Analysis of NNRTIs

3.1
Why Conformational Analysis?

The conformation of a ligand is important for binding to the protein. The complementarity of molecular surfaces and electrostatic potentials between ligand and the receptor site is essential for binding and stability of the complexes, as the total binding energy results from the local interactions between each part of the ligand and the surrounding moiety. To obtain a better agreement between the ligand surface and the complementary molecular surface of the enzyme, conformational changes of the backbone as well as of the side chains of distinct residues take place, which in turn leads to the loss of the activity of the enzyme.

The NNRTIs are a class of compounds that present a common butterfly-like conformation. Parreira et al. [56] investigated the intramolecular factors that contribute to this conformation. Hydrogen bonds were analyzed by geometric and electrostatic criteria. Only the former allows the elucidation of the relative intensity of hydrogen bonds. The interaction between aromatic rings may contribute to the preferential conformation. Novel 1H,3Hnaphtho[2′,3′:4,5]imidazo[1,2-c]thiazoles were studied by Chimirri et al. [57]. It was found that, in spite of a butterfly-like conformation as evident in X-ray analysis, the results of in vitro screening suggest that replacement of the benzene fused ring by the naphthyl moiety negatively influences the activity of this analog. Calculations of the potential energy surfaces of some derivatives of HEPT, TIBO, and nevirapine [58–63] have shown that the geometry of the ligands in the inhibition pocket is rather close to the absolute or to a local conformational energy minimum. A butterfly-like molecular shape has been proposed to be common for inhibitor molecules. For the more recent second generation inhibitors, however, this notion becomes less stringent.

As the inhibitor molecules are relatively small, their geometries can be calculated by rather accurate quantum chemical methods like ab initio (HF/6-31G(d,p), MP2/6-31G(d,p), MP2/cc-pVDZ, MP2/cc-pVTZ) or DFT (B3LYP/6-31G(d,p), B3LYP/6-311++G(d,p), B3LYP/cc-pVTZ, B3LYP/cc-pVTZ) methods. Moreover, the potential energy surfaces obtained can be compared to those calculated using various empirical molecular force fields (UFF, MM3, AMBER) in order to check the quality of these methods, which are then used for molecular calculations and simulations on the larger association complexes. The results of these conformational analyses were found to be correlated with experimental data, like X-ray crystal structures, NMR- or vibrational spectra, showing the reliability of the different methods. As suggested by Das et al. [64], a good inhibitor is expected to occupy much of the available volume in the binding pocket of a drug target and to have favorable interactions with the pocket. For a rapidly mutating target, where the mutations affect the binding of an inhibitor, an effective drug candidate should have the ability to bind to the multiple related pocket conformations found in resistant mutants. This type of problem requires that an inhibitor be optimized against multiple related binding pockets, in contrast to the classical approach where the optimization is carried out against a single binding pocket.

3.2
Flexible Side Chain of 9-Cl TIBO

The conformational analysis of one of the flexible NNRTIs, 9-Cl TIBO, was investigated by Saen-oon et al. [63] using a high level of calculations, ab initio, and DFT theory. The potential energy surface (PES) as the function of two important rotatable dihedral angles of the 9-Cl TIBO side chain was generated by the Hartree–Fock method at 3-21G basis set. Eight pronounced local minima were found to exist within an energy range of less than 2.4 kcal/mol (10 kJ/mol). The energy barriers between the different local minima were less than 3.3 kcal/mol (15 kJ/mol). A second derivative (frequency) analysis showed that all conformations were stable at this level of theory. These structures were used as starting points for full geometry optimizations at the HF/6-31G(d,p) and B3LYP/6-31G(d,p) levels of theory to obtain the absolute geometries and structural information. The comparisons of calculated conformations with the bound conformation in the X-ray structure were sequentially considered. Additionally, to obtain some structural information and to correlate between calculated structures and the structure in solution, NMR chemical shift calculations were also performed on all eight local minimum structures at B3LYP/6-311++G(d,p) level, using the GIAO approach. The calculated ^{1}H- and ^{13}C-NMR chemical shifts for the lowest energy conformation gave the best match to experimental results.

3.3
Potential Energy Surfaces

Due to multiple rotatable bonds in the molecules, potential energy surface analysis is a useful technique to find the local minimum energy structures. The conformational performance of the 9-Cl TIBO compound was examined by the rotation and orientation in the space of the highly flexible DMA side chain. The potential energy surface or the hypersurface of the 9-Cl TIBO compound is shown in Fig. 3 by varying two sensitive dihedral angles of the DMA side chain defined as alpha (α) and beta (β). The graphical presentation of

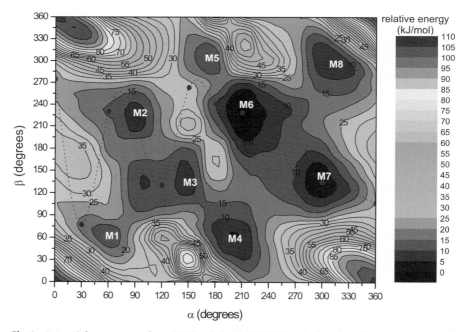

Fig. 3 Potential energy surface (PES) of the 9-Cl TIBO calculated at the HF/3-21G level as a function of two variables: α and β, the C5-N6-C15-C16 and N6-C15-C16=C17 dihedral angles, respectively. Energy values (kJ/mol) refer to differences with respect to the absolute minimum

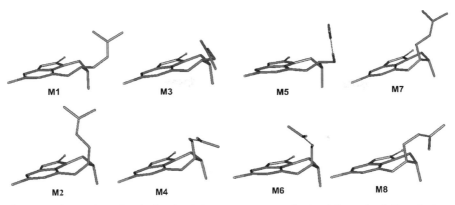

Fig. 4 Conformation of eight local minimum structures obtained from the full optimization at the B3LYP/6-31G(d,p) level of theory

the PES shows the high flexibility of the 9-Cl TIBO structure which is mainly due to the influence of the rotatable DMA side chain. The range of energy fluctuation on the PES, which is relative to the absolute energy, is approximately 0 to 110 kJ/mol calculated at HF/3-21G level. The large flat area for the energy less than 30 kJ/mol covers the range of α dihedral angle from 0 to 360 degrees and is restricted between approximately 60 to 300 degrees for β dihedral angle. Within this area, eight pronounced local minima exist on the PES as depicted in Fig. 4 and defined as M1 to M8. The range of α and β dihedral angles of these eight local minima is determined graphically from the PES within the energy range 0–10 kJ/mol. From these eight local minima, the lowest energy grids (α, β) at each local minimum are selected and given with the relative energy, with respect to the structural energy of local minimum M6.

Based on the conformational space of the 9-Cl TIBO structure, the lowest energy conformation corresponds well to the geometry of the bound conformation in the complex structure with HIV-1 RT. From the structural information of the HIV-1 RT interacting with TIBO derivatives, it was found that the inhibitor molecule should contain a flexible section in order to easily allow the conformational change required to fit into the binding cavity. Based on the adapted conformation bound in the complex, the inhibitor should contain 2π-electron moieties or "wings", connected by the highly flexible part so as to easily rotate or adapt its structure to allow interactions with residues in the binding pocket. The N6 should be retained as S-configuration to orient the DMA side-chain to preserve the van der Waals interaction with the binding pocket. The different structural information of each inhibitor from quantum chemical calculations can be used to detail differences in biological activity, as in the cases of 8-Cl TIBO and 9-Cl TIBO, and will be explained in the next section.

3.4
Influence of the Side Chain Motion on the Conformation of the Heterocyclic Ring

Several different biological activity studies have shown that 8-Cl TIBO is somewhat more potent and less toxic than 9-Cl TIBO. Therefore, it is interesting to investigate the influences of the chlorine substitution on C8 and

Table 1 Monitoring of the seven-membered ring motions

Dihedral angles	Range of dihedral angle changes (in degrees)	
	8-Cl TIBO	9-Cl TIBO
N (C5-N6-C7-C15)	120–240	120–240
A1 (C4-C5-N6-C15)	170–270	210–320
A2 (C14-C5-N6-C15)	50–150	80–190
A3 (C15-N6-C7-C13)	40–150	60–180
A4 (N3-C4-C5-C14)	158–178	55–70
A5 (C12-N3-C4-C5)	289–299	28–48

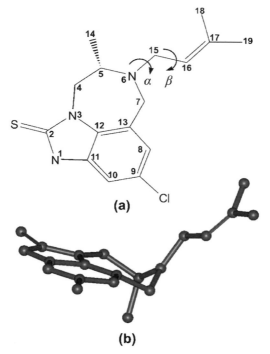

(a)

(b)

Scheme 1 a Chemical structure of 9-Cl TIBO and atomic numbering. **b** 3D structure of the butterfly-like shape of 9-Cl TIBO

C9 of TIBO that cause the large change of HIV-1 RT inhibitory activity on the MT-4 cells. Evidently, there are more degrees of freedom for the conformation of 8-Cl and 9-Cl TIBO, determined by the structural diversity of the seven-membered ring. As a complete conformational analysis of the heterocyclic ring is rather difficult, only geometries of the calculated energy hypersurface were considered. In Table 1, the response of the heterocyclic ring system to changes in the position of the side chain is given as the range of dihedral angles changes, monitored by selected dihedral angles in the seven-membered ring. The dihedral angle A1 (C4-C5-N6-C15) (see Scheme 1) monitors the motion inside the ring system, A2 (C14-C5-N6-C15) describes the relative position of the side chain and the methyl group. A3 (C15-N6-C7-C13) and A4 (N3-C4-C5-C14) depict the position of the side chain and the methyl group relative to the aromatic ring system. A5 (C12-N3-C4-C5) shows the conformation of the ring close to the methyl group. The dihedral angle N defined by C5-N6-C7-C15 does not describe a rotatable bond, but it is a measure for the conformation at the nitrogen atom. The position of the methyl group remains the same as in the crystal structure, and there is no inversion of the heterocyclic ring (the dihedral angle A4 varies only for about 20°). Since three atoms (N3, C12, C13) of the seven-membered ring are part of the planar benzoimidazole system, the conformational flexibility of the ring is determined by the remaining four saturated atoms (C4, C5, N6, C7). As there is no change in the methyl group position, the inversion of atom N6 leads to the predominant conformational change of both 8-Cl and 9-Cl TIBO structures with respect to the orientation of the dimethylallyl side chain. This nitrogen inversion is confirmed by considering the structure along the conformational change of the dihedral angles N, A1, A2, and A3 as shown in Fig. 5.

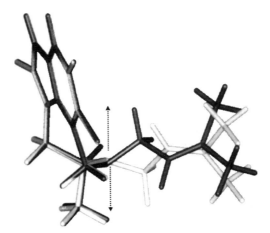

Fig. 5 Influence of the nitrogen inversion on the conformational change of TIBO structure

4
Particular Interaction between HIV-1 RT/NNRTIs

4.1
Hybrid QM/MM

Theoretical investigation has been an alternative method to understand the enzyme–inhibitor interaction in detail for large molecular systems. However, such investigation of larger molecular systems is limited by the computer effort required and the accuracy of the method used. Recently, accurate molecular modeling for larger molecules, such as in molecular biology and surface chemistry, has become feasible due to recent developments in computational chemistry. Hybrid quantum mechanics/molecular mechanics (QM/MM) techniques are becoming increasing popular for modeling large molecular systems [65–67]. In this approach, it is assumed that a large molecular system can be partitioned into a small model system for the chemically active site where a reaction will occur, and a large model system for the chemically inactive part. The chemically active site is treated at a high level of quantum calculation with more accurate methods, while the chemically inactive part is modeled using only a molecular mechanics level. The influence of the inactive part using MM constrains the geometry or reaction of the active model. The key to the success of a hybrid QM/MM technique is the manner in which the influence of the MM part is communicated to the QM part. The problem in defining a suitable QM/MM representation is the construction of an accurate interface between the QM and MM parts of the system. A number of different approaches has led to the development of an automated methodology that reproduces fully quantum mechanical calculations, and also makes significant progress in the reduction of errors. This problem is a difficult one to overcome. Consequently, the QM/MM methodology remains an important active area of computational chemistry research.

4.2
ONIOM Method

Recently, a number of hybrid QM/MM methods were developed and applied to the study of reaction mechanisms in biological systems and material sciences. Guo et al. [68] studied the chorismate mutase acting at the first branchpoint of aromatic amino acid biosynthesis, and catalyzing the conversion of chorismate to prephenate, by using the QM/MM method to understand the enzyme reaction mechanism at the atomic level. In 1995, Maseras and Morokuma [69] developed a new hybrid QM/MM method called the "integrated molecular orbital molecular mechanics" (IMOMO) method. In IMOMO, a hybrid gradient is derived from the energy gradients exerted by the MM part onto the QM part, and this is used to optimize the QM part. Consequently, the ONIOM (our own n-layered integrated molecular orbital and molecular mechanics) method has been introduced and its efficiency has been improved over the year. In the ONIOM approach, a small part of a system (such as the inhibitor and the reacting amino acids in the binding site of an enzyme or chemically active site where a reaction will occur) is treated at a high level of calculation whereas the larger surroundings are modeled using a lower level of calculation. Schematic representation of ONIOM model is demonstrated in Scheme 2. Extrapolated energy can be calculated using the following equations.

The basic idea behind the ONIOM approach can be explained most easily when it is considered as an extrapolation scheme in 2D space; spanned by the size of the system on one axis and the level of theory on the other axis. In the two-layered ONIOM method, the total energy of the system is obtained from three independent calculations as shown in Eq. 1:

$$E^{\text{ONIOM2}} = E_{\text{model}}^{\text{high}} + E_{\text{real}}^{\text{low}} - E_{\text{model}}^{\text{low}} \tag{1}$$

where *real* denotes the full system, which is treated at the *low* level, while *model* denotes the part of the system for which the energy is calculated at both *high* and *low* levels. One can see that the method can be regarded as an extrapolation scheme. Beginning at $E_{\text{model}}^{\text{low}}$, the extrapolation to the high-level calculation ($E_{\text{model}}^{\text{high}} - E_{\text{model}}^{\text{low}}$) and the extrapolation to the real system ($E_{\text{real}}^{\text{low}} - E_{\text{model}}^{\text{low}}$) are assumed to produce an estimation for $E_{\text{real}}^{\text{high}}$. For a three-layer ONIOM scheme, the energy expression can be written as shown in Eq. 2:

$$E^{\text{ONIOM3}} = E_{\text{small model}}^{\text{high}} + E_{\text{intermediate model}}^{\text{medium}} - E_{\text{small model}}^{\text{medium}} \tag{2}$$
$$+ E_{\text{real}}^{\text{low}} - E_{\text{intermediate model}}^{\text{low}}$$

The ONIOM approach provides an extrapolated energy (E_{ONIOM}) for a system partitioned in this manner.

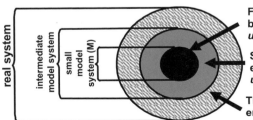

First Layer :
bond-formation/breaking takes place.
use the "high-level" (H) method

Second Layer :
electronic effect on the first layer
use the "medium-level" (M) method

Third layer :
environmental effects on the first layer
use the "low-level" (L) method

Scheme 2 Representation of ONIOM model set up for small model, intermediate model, and real model systems, which will be treated by high-level, medium-level, and low-level methods, respectively [source Morokuma K (2003) Bull Korean Chem Soc 24:797]

4.3
Particular Interaction of HIV-1 RT/NNRTIs

Recently, applicability of the ONIOM2 and ONIOM3 methods to NNRTIs and the HIV-1 reverse transcriptase binding site were studied. Schematic representation of the adopted model systems of HIV-1 RT binding site and nevirapine [70], 8-Cl TIBO [71], and efavirenz [72] in ONIOM calculations are demonstrated in Fig. 6. Kuno et al. [70] investigated the isolated complex of pyridine (part of nevirapine) and methyl phenol (part of Tyr181), which shows a stacking interaction with 8.8 kcal/mol binding energy at MP2/6-31+G(d) level. Optimization of the nevirapine and Tyr181 geometry in the pocket of 16 amino acid residues at the ONIOM3(MP2/6-31G(d):HF/3-

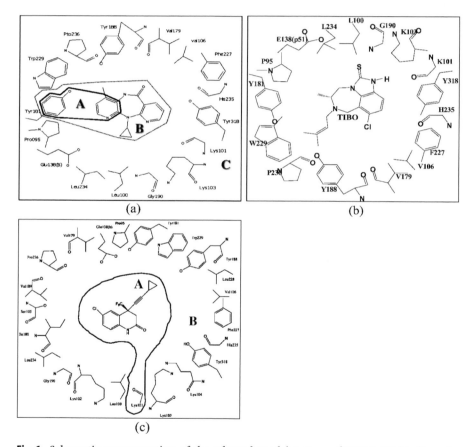

(a) (b)

(c)

Fig. 6 Schematic representation of the adopted model system of HIV-1 RT binding site and bound **a** nevirapine [70], **b** 8-Cl TIBO [71], and **c** efavirenz [72] in ONIOM calculations

21G:PM3) level gave a complex structure with weak hydrogen bonding but without stacking interactions. The binding energy of 8.9 kcal/mol comes almost entirely from the interaction of nevirapine with residues other than Tyr181. Generally, the results indicate that despite using different ONIOM methods, the intermolecular distances and the binding energies calculated are quite similar.

Additionally, the complex structures of HIV-1 RT, both wild type and Tyr181Cys mutant type, with 8-Cl TIBO inhibitor were investigated by Saenoon et al. [71] using the same methods. Two-layered and three-layered ONIOM calculations were carried out to determine the binding energy of 8-Cl TIBO inhibitor bound into the HIV-1 RT binding pocket, consisting of 20 amino acid residues within a radius of 15 Å. The combination of different methods, MP2/6-31G(d,p), B3LYP/6-31G(d,p), and PM3 were performed and different models set up for analysis. The results obtained from different combined QM/QM models clearly indicated that the Tyr181Cys mutation reduces the binding affinity of the inhibitor, and the stability of the complex, by approximately 7–8 kcal/mol. The analysis of components of the interaction energy and deformation energy for 8-Cl TIBO inhibitor upon binding was also examined in more detail. Additionally, the calculations of the interaction energy between 8-Cl TIBO with individual amino acid residues surrounding the binding pocket were performed at MP2/6-31G(d,p) and B3LYP/6-31G(d,p) levels to give energy differences and strengths of binding comparisons between wild type and Tyr181Cys mutant type RT.

The particular interaction between efavirenz and the HIV-1 RT binding site was tackled by Nunrium et al. [72]. The interaction between efavirenz and Lys101 was found to be the strongest, typically 11.29 kcal/mol. The stability of this complex system leads to the estimation of binding energy of approximately 23 kcal/mol. Moreover, hydrogen bonds between benzoxazin-2-one and the backbone carbonyl oxygen and the backbone amino hydrogen of Lys101 were observed. These hydrogen bond interactions play an important role in binding the efavirenz/HIV-1 RT complex.

4.4
Some Remarks

The ONIOM method, as implemented in the Gaussian03 program package [73], has been proven to be applicable to large molecular systems, where different levels of theory are applied on various parts of the molecular cluster [74, 75]. The interactions of non-nucleoside RT inhibitors with the amino acids of the inhibition pocket of the enzyme, and the induced conformational changes, are the results of several sensitive phenomena which need careful and accurate treatment. Calculations of the whole complex, however, is rather difficult because of the large size of the system. Therefore, an approach combining QM/MM or ONIOM methods for such HIV-1

RT/NNRTIs systems would be an exciting area of research for examining the particular interactions between the inhibitors and residues in the enzyme binding site. The information about such interactions will be beneficial for the design of appropriate inhibitors, in particular, active against mutant enzymes.

5
Molecular Dynamics Simulations and Monte Carlo Simulations of NNRTIs

5.1
Molecular Dynamics Simulations

Molecular dynamics (MD) simulations measure (time) averages in an ensemble that is very similar to the microcanonical ensemble. This method is an alternative procedure for the investigation of ligand–receptor interactions [76]. These simulations are used to study the flexibility of large molecules, particularly biomolecules. The total movement of a protein, together with a ligand, can be resolved by integration of Newton's equations of motion. In such a system, for realistic conditions, the surrounding water molecules have to be taken into account. This, however, leads to very large molecular assemblies, which require large amounts of computer resources. According to some recent calculations [77], the movements of some of the most important residues in the active site of HIV-1 RT (Asp110, Asp185, Asp186) are different for free and inhibitor-bound enzymes.

The currently developed steered molecular dynamics (SMD) initiated by Shen et al. [78] investigate the biochemical processes occurring at microsecond or second time scale. Thus SMD provide dynamical and kinetic processes of ligand–receptor binding and unbinding which cannot be accessed by experiment. The binding and unbinding processes of α-APA, an NNRTI, were studied using nanosecond conventional MD and SMD simulations. The simulation results showed that the unbinding process of α-APA consists of three phases based on the position of α-APA in relation to the entrance of the binding pocket. When α-APA is bound in the binding pocket, the hydrophobic interactions between HIV-1 RT and α-APA dominate the binding; however, the hydrophilic interactions (both direct and water-bridged hydrogen bonds) also contribute to the stabilizing forces. Whereas Tyr181 has significant hydrophobic interactions with α-APA, Tyr188 forms a strong hydrogen bond with the acylamino group (N14) of α-APA. These two residues have very flexible side chains and appear to act as two "flexible clamps" discouraging α-APA from dissociating from the binding pocket. At the pocket entrance, two relatively inflexible residues, Val179 and Leu100, gauge the openness of the entrance and form the bottleneck of the inhibitor-unbinding pathway. Two special water molecules at the pocket entrance appear to play important roles in inhibitor recognition of binding and unbinding. These water molecules form water bridges between the polar groups of the inhibitor and the residues around the entrance, and between the polar groups of the inhibitor themselves. The water-bridged interactions not only induce the inhibitor to adopt an energetically favorable conformation so the inhibitor can pass through the pocket entrance, but also stabilize the binding of the inhibitor in the pocket to prevent the inhibitor dissociating. The results from complementary SMD and conventional MD simulation strongly support the hypothesis that NNRTIs inhibit HIV-1 RT polymerization activity by enlarging the DNA-binding cleft and restricting the flexibility and mobility of the p66 thumb subdomain that are believed to be essential during DNA translocation and polymerization.

A series of targeted MD simulations have been carried out in an attempt to assess the effect that the common Lys103Asn mutation in HIV-1 RT has on the binding of three representative NNRTIs: nevirapine, efavirenz, and etravirine. Rodriguez-Barrios et al. [79] have shown the conformation of the enzyme in the unbound state and the drug-bound state in the presence of the NNRTIs. The location of each drug outside the binding pocket was determined by an automated docking program, and steering into the binding pocket followed a route that is likely to represent the actual entrance pathway. The additional hurdle to inhibitor entry imposed by the extra Asn103–Tyr188 hydrogen bond is seen to affect each NNRTI differently, with the ability to disrupt this interaction increasing in the order etravirine and efavirenz $\geq$ nevirapine, in good accordance with experimental findings.

Weizinger and coworkers [80] evaluated the properties of efavirenz and a set of its derivatives (benzoxazinones) by placing them into the binding site of the HIV-1 RT by molecular docking. The resulting geometries were used for a MD simulations and binding energy calculations. The enzyme–inhibitor binding energies were estimated from experimental inhibitory activities. Based on MD simulations, the obtained results indicate that the tight association of the ligand to the HIV-1 RT binding pocket was based on hydrogen bonding between the efavirenz N1 and the oxygen of the backbone of Lys101, with an estimated average distance of 1.88 Å. Moreover, electrostatic interaction was mainly contributed by two amino acids in the binding site: Lys101 and His235. MD simulations open the possibility of studying the reaction of the flexible enzyme to those substances as well as the overall affinity.

These coherent pictures strongly suggest that attempts to overcome resistance through structure-based drug design may be considerably more successful if dynamic structural aspects of the type discussed here are considered, particularly in cases where binding energy-based structure–activity relationship methods are unable to provide the required information.

5.2
Monte Carlo Simulations of NNRTIs

In contrast to MD simulations, Monte Carlo (MC) simulations probe the canonical (i.e., constant-NVT) ensemble [76], which leads to observable differences in the statistical averages computed in MD and MC simulations. Most of these differences disappear in the thermodynamic limit and are already relatively small for systems of a few hundred particles. MC simulations have been used extensively for drug–enzyme interaction studies. For example, the predictions for NNRTIs activities as demonstrated by Rizzo et al. [81]. More than 200 NNRTIs representing eight diverse chemotypes were correlated with their anti-HIV activities in an effort to establish simulation protocols and methods that can be used in the development of more effective drugs. Each inhibitor was modeled in a complex with the protein and by itself in water, and potentially useful descriptors of binding affinity were collected during the MC simulations. A viable regression equation was obtained for each data set using an extended linear response approach, which yielded r^2 values between 0.54 and 0.85 and an average unsigned error of only 0.50 kcal/mol. The most common descriptors confirmed that a good geometric match between the inhibitor and the protein is important and that a net loss of hydrogen bonds with the inhibitor upon binding is unfavorable. Other physically reasonable descriptors of binding are needed on a chemotype case-by-case basis. By including descriptors in common from the individual fits, combination regressions that include multiple data sets were also developed. This procedure led to a refined "master" regression for 210 NNRTIs with an r^2 of 0.60 and a cross-validated q^2 of 0.55. The computed activities showed a root-mean-square error of 0.86 kcal/mol in comparison with experiments and an average unsigned error of 0.69 kcal/mol. Encouraging results were obtained for the predictions of 27 NNRTIs representing a new chemotype not included in the development of the regression model. Predictions for this test set using the master regression yielded a q^2 value of 0.51 and an average unsigned error of 0.67 kcal/mol. Finally, additional regression analysis revealed that use of ligand-only descriptors leads to models with much diminished predictive ability.

To tackle the problem of mutant enzyme inhibition and design of inhibitor, MC simulations were applied on the effects of the Leu100Ile mutation for NNRTIs binding. The investigation as reported by Wang et al. [82] indicated that Monte Carlo with complementary free energy perturbation (MC/FEP) calculations were used to evaluate the binding free energy change for HIV-RT/inhibitor complexes upon Leu100Ile mutation. Inhibitor size and flexibility adjacent to hydrogen-bonding sites were evident as important considerations for antiviral drug design.

In this respect, contribution of the Leu100Ile and Tyr181Cys variants to the enzyme stability and biological activity was investigated by Smith et al. [83].

Using MC simulations of 8-Cl TIBO, both free energy perturbation and linear response calculations were carried out for the transformation of wild-type RT to these two mutations. The newer linear response method estimates binding free energies based on changes in electrostatic and van der Waals energies and solvent-accessible surface areas. In addition, the change in stability of the protein between the folded and unfolded states was estimated for each of these mutations, which are known to emerge upon treatment with the inhibitor. Results from the calculations revealed that there is a large hydrophobic contribution to protein stability in the native folded state. The calculated free energies of binding from both the linear response, and also the more rigorous free energy perturbation method, gave excellent agreement with the experimental differences in activity. The success of the relatively rapid linear response method in predicting experimental activities holds promise for the activity of the inhibitors, not only against the wild-type RT, but also against key protein variants whose emergence undermines the efficacy of the drugs.

6
Protein-based Drug Design and Virtual Screening of NNRTIs

Protein-based drug design has become one of the mature disciplines of medicinal chemistry [84, 85]. Molecular docking calculations have been used as an important method to study orientation and estimation of binding energy between the ligand and the enzyme. This method was originated by Kuntz et al. [86]. The concept of this method is based on the observation that drugs bind to clearly defined molecular targets. A strong and selective binding can be obtained from a high structural and chemical complementary between the macromolecular target and the ligand. A number of computational tools have been described to select putative ligands and to predict their interactions with the protein [87].

6.1
Ligand Docking

Most of the applications of molecular docking in drug design are used to obtain structural information in terms of orientation of the ligand in the binding site, as well as binding energy, in order to rank the activities of proposed compounds prior to synthesis. There are many reports of molecular docking being used as a tool in QSAR and inhibitor design. Examples include Ranise et al. [88], who identified a novel class of NNRTIs, thiocarbamates (TCs) as isosteres of phenethylthiazolylthiourea (PETT) derivative. The lead optimization strategy led to para-substituted TCs, which were active against wild-type HIV-1 in MT-4-based assays at nanomolar concentration. One of the most potent congener bears a methyl group at position 4 of the phthalimide moiety and a nitro group at the para position of the N – Ph ring. Most of the TCs showed good selectivity indexes. Some TCs significantly reduced the multiplication of the Tyr181Cys mutant, but they were inactive against Lys103Arg and Lys103Asn + Tyr181Cys mutants. However, the 41-fold increase in resistance was not greater than that of efavirenz against the Lys103Arg mutant in enzyme assays. The docking model predictions were consistent with in vitro biological assays of the anti-HIV-1 activity of the TCs and related compounds synthesized.

Using structure-based computational methods, the group of Silvestri [89] proposed three pyrrolyl heteroaryl sulfones, (Et 1-[(1H-benzimidazol-2(3H) one-5-yl)sulfonyl]-1H-pyrrole-2-carboxylate, Et 1-[(1H-benzimidazol-5(6)-yl)sulfonyl]-1H-pyrrole-2-carboxylate and Et 1-[(1H-benzotriazol-5(6)-yl) sulfonyl]-1H-pyrrole-2-carboxylate), as novel NNRTIs. Although these compounds were inactive in the cell-based assay, they inhibited the target enzyme with micromolar potency.

Also interesting are pyrrolyl aryl sulfones (PASs), reported as a new class of HIV-1 RT inhibitors by Artico et al. [90]. Pyrrolyl aryl sulfone I (R = H), the most potent inhibitor within the series (EC_{50} = 0.14 µM, IC_{50} = 0.4 µM, and SI > 1429), was then selected as a lead compound for a synthetic project based on molecular modeling studies. Using the 3D structure of RT co-crystallized structures with the α-APA derivative, R95 845, a model of the RT/3 complex was derived by taking into account previously developed structure–activity relationships. Inspection of this model and docking calculations on virtual compounds prompted the design of novel PAS derivatives and related analogs. This computational approach proved to be effective in making qualitative predictions that discriminated active from inactive compounds. Compared with the lead compound, these values represent a three- and eight-fold improvement in the cell-based and enzyme assays, respectively, together with the highest selectivity achieved so far in the PAS series.

6.2
Virtual Screening of NNRTIs

Virtual screening uses computational methods to evaluate virtual libraries (databases) "in silico", against virtual receptors (targets) aiming to increase the speed of the drug discovery process. This method is designed for searching large scale hypothetical databases of chemical structures or virtual libraries by using computational analysis, and for selecting a limited number of candidate molecules likely to be active against a chosen biological receptor [91]. Therefore, virtual screening is a logical extension of 3D pharmacophore-based database searching (PBDS) or molecular docking, capable of automatically evaluating very large databases of compounds. There are several excellent reviews about the application of virtual screening in drug discovery [92, 93]. Some recent success stories are also illustrated. For instant,

Hemmateenejad et al. [94] designed TIBO-like derivatives by contraction of the β-ring in the TIBO structure from a seven- to a six-membered ring. The Autodock program [95] was used to study the binding of the molecule to HIV-1 RT. A total of 16 TIBO derivatives were examined. They found that both the seven- and six-membered ring compounds could dock into the enzyme. A linear relationship was obtained between the total free energy of docking and IC_{50} of the conventional TIBO derivatives. The total free energy of docking indicated that some of these derivatives were bonded to the receptor more strongly than the seven-membered derivatives. Meanwhile, in comparison with the 8-Cl TIBO molecule, which is currently used in HIV treatment, one six-membered ring derivative showed greater binding affinity toward the reverse transcriptase enzyme.

The group of Varnek et al. [96] applied a substructural molecular fragments (SMF) method to design new TIBO and HEPT derivatives. Using available experimental data, the SMF method was first applied to build QSAR models based on fragment descriptors (atom/bond sequences and "augmented atoms"). Virtual combinatorial libraries containing 891 TIBO derivatives and 2640 HEPT derivatives were then generated by systematically attaching selected substituents to corresponding Markush structures. Finally, a screening of those libraries using developed QSAR models led to several hits which potentially possess high anti-HIV activity.

The application of molecular docking to the search for bioactive compounds from libraries of natural products has been introduced by the group of Jiang [97]. Natural products containing inherently large-scale structural diversity, more so than synthetic compounds, have been major resources of bioactive agents and will continue to play as protagonists in the discovery of new drugs. However, accessing this diverse chemical space efficiently and effectively is an exciting challenge for medicinal chemists and pharmacologists. Virtual screening, which has shown great promise in drug discovery, will play an important role in finding lead and active compounds. Moreover, Jiang and coworkers demonstrated that chemogenomics is also a new technology for initiating target discovery by using active compounds as probes to characterize proteome functions.

In a similar approach, Sangma et al. [98] used the combination of molecular docking with neural networks using a self-organizing map (SOM) through pharmacophore generation for screening of anti-HIV-1 RT and anti-HIV-1 PR inhibitors from the Thai Medicinal Plants Database. Based on nevirapine and calanolide A as reference structures in the HIV-1 RT binding site and XK-263 in the HIV-1 PR binding site, 2684 compounds in the database were docked into the target enzymes. SOMs were then generated with respect to three types of pharmacophoric groups. The map of the reference structures were then superimposed on the feature maps of all screened compounds. By using the SOMs, the number of candidates for HIV-1 RT was reduced to less than ten compounds. For the HIV-1 PR target, there are 135 screened compounds that showed good agreement with the XK-263 feature map. The obtained results indicate that this combined method is clearly helpful in performing successful screening and in reducing the analysis step from AutoDock and scoring procedures.

Many attempts have been made to enhanced virtual screening capabilities such as by using combinations of two docking methods. This can be seen in the work of the group of Maiorov [99]. The aim of such an approach is to obtain the most results on a limited budget when screening large databases (hundreds of thousands to millions of compounds). The authors present a combination of two methods: their "fast-free-approximate" in-house docking program, and the "slow-costly-accurate" [100] as an example of one solution to the problem. The proposed protocol is illustrated by a series of virtual screening experiments aimed at identifying active compounds in the database. In more than half of the 20 cases examined, at least several active compounds per protein target were identified in approximately 24 h per target.

Similarity search profiles as a diagnostic tool for the analysis of virtual screening calculations were reported by Xue et al. [101]. The analysis is based on systematic similarity search calculations using multiple template compounds over the entire value range of similarity coefficients. Moreover, a new approach for rapid flexible docking based on an improved multipopulation genetic algorithm was developed on GAsDock by Li et al. [102]. The docking time is approximately in proportion to the number of rotatable bonds of ligands, and GAsDock can complete a docking simulation within 60 s for a ligand with up to 20 rotatable bonds. Results indicate that GAsDock is an accurate and remarkably faster docking program in comparison with other docking programs, which is much appreciated in the application of virtual screening.

7
Outlook and Conclusion

QSAR and computer-aided drug design are currently being used for understanding of structural and energetic properties of NNRTIs. As HIV-1 RT structures have been solved increasingly rapidly, opportunities to apply a protein-based design approach in order to obtain orientation, estimation of binding energies, and ranking of the potent designed compounds prior to synthesis have been developed. Both analog-based and protein-based design methods can be used to support each other. Moreover, molecular simulations such as MD and MC simulations are important tools for predicting the dynamics properties of the enzyme and inhibitor in the solutions.

Also, a number of mutations in the enzyme cause severe problems in anti-AIDS treatment. Therefore, in the case of NNRTIs, new potent inhibitors are still under development for activity against mutant enzymes. In this respect, structural information in terms of particular interactions of the inhibitor complexes with single and double mutations in the enzymes will be of great importance. These can be treated by the application of a variety of methods, as reviewed above. Explanation of the resistance profile for the second generation NNRTIs will be especially interesting, and should lead to suggestions for newer and improved candidates for future drug therapies.

Acknowledgements The author thanks Prof. Gupta for his encouragement in writing this review and Prof. Wolschann for many useful suggestions. Financial support from the Thailand Research Fund (BRG4780007) and National Research Council of Thailand (1.AU 49/2547), Postgraduates on Education and Research on Petroleum and Petrochemical Technology and KURDI are gratefully acknowledged. Thanks are due to Patchareenart Saparpakorn for excellent assistance in preparation of the manuscript and to W.J. Holzschuh for reading the manuscript.

References

1. Jonckheere H, Annè J, De Clercq E (2000) Med Res Rev 20:129
2. De Clercq E (2001) J Clin Virol 22:73
3. De Clercq E (2002) Biochim Biophys Acta 1587:258
4. De Clercq E (2004) Chem Biodiver 1:44
5. De Clercq E (1998) Antiviral Res 38:153
6. Lawtrakul L, Beyer A, Hannongbua S, Wolschann P (2004) Monatsh Chem 135:1033
7. Beyer A, Lawtrakul L, Hannongbua S, Wolschann P (2004) Monatsh Chem 135:1047
8. Ren J, Esnouf R, Garman E, Somers D, Ross C, Kirby I, Keeling J, Darby G, Jones Y, Stuart D, Stammers D (1995) Nat Struct Biol 2:293
9. Buckheit RW, Fliakas-Boltz V, Yeagy-Bargo S, Weislow O, Mayers DL, Boyer PL, Hughes SH, Pan BC, Chu SH, Bader JP (1995) Virology 210:186
10. Richman D, Shih CK, Lowy I, Rose J, Prodanovich P, Goff S, Griffin J (1991) Proc Natl Acad Sci USA 88:11241
11. Berman HM, Westbrook J, Feng Z, Gililand G, Bhat TN, Weissing H, Shindyalov IN, Bourne PE (2000) Nucleic Acids Res 28:235
12. Kohlstaedt LA, Wang J, Friedman JM, Rice PA, Steitz TA (1992) Science 256:1783
13. Rodgers DW, Gamblin SJ, Harris BA, Ray S, Culp JS, Hellmig B, Woolf DJ, Debouck CD, Harrison SC (1995) Proc Natl Acad Sci USA 92:1222
14. Wang J, Smerdon SJ, Jager J, Kohlstaedt LA, Rice PA, Friedman JM, Steitz TA (1994) Proc Natl Acad Sci USA 91:7472
15. Ren J, Nichols C, Bird L, Chamberlain P, Weaver K, Short S, Stuart DI, Stammer DK (2001) J Mol Biol 312:795
16. Lindberg J, Sigurdsson S, Löwgren S, Andersson HO, Sahlberg C, Noreen R, Fridborg K, Zhang H, Unge T (2002) Eur J Biochem 269:1670
17. Shafer RW, Kantor R, Gonzales MJ (2000) AIDS Rev 2:211
18. Carloni P, Alber F (eds) (2003) Quantum medicinal chemistry. Wiley, Weinheim
19. Höltje H-D, Sippl W, Rognan D, Folkers G (eds) (2003) Molecular modelling: basic principles and applications, 2nd edn. Wiley, Weinheim
20. Graves JD, Krebs EG (1999) Pharmacol Ther 82:111
21. Varnai P, Richards WG, Lyne PD (1999) Proteins 37:218

22. Ridder L, Rietjens IMCM, Vervoort J, Mulholland AJ (2002) J Am Chem Soc 124:9926
23. Laio A, Vandevondele J, Röthlisberger U (2002) J Chem Phys 116:6941
24. Laio A, Vandevondele J, Röthlisberger U (2002) J Phys Chem B 106:7300
25. Davies MS, Berners-Price SJ, Hambley TW (2000) Inorg Chem 39:5603
26. Legendre F, Bas V, Kozeika J (2000) Chem Eur J 6:2002
27. Davies MS, Berners-Price SJ, Hambley TW (1998) J Am Chem Soc 120:11380
28. Leach AR (1996) Molecular modelling: principles and applications. Longman, Essex
29. Kubinyi H (1993) QSAR:Hansch analysis and related approaches. In: Mannhold R, Krogsgarrd Larsen P, Timmerman H (eds) Methods and principles in medicinal chemistry, vol 1. Wiley, Weinheim
30. Kubinyi H (ed) (1993) 3D QSAR in drug design: theory, methods and applications. ESCOM, Leiden
31. Gupta SP, Garg RJ (1996) Enz Inh 11:23
32. Barreca ML, Carotti A, Rao A (1999) Bioorg Med Chem 7:2283
33. Silverman BD, Platt DE (1996) J Med Chem 39:2129
34. Luco JM, Ferritti FH (1997) J Chem Inf Comp Sci 37:392
35. Tronchet JMJ, Grigorov M, Dolatshahi N, Moriaud F, Weber JA (1997) Eur J Med Chem 32:279
36. Jalali-Heravi M, Parastar F (2000) J Chem Inf Comp Sci 40:147
37. Gussio R, Pattabiraman N, Zaharevitz DW, Kellogg GE, Toplo IA, Rice WG, Schaeffer CA, Erickson JW, Burt SK (1996) J Med Chem 39:1645
38. Hopfinger AJ, Duraiswami C (1997) J Am Chem Soc 119:10509
39. Hannongbua S, Lawtrakul L, Limtrakul J (1996) J Comput Aided Mol Des 10:145
40. Hannongbua S, Pungpo P, Limtrakul J, Wolschann P (1999) J Comput Aided Mol Des 13:563
41. Hall LH, Kier LB (1992) Rev Comput Chem 2:367
42. Randic M (1975) J Am Chem Soc 97:6609
43. Balaban AT (1982) Chem Phys Lett 89:399
44. Prasithichokekul S, Pungpo P, Hannongbua S, Ecker G, Wolschann P (2001) In: Proceedings of the 5th annual national symposium on computational science and engineering. Academic, Bangkok, p 351
45. Hannongbua S, Nivasanond K, Lawtrakul L, Pungpo P, Wolschann P (2001) J Chem Inf Comput Sci 41:848
46. Pungpo P, Hannongbua S (2000) J Mol Graph Mod 18:581
47. Pungpo P, Hannongbua S, Wolschann P (2003) Curr Med Chem 10:1661
48. Medina-Franco JL, Rodriguez-Morales S, Juarez-Gordiano C, Hernandez-Campos A, Castillo R (2004) J Comp Aided Mol Des 18:345
49. Klabe G, Abraham U, Mietzner T (1994) J Med Chem 37:4130
50. Samee W, Ungwitayatorn J, Matayatsuk C, Pimthon J (2004) Science Asia 30:81
51. Zhou Z, Madura JD (2004) J Chem Inf Comput Sci 44:2167
52. Chen HF, Yao XJ, Li Q, Yuan SG, Panaye A, Doucet JP, Fan BT (2003) SAR QSAR Environ Res 14:455
53. Ishiki HM, Galembeck SE, do Amaral AT (2001) Rational approaches to drug design. In: Proceedings of the 13th European symposium on quantitative structure–activity relationships, Duesseldorf, Germany, 27 Aug–1 Sept, 2000. Prous Science, Barcelona, p 340
54. Kireev DB, Chretien JR, Grierson DS, Monneret C (1997) J Med Chem 40:4257
55. Hopkins AL, Ren J, Esnouf RM, Willcox BE, Jones EY, Ross C, Miyasaka T, Walker RT, Tanaka H, Stammers DK, Stuart DI (1996) J Med Chem 39:1589
56. Parreira RLT, Abrahao O, Galembeck SE (2001) Tetrahedron 57:3243

57. Chimirri A, Grasso S, Monforte A-M, Monforte P, Rao A, Zappala M, Bruno G, Nicolo F, Scopelliti R (1997) Farmaco 52:673
58. Lawtrakul L, Hannongbua S, Beyer A, Wolschann P (1999) Biol Chem 380:265
59. Lawtrakul L, Hannongbua S, Beyer A, Wolschann P (1999) Monatsh Chem 130:1347
60. Hannongbua S, Prasithichokekul S, Pungpo P (2001) J Comput Aided Mol Des 15:997
61. Abrahao O, Nascimento PBD, Galembeck SE (2001) J Comput Chem 22:1817
62. Hannongbua S, Saen-oon S, Pungpo P, Wolschann P (2001) Monatsh Chem 132:1157
63. Saen-oon S, Wolschann P, Hannongbua S (2003) J Chem Inf Comp Sci 43:1412
64. Das K, Lewi PJ, Hughes SH, Arnold E (2005) Prog Biophys Mol Biol 88:209
65. Vreven T, Morokuma K, Farkas O, Schlegel HB, Frisch MJ (2003) J Comput Chem 24:760
66. Zhang Y, Kua J, McCammon JA (2002) J Am Chem Soc 124:10572
67. Decker SA, Cundari TR (2001) J Organomet Chem 635:132
68. Guo QC, Lipscomb WN, Karplus M (2001) Proc Natl Acad Sci USA 98:9032
69. Maseras F, Morokuma K (1995) J Comput Chem 16:1170
70. Kuno M, Hannongbua S, Morokuma K (2003) Chem Phys Letts 380:456
71. Saen-oon S, Kuno M, Hannongbua S (2005) Proteins 61:859
72. Nunrium P, Kuno M, Saen-oon S, Hannongbua S (2005) Chem Phys Letts 405:198
73. Frisch MJ, Trucks GW, Schlegel HB, Scuseria GE, Robb MA, Cheeseman JR, Montgomery JA, Vreven JrT, Kudin KN, Burant JC, Millam JM, Iyengar SS, Tomasi J, Barone V, Mennucci B, Cossi M, Scalmani G, Rega N, Petersson GA, Nakatsuji H, Hada M, Ehara M, Toyota K, Fukuda R, Hasegawa J, Ishida M, Nakajima T, Honda Y, Kitao O, Nakai H, Klene M, Li X, Knox JE, Hratchian HP, Cross JB, Adamo C, Jaramillo J, Gomperts R, Stratmann RE, Yazyev O, Austin AJ, Cammi R, Pomelli C, Ochterski JW, Ayala PY, Morokuma K, Voth GA, Salvador P, Dannenberg JJ, Zakrzewski VG, Daniels AD, Farkas O, Rabuck AD, Raghavachari K, Ortiz JV (2003) GAUSSIAN03. Gaussian, Pittsburgh, PA
74. Dapprich ST, Komaromi I, Byun KS, Morokuma K, Frisch MJ (1999) J Mol Struct THEOCHEM 462:1
75. Kerdcharoen T, Morokuma K (2002) Chem Phys Letts 355:257
76. Frenkel D, Smit B (eds) (1996) Understanding molecular simulation: from algorithms to applications. Academic, Chestnut Hill
77. Zhou Z, Madrid M, Evansak JD, Madura JD (2005) J Am Chem Soc 127:17253
78. Shen L, Shen J, Luo X, Cheng F, Xu Y, Chen K, Arnold E, Ding J, Jiang H (2003) Biophys J 84:3547
79. Rodriguez-Barrios F, Balzarini J, Gago F (2005) J Am Chem Soc 127:7570
80. Weinzinger P, Hannongbua S, Wolschann P (2005) J Enz Inh Med Chem 20:129
81. Rizzo RC, Udier-Blagovic M, Wang DE-P, Watkins EK, Kroeger S, Marilyn B, Smith RH, Tirado-Rives J, Jorgensen WL, Western MC (2002) J Med Chem 45:2970
82. Wang DP, Rizzo RC, Tirado-Rives J, Jorgensen WL (2001) Bioorg Med Chem Lett 11:2799
83. Smith MBK, Lamb ML, Tirado-Rives J, Jorgensen WL, Michejda CJ, Ruby SK, Smith RH Jr (2000) Protein Eng 13:413
84. Weber HP (1994) J Comput Aided Mol Des 8:1
85. Colman PM (1994) Curr Opin Struct Biol 4:868
86. Kuntz ID (1992) Science 257:1078
87. Gubernator K, Böhm H (1997) Structure-based ligand design. In: Mannhold R, Krogsgarrd Larsen P, Timmerman H (eds) Methods and principles in medicinal chemistry, vol 6. Wiley, Weinheim

88. Ranise A, Spallarossa A, Cesarini S, Bondavalli F, Schenone S, Bruno O, Menozzi G, Fossa P, Mosti L, La Colla M, Sanna G, Murreddu M, Collu G, Busonera B, Marongiu ME, Pani A, La Colla P, Loddo R (2005) J Med Chem 48:3858
89. Silvestri R, Artico M, De Martino G, Novellino E, Greco G, Lavecchia A, Massa S, Loi AG, Doratiotto S, La Colla P (2000) Bioorg Med Chem 8:2305
90. Artico M, Silvestri R, Pagnozzi E, Bruno B, Novellino E, Greco G, Massa S, Ettorre A, Loi AG, Scintu F, La Colla P (2000) J Med Chem 43:1886
91. Good AC, Krystek SR, Mason JS (2000) Drug Discov Today 5:S61
92. Shoichet BK, McGovern SL, Wei B, Irwin JJ (2002) Curr Opin Chem Biol 6:439
93. Abagyan R, Totrov M (2001) Curr Opin Chem Biol 5:375
94. Hemmateenejad B, Tabaei S, Mohammad H, Namvaran F (2005) J Mol Struct THEOCHEM 732:39
95. Morris GM, Goodsell DS, Halliday RS, Huey R, Hart WE, Belew RK, Olson AJ (2001) Autodock reference manual, version 3.0.5. Scripps Research Institute, La Jolla, CA
96. Varnek A, Solov'ev VP (2005) Comb Chem High Throughput Screen 8:403
97. Shen J, Zu X, Feng C, Liu H, Luo X, Shen J, Chem K, Zhao W, Shen X, Jiang H (2003) Curr Med Chem 10:1241
98. Sangma C, Chuakheaw D, Jongkon N, Saenbandit K, Nunrium P, Uthayopas P, Hannongbua S (2005) Comb Chem High Throughput Screen 8:417
99. Maiorov V, Sheridan RP (2005) J Chem Inf Comput Sci 45:1017
100. Totrov M, Abagyan R (1997) Proteins 1(Suppl):215
101. Xue L, Godden JW, Stahura FL, Bajorath J (2004) J Chem Inf Comput Sci 44:1275
102. Li H, Li C, Gui C, Luo X, Chen K, Shen J, Wang X, Jiang H (2004) Bioorg Med Chem Lett 14:4671

Top Heterocycl Chem (2006) 4: 85–106
DOI 10.1007/7081_026
© Springer-Verlag Berlin Heidelberg 2006
Published online: 23 March 2006

QSAR Approach in Study of Mutagenicity of Aromatic and Heteroaromatic Amines

Marjan Vračko

Kemijski inštitut/National Institute of Chemistry, Hajdrihova 19, 1000 Ljubljana, Slovenia
Marjan.vracko@ki.si

Abstract In this chapter we give an overview on QSAR models for treating the mutagenicity of cyclic amines. An extensive discussion is focused on the topological, E-state, quantum chemical, and empirical descriptors ($\log P$) that are often used in corresponding models. Two case studies are presented in more detail. The conclusion addresses the OECD principles for validation of models that are used for regulatory purposes.

Keywords Amines · Pyriminoizodiamines · Mutagenicity · Descriptors · Neural network

Abbreviations

(Q)SAR	(Quantitative) structure–activity relationship
E_{HOMO}	Energy of the highest occupied orbital
E_{LUMO}	Energy of the lowest unoccupied orbital
2D, 3D	Two dimensional, Three dimensional
OECD	Organization for economic cooperation and development

1
Introduction

Governmental agencies and society in general have a great interest in setting the strategy for risk assessment of chemicals, which are extensively used in agriculture, industry, cosmetics, food, and for diagnostic and therapeutic purposes. An important toxicological endpoint in assessment of chronic toxicity is mutagenicity, which is associated with rodent carcinogenicity and potentially to human carcinogenicity [1]. In contrast to rodent carcinogenicity, which can be evaluated in time-consuming animal experiments, mutagenicity can be relatively easy determined in different strains of *Salmonella typhimurium* (Ames test) [2]. However, it should be emphasized that the correlation between mutagenicity and carcinogenicity is still under discussion [3]. Aromatic amines represent a very important class of industrial and environmental chemicals. They have been widely used in industry in production of polymers and rubber, as dyes and pigments, in agricultural chemicals, and also in the pharmaceutical industry. Many of them are carcinogens, mutagens, or other toxicants. In addition, many aromatic amines are present in cigarette smoke or they are created during cooking processes. Epidemiological studies indicate that they may be responsible for cancer [4, 5] and this fact was already recognized 70 years ago [6]. Due to their hazard potential the aromatic amines have been the subject of many in vivo and in vitro tests. The data obtained have been extensively used for SAR and QSAR modeling. In developing QSAR models two main strategies can be applied. One selects a set of congeneric compounds and uses the model to get an insight in the mechanism of activity. Alternatively, one takes a broad set of diverse chemicals and studies the general relationship between structure and activity. The second strategy also includes classification and the search for analogs. An overview over QSAR studies related to aromatic amines is given by Benigni et al. [7, 8]. Most of the models reported in [7] grounds on four descriptors: $\log P$, E_{HOMO}, E_{LUMO} and the energy of hydroxylation of the amine group. They support the hypothesis that transformation is necessary for activation of amines. The activation of an amine group runs in two steps. The first step is the hydroxylation of amine group, and the second is the formation of a nitrenium ion. In the following reports, authors used different descriptors and modeling techniques. Klopman et al. [9] used the CASE program to analyze the mutagenicity of about 100 aromatic amines. CASE searches for molecular fragments, which are statistically significantly correlated to the mutagenicity. The work was extended by Zhang et al. who included quantum chemical descriptors in the study [10]. Benigni et al. [1] reported a discriminating model for aromatic amines. Mutagenicity was expressed as number of revertants and as a class (mutagenic or not). They used descriptors $\log P$, E_{HOMO}, E_{LUMO}, molar refractivity, molar refractivity related to *ortho*-substituents, and an indicator variable. A stepwise regression equation was used for modeling and classification. Glende et al. [11] studied 4-aminobiphenyl and 2-aminofluorene and the role of substituents on amino sites in mutagenicity, both experimentally and using QSAR models. Authors suggested introduction of a steric descriptor to include the information on geometry into the models. Debnath et al. report the QSAR models based on $\log P$ and E_{HOMO},

E_{LUMO} descriptors [12]. Garg et al. [13] studied a set of 43 aminoazobenzene derivatives using CODESSA descriptors [14]. Authors compared linear regression and neural network models. Gramatica et al. [15] studied mutagenicity for a set of 146 aromatic amines. Mutagenicity was measured for *Salmonella typhimurium* TA98+S9 and TA100+S9 strains. Molecules were described by descriptors using the DRAGON package, which calculates more than 1400 different descriptors [16]. The data set of 146 compounds was taken from literature [12]. The experimental design program DOLPHIN was used for optimal division of compounds into training set and test set [17]. A genetic algorithm was used to select the most relevant descriptors [18]. Chung et al. [19] studied the correlation between physicochemical parameters and mutagenicity for a set of 11 benzidines and their analogs. No clear correlation between parameters oxidation potential, E_{LUMO}-E_{HOMO}, ionization potential, dipole moment, partition coefficient and mutagenicity was found. The inhibitory activity of flavonoids toward mutagenicity of heterocyclic amines was reported by Hatch [20]. Patlewicz et al. report a review of different models for evaluation of mutagenicity and carcinogenicity [21]. Votano et al. report an extensive study on 3363 diverse compounds, which studied their mutagenicity (Ames test) and applied three QSAR modeling techniques: artificial neural net, k-nearest neighbours, and decision forest [22]. Compounds were described by 148 descriptors classified as molecular connectivity descriptors, electrotopological indices (atom-type, bond-type and group-type E-state indices), and binary indices. Log P was not indicated as an important index. Lozano et al. studied a set of heterocyclic amines present in cooked food [23]. They studied the metabolism by human cytocrome P450 by applying 3D-QSAR methodology. Mattioni et al. [24] studied a set of 334 aromatic amine compounds and their genotoxicity, which was determined in the SOS Chromotest in the presence and absence of an S9 rat liver homogenate, and expressed binary as positive or negative. Genetic algorithm was applied for the selection of most relevant descriptors and the models were selected for their ability to minimize misclassifications and to minimize the false negative predictions. Contrera et al. [25] report several QSAR models for mutagenicity in different *Salmonella typhimurium* strains and *Escherichia coli* gene mutations in terms of OECD principles (see Sect. 5) using MDL QSAR modules.

In Sect. 2 we present an overview of descriptors and computational techniques. The discussion is focused on these descriptors that are often used in models treating the mutagenicity of amines. An extensive comment is dedicated to the quantum chemical descriptors. Orbital energies, particularly E_{HOMO} and E_{LUMO}, are important descriptors measuring the ability of molecules to accept or donate electrons. In Sects. 3 and 4 two case studies are presented. The first study is on a set of 12 pyriminoizodiamine isomers and the second on a diverse set of 95 aromatic and heteroaromatic amines. In Sect. 5 we introduce the OECD principles for validation of (Q)SAR models used for regulatory purposes.

2
Computational Methods

2.1
Descriptors

In QSAR modeling, the question of molecular representation is central. For the modeling, a molecule is represented as a multidimensional vector, i.e., a molecule is a point in multidimensional representational space. An ideal representation should be: unique, uniform, reversible, and invariant on rotation and translation of molecules. Unique means that different structures give different representations, uniform means that the dimension of representation is the same for all structures, reversible means that the structure can be unambiguously reconstructed from the representation vector. Furthermore, invariant means that the representation is not sensitive if a molecule is rotated or translated. It is not expected that we would find a general representation that fulfills all requirements simultaneously [26]. Nowadays, thousands of descriptors and structural representations are in use [26–29]. We will give a short overview, and references about descriptors that often appear in QSAR studies related to mutagenicity of aromatic amines are presented in more detail.

Fragments as Descriptors
A molecular structure can be encoded in terms of its fragments (substructures). In this representation an individual element of a representation vector shows the presence or absence of a particular fragment [10].

Structural Descriptors
Structural descriptors are parameters calculated from molecular structure. The information about chemical structure can be hierarchically ordered into 1D, 2D, 3D, quantum chemical, etc. classes [30].

One-Dimensional (Constitutional) Descriptors
To this class belong the discrete numerical descriptors that describe the basic molecular structure, e.g., molecular weight, number of atoms, number of rings, etc.

Two-Dimensional Descriptors
These are deduced from a topological picture (2D picture) of the molecules. The picture carries information on how the atoms are connected and what is the nature of bonds (structural formula of a molecule). Mathematically, the topology picture is described with the connectivity matrix. Pioneering work in this field was published in 1947 by Wiener on paraffin hydrocarbons [31]. It is defined as a half sum of the off-diagonal elements in the topological distance matrix. In the last few decades dozens of descriptors have been deduced

from the topological picture of molecules; these are known as topological indices. The reminiscent way of topological indices and the views of authors are given in [32–35]. To this class belongs the branching index proposed by Randić [36, 37] and other connectivity indices proposed by Kier and Hall [38], which appear as important descriptors in many QSAR models (see examples below).

Electrotopological (E-state) descriptors

These summarize topological information about a molecule with atomic properties [39]. A molecular E-state index is expressed as a sum of atomic E-state indices, which are composed of two parts. First is the intrinsic atomic part, and second is perturbation, which depends on its neighborhood (other atoms in the structure). The intrinsic part includes information about the σ- and π-orbitals, lone pairs, hydrogen atoms attached to heavy atoms, and the principal quantum numbers of valence electrons. The perturbation part is a sum of all other atomic parts modified by function, which descends with distance.

Geometrical Descriptors

Theses are calculated from the 3D molecular structure, which is defined with coordinates of all atoms in the molecule. The step from 2D to 3D description of molecules is the crucial one [40–42]. The 3D structure is an ambiguously defined quantity, which depends on molecular environment, i.e., it is different in crystal structure, in solutions, or in vacuo. If it is theoretically determined it depends on the computational method. However, the 3D structures form a basis for a broad range of descriptors. Examples are mass distribution descriptors such as moment of inertia or gravitation index, and shape indices such as shadow indices, surface area indices, and van der Waals indices [29, 40, 41].

Quantum Chemical Descriptors

These are derived from the electronic structure of molecules [43, 44]. The electronic structure is considered to be a complete description of a molecule and it determines the physicochemical properties of a molecule and its interaction with the environment. It is clear that these descriptors are often used in modeling of biological properties when specific covalent or electrostatic interactions are postulated. Basically, the results of quantum chemical calculations are orbital energies and molecular orbitals, which are mathematically the eigenvalues and eigenvectors of the corresponding Hamilton operator, respectively. The electronic wave function, which is the antisymmetric product of molecular orbitals, determines the distribution of electrons in a molecule. We must be aware that in calculating an electronic structure several assumptions and approximations are made. In ongoing quantum chemical calculations we construct the Schrödinger equation for electrons in an electrostatic field, which is defined with the spatial positions of atomic nuclei.

In this first step we have already made an assumption. In the so-called Born–Oppenheimer approximation the quantum mechanical description of atomic nuclei has been neglected. The electrostatic interaction between electrons is described with classical Coulomb potential. Here, the second assumption is made while the relativistic character of electrostatic interaction is neglected. In heavy atoms, the relativistic effects play a considerable role and, thus, the interaction should include relativistic corrections. The electron–electron interaction makes the solving of the Schrödinger equation impossible for real-world molecules. The only chemical systems for which an exact solution of a Schrödinger equation exists are the hydrogen atom and H_2^+ ion. To solve many-electron problems we usually apply the Hartree–Fock approximation. In this approximation the electron–electron interaction is replaced by an average electrostatic electron potential and exchange potential. The latter is quantum mechanical potential and has no counterpart in classical mechanics. The Hartree–Fock equation is solved in a self-consistent way, i.e., the solution (wave function) of one step determines the potential for the next step. The steps are repeated until convergence is reached. It has been rigorously proven that the Hartree–Fock potential best approximates the electron–electron interacting potential. Direct integration (solution) of the Hartree–Fock equation is still difficult for complex systems. In quantum chemistry we rather set the solution of the Hartree–Fock equation as a linear combination of atomic orbitals (LCAO) and so the Hartree–Fock equation becomes the Hartree–Fock–Roothaan matrix equation. In the last few decades several sets of atomic orbitals have been developed [45]. It must be emphasized that the results may drastically depend on the selected basis set. The treatment of electron–electron interactions beyond the Hartree–Fock approximation (usually referred to as electron correlation calculations) is possible within configuration interaction approximations or within many-body perturbation theory (for example, the Møller–Plesset perturbation scheme). The electron correlation effects should be considered particularly when the excited states are investigated. The calculations on this level are often referred as ab initio calculations because only fundamental physical parameters like electron charge and mass occur in the equations [45]. Ab initio calculations are time-consuming and are possible only for small or medium-sized molecules. If thousands of molecules have to be treated, a situation which is likely to occur in QSAR research, further approximations are necessary. In recent decades several semi-empirical methods have been developed. In these approximations the one- and two-electron integrals are replaced with empirically obtained parameters. Due to the fact that calculation of these integrals represents the bottle-neck of the calculation, the entire calculation becomes simpler and faster. For QSAR purposes the most-used semi-empirical methods are MNDO, Austin model 1 (AM1), parameterized model 1 (PM3), and parameterized model 5 (PM5) [46]. A scheme of approximations used in quantum chemistry is shown in Fig. 1. The next question is how we use and interpret the calculated results. According to the Koopmans' theorem, the E_{HOMO} and E_{LUMO} are related to ionization potential (IP) and electron affinity (EA), respectively [47]:

$$IP = - E_{HOMO}$$
$$EA = - E_{LUMO}$$

IP and EA express the readiness of a molecule to accept or donate electrons, respectively. Koopmans' theorem is an approximation. It rests on the assumption that the electron wave function of remaining electrons does not change if one electron is removed or added. Indeed, the electron structure of an ionized molecule differs from that of a neutral one. Further quantum chemical descriptors, like those related to electron delocalizability and polarizability, are described in [48]. It should be emphasized that the quantum chemical descriptors depend (sometimes drastically) on the selected method and approximation. An example is shown in [48] where different quantum chemical descriptors were calculated for set of 607 compounds. As a final note on quantum chemical descriptors, we emphasize that the information on electronic structures of molecules can be obtained from spectroscopic measurements. For example, the energies of individual electronic states can be directly measured in photoelectron spectroscopy experiments.

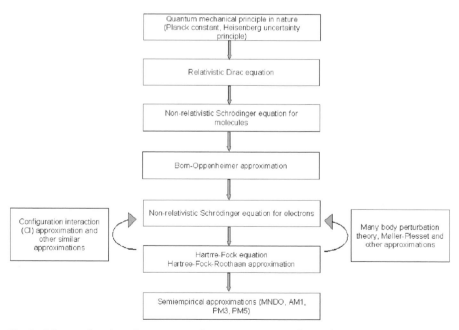

Fig. 1 Scheme showing the structure from quantum mechanical principles to quantum chemical methods

Hydrophobic Descriptors

Hydrophobicity (or lipophilicity) characterizes the readiness of a molecule to escape or to prefer the water environment. It plays a fundamental role in biochemical processes and influences the fate of a molecule in the environment. Thus, hydrophobic descriptors play an important role in QSAR modeling that is used in drug research and for risk characterization. The most widely used hydrophobic descriptor is the octanol–water partition coefficient ($\log P$) proposed by Hansh [49]. P is a quotient between solubilities in octanol and water. It is defined by following equation:

$$P = \frac{[\text{compound}]_{\text{octanol}}}{[\text{compound}]_{\text{water}}(1 - \alpha)} ,$$

where α measures degree of dissociation. For nonionizable compounds the $\log P$ is independent of pH. When a compound is ionizable $\log P$ is often replaced with pH-dependent $\log D$ values [50]. Log P and $\log D$ were introduced on the assumption that transport from the site of application of drug to its site of action depends on the lipophilicity of a molecule. Log P can be precisely measured experimentally. On the other hand, it can be (easily) calculated (which is not always the case). Several commercial and free computer programs that are available for calculation of $\log P$ values mostly used the fragmental approach. In this approach a molecule is broken down into fragments and $\log P$ is the sum of fragmental constants. They are determined from large set of data with statistical analysis (regression). Many publications report the comparison of results obtained by different programs. For example, Medić-Šarić et al. [51] report measured and calculated $\log P$ values for 3,3'-(2-methoxybenzylidene)bis(4-hydroxycoumarin).The measured value was 2.5, while values calculated with six different programs span the range from 2.54 to 4.54. A further comparison study for 22 drugs is reported in [52, 53]. The model proposed by Hansh also includes two further empirical descriptors. The Hammet and Taft substituents constant describes the electronic and steric properties of molecule.

Encoding of Structures

Beside the descriptors, further attempts have been made to encode the 3D molecular structures with functions. Such are 3D-MoRSE code [54] spectrum-like representations [55] and radial distribution functions [56]. Also, experimentally determined infrared, mass, or NMR spectra can be taken to represent a molecule [57]. Another example is comparative molecular field analysis (CoMFA) where the molecular 3D structures are optimized together with the receptor [58]. This approach is often applied in drug design or in specific toxicology studies where the receptor is known. The field of molecular descriptors and molecular representations has exploded in the recent decades. Over 200 programs for calculating descriptors and different QSAR applications are listed on web page [59].

2.2
Modeling Methods

Self-Organizing Maps
When the representation of molecular structure is set, the next question in the building of a model concerns the selection of modeling method. In the most of the "standard" QSAR models the multiple linear regression method has been used to express the relationship between descriptors and property (toxicity). However, mathematically more complex methods are now more and more in use for modeling [60, 61]. These include neural networks, expert systems, genetic algorithm for descriptor selection, fuzzy logic approaches, and other techniques for data mining. In the two case studies presented below we have applied the self-organizing maps (Kohonen neural networks) and the counter propagation neural networks and, therefore, we shall give a short introduction to both methods.

The principle of a self-organizing map is the mapping from a multidimensional descriptor space onto a 2D grid of neurons (Kohonen map). The mapping is done over a nonlinear learning algorithm, which repeatedly passes through the data and adjusts the weights of the network to minimize the error. A final model represents an arrangement of objects (molecules) into a 2D network (self-organizing map) on the basis of similarity among objects. Such a network is a visualization of data. Our mind cannot analyze objects located in multidimensional descriptor space but it can very efficiently analyze 2D pictures. A generalization of the self-organizing maps is the counter propagation neural network, which includes the output values (toxicity values) as well as the input variables (descriptors) into a training. Details about the architecture and learning strategy of both methods can be found in numerous text books and articles [60, 62, 63]. Both methods can be used to treat different problems related to QSAR modeling: to build the predictive models, for analyzing clusters, for classification, and for descriptor selection [63–67].

3
Case Studies – Data Sets

3.1
Data Set 1

During the cooking processes, numerous genotoxic heterocyclic amines are created and some of them have been found in rodent bioassays to be carcinogenic [68]. Felton et al. [69] report the data for a set of 12 pyriminoizodiamine isomers, which are shown in Table 1. The mutagenic potencies, which are reported for ten isomers, were determined for two strains of *Salmonella typhimurium*, YG1024 and TA98 and are expressed as number of revertants per

Table 1 Structures of 12 isomers together with mutagenic activities, which are expressed as natural logarithm of the number of revertants. (For the TA98 the number of revertants is multiplied by a factor of ten to avoid negative numbers)

Isomer no.		ln(R)YG1024	ln(10*R)TA98
1		1.946	1.609
2		1.902	1.609
3		Not measured	Not measured
4		1.792	2.197

Table 1 (continued)

Isomer no.		ln(R)YG1024	ln(10*R)TA98
5		2.219	3.135
6		2.603	2.773
7		4.635	5.635
8		2.460	4.043
9		6.319	8.061

Table 1 (continued)

Isomer no.		ln(R)YG1024	ln(10*R)TA98
10		Not measured	Not measured
11		6.766	8.061
12		2.747	3.434

nanomole. Although the isomers form a congeneric set of compounds they show very different mutagenic potencies. The number of revertants spans a range over two orders of magnitude, from < 0.5 to 300 and from 6 to 860 from TA98 and YG1024, respectively. It is clear that the set is an interesting example for QSAR studies. For the modeling, the number of revertants (R) was converted to the natural logarithm (ln R). (In the reported study the numbers R in modeling of the TA98 were multiplied by a factor of ten to avoid negative numbers).

3.2
Data Set 2

As the second case study we give an overview of QSAR models on set of 95 aromatic and heteroaromatic amines, which was compiled by Debnath et al. [12]. The mutagenic potency was expressed as log R, where R is the number of revertants per nanomole in the strain *Salmonella typhimurium* TA98+S9. The entire data set or subsets have been a target of many QSAR studies. In the original paper authors proposed a linear regression QSAR model with four descriptors: $\log P$, an indicator variable, and orbital energies E_{HOMO} and E_{LUMO}. Cash [70] developed QSAR models using electrotopological state indices as descriptors. Furthermore, Cash et al. [71] report the validation of models using an external set of 29 aromatic amines. Maran et al. [72] used atomic surface areas, Coulombic and exchange energies as descriptors. Karelson et al. [73] used an error-back propagation method and reported an improvement in comparison to the six-parameter linear model. Basak and coauthors published several articles on this data set exploring different descriptors and different modeling techniques [74–76]. In the description of molecules they used hierarchical ordering of descriptors: topological, topochemical, 3D and quantum chemical descriptors [76].

4
Results

4.1
Set of 12 Pyriminoizodiamine Isomers

Ten isomers were used for building the models, which were further used for prediction of two isomers with no experimental data. In the first modeling experiments the program package CODESSA was used for considering the topological and geometrical descriptors. For the selection of descriptors and for modeling the *heuristic option* was applied. In this option the selection algorithm is as follows. The program calculates all correlations between individual descriptors and property (mutagenic potency) and eliminates descriptors that not fulfill following criteria:

1. F-test's value is below 1
2. Correlation coefficient is less than $r_{\min}$, which was set to 0.1
3. The t value is less than t_1, which was set to 0.1

In this way 12 descriptors were selected, which are shown in Table 2. The correlation coefficients for the best models built with selected descriptors were: for mutagenicity YG1024 $r^2 = 0.713$ for model and $r^2_{cv} = 0.506$ for leave-one-out cross validation test; and for mutagenicity TA98 $r^2 = 0.775$ and $r^2_{cv} = 0.503$, respectively.

In the second modeling experiment the orbital energies were considered as descriptors. The orbital energies were calculated with 6-311 G basis set whereas the geometries were optimized with AM1 approximation. Program GAUSSIAN 94 was using to perform all calculations [77]. Twenty seven occupied molecular orbital energies of valence region and five unoccupied

Table 2 Twelve selected descriptors and the squares of correlation coefficients (r^2) to YG1024 and TA98

	Descriptor	r^2 YG1024	r^2 TA98
1	Molecular volume/XYZ box	0.6751	0.5804
2	ZX shadow/ZX rectangle	0.4544	0.3106
3	Average information content (order 1)	0.3121	0.3370
4	Kier&Hall index (order 2)	0.2608	0.4194
5	Wiener index	0.2551	0.2451
6	Moment of inertia C	0.2483	0.2195
7	YZ shadow/YZ rectangle	0.2420	0.1712
8	Moment of inertia B	0.1866	0.1521
9	XY shadow/XY rectangle	0.1837	0.2369
10	Molecular surface area	0.1698	0.1415
11	Moment of inertia A	0.1619	0.1186
12	Gravitation index (all pairs)	0.1543	0.1473

Table 3 Orbital energies with the highest correlation coefficients to mutagenicities

	Orbital energy	r^2 YG1024	r^2 TA98
1	LUMO+1	0.7204	0.7082
2	HOMO-8	0.4827	0.4342
3	HOMO-6	0.3974	0.3855
4	LUMO+2	0.3784	0.2736
5	HOMO-3	0.3727	0.3078
6	HOMO	0.2855	0.2599
7	HOMO-9	0.2314	0.2718
8	HOMO-8	0.2025	0.3010
9	LUMO+4	0.2002	0.1806
10	HOMO-4		0.1748

molecular orbital energies were using as descriptors. Program CODESSA was used to correlate the orbital energies with mutagenic potencies and to find the most correlated multiparametric models. The correlation coefficients between orbital energies and mutagenic potencies are shown in Table 3. The correlation coefficients for the best models obtained with three orbital energies were: for YG1024 $r^2 = 950$ and $r^2_{cv} = 0.847$, and for TA98 $r^2 = 0.968$ and $r^2_{cv} = 0.843$.

In the following study, selected descriptors were used for counter propagation neural network modeling. Figures 2a and b show the positions of molecules in a Kohonen network. The two molecules and their neighbors are indicated. The predictions for two molecules with unknown mutagenicity

Table 4 Predictions for mutagenicities expressed as number of revertants for molecules 3 and 10

Model	Predictions			
	YG1024		TA98	
	Mol 3	Mol 10	Mol 3	Mol 10
Linear CODESSA	1.61	7.47	0.38	2.63
Linear MO	10.28	9.94	0.87	0.26
Neural net MO	7.99	13.95	1.04	4.15

values are shown in Table 4. It is evident that the predictions show considerable variation although they were obtained by models with acceptable statistical parameters. Further details on this study can be found in [78].

4.2
Set of 95 Aromatic and Heteroaromatic Amines

This is an overview of work reported by Valkova, Basak, Jezierska, and Vračko [67, 74–76, 79, 80]. The counter propagation neural networks were used to address important problems in QSAR modeling: selection of descriptors, division of data set into training and test set, determination of outliers, and analysis of clusters in the training set. As a starting point we had a pool of 375 calculated descriptors, which were hierarchically organized in topological, topochemical, geometrical, and quantum chemical classes. Some 359 topological and topochemical descriptors were calculated with programs POLLY, Triplet, and Molconn-Z 3.50 [81]. Ten geometrical descriptors were calculated with Molconn-Z and Sybyl 6.2 programs. Six quantum chemical descriptors were calculated within AM1 approximation using standard geometry optimization procedure. A set of 359 descriptors was reduced using the VARCLUS procedure of the SAS statistical package [82]. The reduced set of descriptors used for modeling consisted of five topological, 16 topochemical, ten geometrical, and six quantum chemical descriptors. The comparison of multiple linear regression models and counter propagation models is shown in [74, 80]. The models were built considering different classes of descriptors. The Kohonen network and the counter propagation neural methods were applied to analyze to similarity relationships among objects of the data set. Table 4 from [80] shows the compounds located on the same neurons. These compounds are recognized by the model as identical or very similar. It is expected that different classes of descriptors indicate different similarity relationships among compounds. The first example shows that the model considers as identical p-cresidine and m-cresidine when topological/topochemical or 3D descriptors are used, and p-cresidine and o-cresidine when quantum chemical descriptors are used.

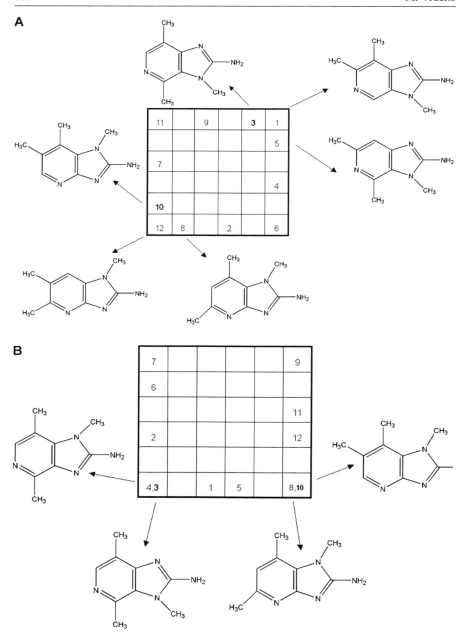

Fig. 2 Kohonen maps built with orbital energies as descriptors (**a**) and CODESSA descriptors (**b**)

The second example shows that 4-aminopyrene and 3-aminofluoranthene are considered as identical when topological/topostructural indices are used; to them the 3,3'-dichlorobenzidine is attached when 3D descriptors are used. When quantum chemical descriptors are used the 4-aminopyrene is similar to 1-aminopyrene. In the next example the counter propagation model was used to predict the mutagenicity for two highly active compounds: 4-aminopyrene and 2,7-diaminophenazine. The prediction for 4-aminopyrene was made on the basis of highly active 1-aminophenantrene and 3-aminofluoranthene. It should be emphasized that all three compounds express a substantial degree of structural similarity. On the other hand, the highly active compound 2,7-diaminophenazine was situated in the network closely to six phenanzines (1,7-diaminophenazine, 2-aminophenazine, 2,8-diaminophenazine, 1,6-diaminophenazine, 1,9-diaminophenazine, and 1-aminophenazine) showing mutagenic potency between – 2.1 and 1.12. The 2,7-diaminophenazine with an activity of 3.97 is an outlier in this subset.

Jezierska et al. [67] applied the Kohonen neural network to select the most relevant descriptors. Here a Kohonen network is built with the "transposed" matrix, i.e., with the matrix where the roles of descriptors and molecules are exchanged. From the map of descriptors, 36 descriptors were selected. This number of descriptors was further reduced to six, five, four, or three descriptors. Statistical parameters of compared models are reported in Table 3 of [67]. It is evident from the table that the model built with four selected descriptors show comparable parameters to the model built with 36 descriptors. The selected descriptors belong to topostructural and topochemical classes.

Valkova et al. [79] studied the same set with the same descriptors. The question addressed was how robust is the data set, or can we divide the entire set into two subsets, training and test set. The set of 95 compounds with descriptors as described above was divided into a training set and a test set. The program SphereExcluder was used for division of the set into training and test set [83]. The division was determined with two indices: the first is the diversity index of training set $I_{tr,tr}$ and the second is the diversity index of the test set with respect to training set $M_{te,tr}$. The precise definition of both indices is given in [84]. Considering the optimal values for indices ($I_{tr,tr} = 1$ and $M_{te,tr} = 0$) the set was divided into 29 (31%) compounds for the training set and 66 (69%) compounds for the test set. Figure 1 in [79] shows the distribution of compounds in the Kohonen map. One recognizes equivocal distribution of training and test compounds over entire informational space. After analyzing the individual predictions, the 2,2'-diaminobiphenyl and 2-amino-3'-nitrobiphenyl were deleted as outliers. The final model was built with 31 training compounds and tested with 62 compounds. The correlation coefficients for training set and test set were $r^2 = 0.986$ and $r^2_{cv} = 0.816$, respectively. To test the model, two additional tests were performed. First, the roles of the training and test sets were exchanged, i.e., the model was trained with the test set and tested with the training set. The corresponding correlation coefficient for the prediction for the "new" test set was $r^2 = 0.728$. The performance of this model was lower than for the original model, however, it is still acceptable. In the second test the mutagenicity values were randomly permuted for the compounds of the training set (scrambling test). A model built with scrambled mutagenicity values should lose its predictive ability because the relationship between structures and mutagenicity values is deliberately broken. The predictions for the test set confirm this (correlation coefficient $r^2 = 0.25$). Further details of this study are reported in [79].

5
Conclusions

Mutagenicity is an important biological endpoint in assessing the potential risk of chemicals [85]. There are more than 82 000 chemicals in current use and the list in the the Toxic Substances Control Act (TSCA) is growing by nearly 2000 to 3000 new compounds per year. Only about 15% of TSCA chemicals have any mutagenicity data. It is of general interest to develop and use the QSAR models for evaluation of the mutagenic potency of chemicals.

For the QSAR models, which are used for regulatory purposes, the OECD adopted five principles for validation of QSAR models [86]. The principles may ensure the statistical reliability and transparency of models. Briefly, a (Q)SAR model considered for regulatory purposes should be associated with the following information:

- *A defined endpoint.* A (Q)SAR should be associated with a "defined endpoint", which refers to any physicochemical, biological, or environmental effect. The intent of this principle is to ensure transparency of the modeled data.
- *An unambiguous algorithm.* The intent of this principle is to ensure the transparency of the modeling algorithm. Sometimes, it is a difficult task to satisfy this principle, particularly when complex methods like neural networks or fuzzy logic techniques are used for modeling.
- *A defined domain of applicability.* QSAR models are limited in terms of the types of chemical structures, physicochemical properties, and mechanisms of actions.
- *Appropriate measures of goodness-of-fit, robustness and predictivity.* This principle expresses the need to provide two types of information: the internal performance of a model (as expressed by goodness-of-fit and robustness) and the predictivity of a model using an appropriate test set.
- *A mechanistic interpretation, if possible.* The intent of this principle is to ensure that there is an assessment of the possibility of a mechanis-

tic association between the descriptors used in a model and the endpoint being predicted, and that any association is documented. Unfortunately, such interpretation is not always possible from a scientific point of view.

In our contribution we have focused the discussion on descriptors. The understanding of descriptors is essential for transparency of models and can also lead to mechanistic interpretation of models. Several questions are associated with descriptors. First of all, nowadays thousand of descriptors are defined and can be easily calculated with available software and the first question is how to the select the most relevant descriptors. The topological descriptors are sometimes promising, but there is no clear physicochemical interpretation for them. 3D molecular structure is a problematic quantity as it depends on the media where the molecule is, or on the method of determination. Quantum chemical descriptors, which have a clear physicochemical interpretation, are difficult to calculated. In the cases studies we have addressed some of those questions. We have discussed the sensitivity of the models, and particularly predictions, to descriptors used. According to the critical review of Snyder and Smith [87] on QSAR models for mutagenicity prediction a lot of work still remains to be done.

References

1. Benigni R, Giuliani A, Franke R, Gruska A (2000) Chem Rev 100:3697
2. McCann J, Choi E, Yamasaki E, Ames BN (1975) Proc Natl Acad Sci USA 72:5135
3. Zeiger E (2001) Mutat Res 492:29
4. Dolin PJ (1992) Br J Cancer 65:476
5. Vincis P, Pirastu R (1997) Cancer Cause Control 8:346
6. Kinosita R (1937) Tr Soc Path Jap 27:665
7. Benigni R, Giuliani A, Gruska A, Franke R (2003) QSARs for the mutagenicity and carcinogenicity of the aromatic amines. In: Benigni R (ed) Quantitative structure–activity relationship (QSAR) models of mutagens and carcinogens. CRC, Boca Raton FL, p 125
8. Benigni R (2005) Chem Rev 105:1767
9. Klopman G, Frierson MR, Rosenkranz HS (1985) Environ Mutagen 7:625
10. Zhang YP, Klopman G, Rosenkranz HS (1993) Environ Mol Mutagen 21:100
11. Glende C, Klein M, Schmitt H, Erdinger L, Boche G (2002) Mutat Res 515:15
12. Debnath AK, Debnath G, Shusterman AJ, Hansh C (1992) Environ Mol Mutagen 19:37
13. Garg A, Bhat KL, Bock CW (2002) Dyes Pigments 55:35
14. CODESSA^{TM}, v20, Semichem, 7204 Mullen, Shawnee KS 66216, USA
15. Gramatica P, Connsonni V, Pavan M (2003) SAR QSAR Environ Res 14:237
16. Todeschini R, Consonni V, Mauri A, Pavan M (2002) DRAGON rel 2.1 for Windows, Milano, Italy
17. Todeschini R, Mauri A (2000) DOLPHIN rel 2.1 for Windows, Milano, Italy
18. Leardi R, Boggia R, Terrile M (1992) J Chemom 6:267

19. Chung KT, Chen SC, Wong TY, Li YS, Wei CI, Chou MW (2000) Toxicol Sci 56:351
20. Hatch FT, Lightstone, Colvin ME (2000) Environ Mol Mutagen 35:279
21. Patlewicz G, Rodford R, Walker JD (2003) Environ Toxicol Chem 22:1885
22. Votano JR, Parham M, Hall LH, Kier LB, Oloff S, Tropsha A, Xie Q, Tong W (2004) Mutagenesis 19:365
23. Lozano JJ, Pastor M, Cruciani G, Gaedt K, Centeno NB, Gago F, Sanz F (2000) J Comput Aid Mol Des 14:341
24. Mattioni BE, Kauffman GW, Jurs PC, Custer LL, Durham SK, Pearl GM (2003) J Chem Inf Comput Sci 43:949
25. Contrera JF, Matthews EJ, Kruhlak NL, Benz RD (2005) Regul Toxicol Pharmacol 43:313
26. Zupan J, Vračko M, Novič M (2000) Acta Chim Slov 47:11
27. Todeschini R, Consonni V (2000) The handbook of molecular descriptors. Wiley, New York
28. Schuur JH, Selzer P, Gasteiger J (1996) J Chem Inf Comput Sci 36:334
29. Diudea MV (ed) (2001) QSPR/QSAR studies by molecular descriptors. Nova Science, Hungtington, New York
30. Basak SC, Mills D (2001) SAR QSAR Environ Res 12:481
31. Wiener H (1947) J Am Chem Soc 69:2636
32. Balaban A (2001) A personal view about topological indices for QSAR/QSPR. In: Diudea MV (ed) QSPR/QSAR studies by molecular descriptors. Nova Science, Hungtington, New York
33. Devillers J, Balaban AT (eds) (1999) Topological indices and related descriptors in QSAR and QSPR. Gordon and Breach, Reading, UK
34. Randić M (1998) Topological indices. In: Schleyer PvR, Allinger NL, Clark T, Gasteiger J, Kollman PA, Schaefer III HF, Schreiner PR (eds) Encyclopedia of computational chemistry. Wiley, Chichester
35. Netzeva TI (2004) Whole molecule and atom-based topological descriptors. In: Cronin MTD (ed) Predicting chemical toxicity and fate. CRC, Boca Raton FL
36. Randič M (1975) J Am Chem Soc 97:6609
37. Randič M (2001) J Mol Graphics Modelling 20:19
38. Kier LB, Hall LH (1976) J Pharm Sci 65:1806
39. Rose K, Hall LH, Kier LB (2002) J Chem Inf Comput Sci 42:651
40. Katritzky AR, Lobanov VS, Karelson M (1994) CODESSA Reference manual 2.0, Gainesville
41. Rohrbaugh RH, Jurs PC (1987) Anal Chim Acta 199:99
42. Randić M, Razinger M (1997) On characterization of 3D molecular structure. In: Balaban AT (ed) From chemical topology to three-dimensional structure. Plenum, New York
43. Thouless DJ (1972) The quantum mechanics of many-body systems. Academic, New York
44. Schaefer III HF (1977) Methods of electronic structure theory. Plenum, New York
45. Hehre WJ, Radom L, Schleyer PR, Pople JA (1986) Ab initio molecular orbital theory. Wiley, New York
46. Murrell JN, Herget AJ (1972) Semi-empirical self-consistent-field-molecular theory of molecules. Wiley, New York
47. Koopmans T (1934) Physica 1:104
48. Schüürmann G (2004) Quantum chemical descriptors in structure–activity relationships – calculation, interpretation, and comparison of methods. In: Cronin MTD (ed) Predicting chemical toxicity and fate. CRC, Boca Raton FL
49. Hansch C, Maloney PP, Fujita T, Muir RM (1962) Nature 194:178

50. Silverman RB (2004) The organic chemistry of drug design and drug action. Elsevier, Amsterdam, p 55
51. Medić-Šarić M, Mornar A, Badovinac-Črbjević T, Jasprica I (2004) Croat Chem Acta 1–2:367
52. Petrauskas AA, Kolovanov EA (2000) Perspect. Drug Discovery and Design 19:99
53. Eros D, Kovesdi I, Orfi L, Takacs-Novak K, Acsady G, Keri G (2002) Current Med Chem 9:1819
54. Schuur JH, Selzer P, Gasteiger J (1996) J Chem Inf Comput Sci 36:334
55. Zupan J, Vračko M, Novič M (2000) Acta Chim Slov 47:19
56. Hemmer CM, Gasteiger J (2000) Anal Chim Acta 420:145
57. Bursi R, Dao T, Wijk Tv, Gooyer Md, Kellenbach E, Verwer P (1999) J Chem Inf Comput Sci 39:861
58. Cramer RD, DePriest SA, Patterson DE, Hecht P (1993) The developing practice of comparative molecular field analysis. In: Kubinyi H (ed) 3D QSAR in drug design theory, methods and applications, vol 1. ESCOM Leiden 443–485
59. http://www.ndsu.nodak.edu/qsar_soc/resource/software.htm
60. Leardi R (ed) (2003) Nature-inspired methods in chemometrics: Genetic algorithms and artificial neural networks. Elsevier, Amsterdam
61. Helma C (ed) (2005) Predictive toxicology. Taylor Francis, Boca Raton FL
62. Zupan J, Gasteiger J (1999) Neural networks in chemistry and drug design. Wiley, Weinheim
63. Vračko M (2005) Curr Comput-Aided Drug Des 1:73
64. Panek JJ, Jezierska A, Vračko M (2005) J Chem Inf Model 45:264
65. Spycher S, Pellegrini E, Gasteiger J (2005) J Chem Inf Model 45:200
66. Roncaglioni A, Novič M, Vračko M, Benfenati E (2004) J Chem Inf Comput Sci 44:300
67. Jezierska A, Vračko M, Basak SC (2004) Mol Divers 8:371
68. Shirai T, Sano M, Tamano S, Takahashi S, Hirose T, Futakuchi M, Hasegawa R, Imaida K, Matsumoto K-I, Wakabayashi K, Sugimura T, Ito N (1997) Cancer Res 57:195
69. Felton JS, Knize MG, Hatch FT, Tanga MJ, Colvin ME (1999) Cancer Letters 143:127
70. Cash GG (2001) Mutat Res Genet Toxicol Environ Mutagen 491:31
71. Cash GG, Anderson B, Mayo K, Bogaczyk S, Tunkel J (2005) Mutat Res 585:170
72. Maran U, Karelson M, Katritzky AR (1999) Quant Struct-Act Relat 18:3
73. Karelson M, Sild S, Maran U (2000) Mol Simul 24:229
74. Basak SC, Mills D (2001) SAR QSAR Environ Res 12:481
75. Basak SC, Mills D, Balaban AT, Gute BD (2001) J Chem Inf Comput Sci 41:671
76. Basak SC, Gute BD, Grunwald GD (1998) Relative effectiveness of topological, geometrical, and quantum chemical parameters in estimating mutagenicity of chemicals. In: Chen F, Schüürmann G (eds) Proceedings of the quantitative structure–activity relationships in environmental sciences. VII SETAC, Pensacola FL, p 245
77. Frisch MJ, Trucks GW, Schlegel HB, Gill PMW, Johnson BG, Robb MA, Cheeseman JR, Keith T, Petersson GA, Montgomery JA, Raghavachari K, Al-Laham MA, Zakrzewski VG, Ortiz JV, Foresman JB, Peng CY, Ayala PY, Chen W, Wong MW, Andres JL, Replogle ES, Gomperts R, Martin RL, Fox DJ, Binkley JS, Defrees DJ, Baker J, Stewart JP, Head-Gordon M, Gonzalez C, Pople JA (1995) GAUSSIAN 94. Gaussian Inc, Pittsburgh, PA
78. Vračko M, Szymoszek A, Barbieri P (2004) J Chem Inf Comput Sci 44:352
79. Valkova I, Vračko M, Basak SC (2004) Anal Chim Acta 509:179
80. Vračko M, Mills D, Basak SC (2004) Environ Toxicol Pharmacol 16:25
81. MOLCONN-Z (2000) Version 3.5. Hall Associates Consulting, Quincy, MA

82. SAS Institute(1988) Release 6.03. Cary, NC
83. Golbraikh A, Tropsha A (2003) J Comput-aided Mol Des 16:357
84. Golbraikh A (2000) J Chem Inf Comput Sci 40:414
85. OECD series on testing and assessment, Number 12. ENV/JM/MONO(99)2
86. The principles for establishing the status of development and validation of (quantitative) structure–activity relationships [(Q)SARs]. OECD document ENV/JM/TG(2004)27
87. Snyder RD, Smith MD (2005) Drug Discov Today 10:1119

Top Heterocycl Chem (2006) 4: 107–159
DOI 10.1007/7081_024
© Springer-Verlag Berlin Heidelberg 2006
Published online: 24 March 2006

Modeling Reaction Mechanism
of Cocaine Hydrolysis and Rational Drug Design
for Therapeutic Treatment of Cocaine Abuse

Chang-Guo Zhan

Department of Pharmaceutical Sciences, College of Pharmacy, University of Kentucky,
725 Rose Street, Lexington, KY 40536, USA
zhan@uky.edu

Abstract Cocaine is a widely abused heterocyclic drug and there is no available anti-cocaine therapeutic. The disastrous medical and social consequences of cocaine addiction have made the development of an effective pharmacological treatment a high priority. An ideal anti-cocaine medication would accelerate cocaine metabolism producing biologically inactive metabolites. The main metabolic pathway of cocaine in the body is hydrolysis at its benzoyl ester group. State-of-the-art molecular modeling of the reaction mechanism for the hydrolysis of cocaine and the mechanism-based design of anti-cocaine therapeutics will be discussed. First of all, competing reaction pathways and the transition state stabilization of the spontaneous hydrolysis of cocaine in solution will be examined. It will be demonstrated that the information obtained about the transition states and their stabilization has been very useful in the rational design of stable analogs of the transition states of cocaine hydrolysis, in order to elicit anti-cocaine catalytic antibodies. Detailed molecular modeling of the reaction mechanism for cocaine hydrolysis catalyzed by human butyrylcholinesterase (BChE), the primary cocaine-metabolizing enzyme in body, will be examined. Then, we will describe the application of these mechanistic insights to the rational design of human BChE mutants as a new therapeutic treatment of cocaine abuse. Finally, future directions of the mechanism-based design of anti-cocaine therapeutics will be discussed.

Keywords Cocaine · Hydrolysis mechanism · Transition-state simulation · Rational enzyme redesign · Catalytic antibody

Abbreviations

ACh	Acetylcholine
AChE	Acetylcholinesterase
BCh	Butyrylcholine
BChE	Butyrylcholinesterase
QM	Quantum mechanics
MM	Molecular mechanics
QM/MM	Quantum mechanics/molecular mechanics
MD	Molecular dynamics
BE	Benzoylecgonine
EME	Ecgonine methyl ester
CNS	Central nervous system
PET	Positron emission tomography
$B_{AC}2$	Base-catalyzed, acyl-oxygen cleavage, bimolecular
IRC	Intrinsic reaction coordinate
TSA	Transition state analog
TS	Transition state
TS1	Transition state for the first reaction step
TS2	Transition state for the second reaction step
TS3	Transition state for the third reaction step
TS4	Transition state for the fourth reaction step
INT	Intermediate
INT1	First intermediate
INT2	Second intermediate
INT3	Third intermediate
ES	Prereactive enzyme–substrate complex
SCRF	Self-consistent reaction field

SVPE   Surface and volume polarization for electrostatic interactions
FPCM   Fully polarizable continuum model
PCM   Polarizable continuum model
HBR   Hydrogen-bonded reactant complex
NPA   Natural population analysis
HBE   Hydrogen bonding energy
3D   Three-dimensional
ZPVE   Zero-point vibration energy

1
Introduction

1.1
Cocaine Abuse and Its Pharmacological Treatment

Cocaine is a widely abused heterocyclic drug (Fig. 1). Addiction and overdose of cocaine are major medical and public health problems that continue
to defy treatment [1–3]. This drug molecule reinforces self-administration
in relation to the peak serum concentration of the drug, the rate of rise to
the peak, and the degree of change of the serum level. Potent central nervous system stimulation is followed by depression [4]. With overdose of the
drug, respiratory depression, cardiac arrhythmia, and acute hypertension are
common effects. The disastrous medical and social consequences of cocaine
addiction (such as violent crime, loss in individual productivity, illness, and
death) have made the development of an effective pharmacological treatment
a high priority [5, 6].

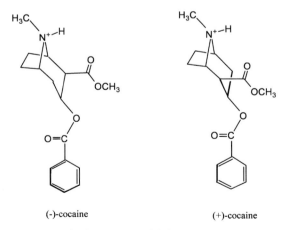

(-)-cocaine   (+)-cocaine

Fig. 1 Molecular structures of (−)-cocaine and (+)-cocaine

Pharmacological treatment for cocaine abuse and dependence can be either pharmacodynamic or pharmacokinetic. Most previously employed anti-addiction strategies use the classical pharmacodynamic approach, (i.e., developing small molecules that interact with one or more neuronal binding sites) with the goal of blocking or counteracting a drug's neuropharmacological actions. However, despite of decades of effort, existing pharmacodynamic approaches to cocaine abuse treatment have not yet proven successful [5–8].

1.2
Pharmacokinetic Approach

The inherent difficulties in antagonizing a blocker like cocaine have led to the development of a pharmacokinetic approach that aims at acting directly on the drug itself to alter its distribution or accelerate its clearance [7–14]. Pharmacokinetic antagonism of cocaine could be implemented by administration of a molecule, such as an anti-cocaine antibody, that binds tightly to cocaine so as to prevent it from crossing the blood–brain barrier [15–20].

The blocking action could also be implemented by administration of an enzyme or a catalytic antibody (regarded as an artificial enzyme) that not only binds but also accelerates cocaine metabolism and thereby frees itself for further binding [16–25]. Usually, a pharmacokinetic agent would not be expected to across the blood–brain barrier and thus would itself have no direct pharmacodynamic action, such as abuse liability [5].

1.3
Cocaine Metabolism

The primary pathway for metabolism of cocaine in primates is hydrolysis at the benzoyl ester or methyl ester group [5, 26]. Benzoyl ester hydrolysis generates ecgonine methyl ester (EME), whereas the methyl ester hydrolysis yields benzoylecgonine (BE). The major cocaine-metabolizing enzymes in humans are butyrylcholinesterase (BChE), which catalyzes benzoyl ester hydrolysis, and two liver carboxylesterases (denoted by hCE-1 and hCE-2), which catalyze hydrolysis at the methyl ester and the benzoyl ester, respectively. Of the three, BChE is the principal cocaine hydrolase in human serum. Hydrolysis accounts for about 95% of cocaine metabolism in humans. The remaining 5% is deactivated through oxidation by the liver microsomal cytochrome P450 system, producing norcocaine [5, 27]. EME appears the least pharmacologically active of the cocaine metabolites and may even cause vasodilation, whereas both BE and norcocaine appear to cause vasoconstriction and lower the seizure threshold, similarly to cocaine itself. Norcocaine is hepatotoxic and a local anesthetic [28]. Thus, hydrolysis of cocaine at the benzoyl ester by BChE is the pathway most suitable for amplification.

1.4
Butyrylcholinesterase

BChE, designated in older literature as pseudo-cholinesterase or plasma cholinesterase to distinguish it from its close cousin acetylcholinesterase (AChE), is synthesized in the liver and widely distributed in the body, including plasma, brain, and lung [5, 29]. Studies in animals and humans demonstrate that enhancement of BChE activity by administration of exogenous enzyme substantially decreases cocaine half-life [30–34]. For example, the addition of human BChE (extracted from donated blood) to human plasma containing cocaine (2 µg/mL) decreased the cocaine half-life in vitro from 116 min at a BChE concentration of 3.02 µg/mL to 10 min at a BChE concentration of 37.6 µg/mL. In vivo studies in animals have also revealed significant enhancement of BChE activity on cocaine's effects. Further, a single injection of the enzyme may increase plasma BChE activity for several days [5]. Clinical studies suggest that BChE has unique advantages. First, human BChE has a long history of clinic use, and no adverse effects have been noted with increased BChE plasma activity. Second, about 20 different naturally occurring mutants of human BChE have been identified [35], and there is no evidence that these mutants are antigenic. BChE also has potential advantages over active immunization since BChE administration would immediately enhance cocaine metabolism and would not require an immune response to be effective. For these reasons, enhancement of cocaine metabolism by administration of BChE is considered a promising pharmacokinetic approach for treatment of cocaine abuse and dependence [5, 6].

However, the catalytic activity of this plasma enzyme is three orders-of-magnitude lower against the naturally occurring (–)-cocaine than that against the biologically inactive (+)-cocaine enantiomer [36–39]. (+)-Cocaine can be cleared from plasma in seconds and prior to partitioning into the central nervous system (CNS), whereas (–)-cocaine has a plasma half-life of ∼ 45–90 min, long enough for manifestation of the CNS effects, which peak in minutes [21]. Thus, positron emission tomography (PET) applied to mapping of the binding of (–)-cocaine and (+)-cocaine in baboon CNS showed marked uptake corresponding to (–)-cocaine at the striatum along with other areas of low uptake, whereas no CNS uptake corresponding to (+)-cocaine was observed [37]. (+)-Cocaine was hydrolyzed by BChE so rapidly that it never reached the CNS for PET visualization. One may expect great progress in pharmacological treatment if a BChE mutant capable of hydrolyzing (–)-cocaine with the rate of (+)-cocaine hydrolysis by wild-type BChE is developed. Hence, BChE mutants with a significantly improved catalytic efficiency against (–)-cocaine are highly desirable for use as an exogenous enzyme in humans. As discussed below, encouraging progress has been made in recent computational design of high-activity mutants of human BChE.

2
Mechanism for Non-enzymatic Hydrolysis of Cocaine

Anti-cocaine catalytic antibodies are a novel class of artificial enzymes with unique potential as therapeutic agents for cocaine overdose and addiction [21, 22]. This novel class of artificial enzymes, elicited by immunization with transition-state analogs of cocaine benzoyl-ester hydrolysis, have unique potential as therapeutic artificial enzymes due to their biocompatibility and extended plasma half-life. The design of a transition-state analog that would elicit a catalytic antibody [40] is based on the mechanism of the corresponding non-enzymatic reaction, specifically the transition-state structure for the rate-determining step. Hence, a more complete understanding of the mechanism of cocaine hydrolysis in aqueous solution could provide additional insights into the rational design of more effective transition-state analogs. This is why computational studies for development of anti-cocaine catalytic antibodies have been focused on the reaction coordinate calculations on the detailed mechanisms for non-enzymatic hydrolysis of cocaine in water.

2.1
Hydrolysis of Cocaine Free Base

As one can see from Fig. 1, cocaine has two carboxylic acid ester groups: benzoyl ester and methyl ester. Hence, the fundamental reaction pathway for non-enzymatic hydrolysis of cocaine at both benzoyl ester and methyl ester groups is expected to be similar to that for the usual non-enzymatic hydrolysis of a carboxylic acid ester. The hydrolysis of the majority of common alkyl esters, RCOOR′, in neutral solution occurs through attack at the hydroxide ion at the carbonyl carbon [41–48]. This mode of hydrolysis has been designated as $B_{AC}2$ (base-catalyzed, acyl-oxygen cleavage, bimolecular) [43], and is believed to occur by a two-step mechanism, although a concerted pathway can arise in the case of esters containing very good leaving groups (corresponding to a low pK_a value for R′OH) [49–56]. The generally accepted two-step mechanism consists of the formation of a tetrahedral intermediate (first step), followed by decomposition of the tetrahedral intermediate to yield products RCOO⁻ + R′OH (second step) [43]. Degradation of cocaine may take place through the $B_{AC}2$ route of hydrolysis of either the benzoyl ester group or the methyl ester group.

The earliest theoretical calculations of cocaine hydrolysis focused on the first step of the hydrolysis of the benzoyl ester [57, 58]. In these computational studies [57, 58], MNDO, AM1, PM3, and SM3 semiempirical molecular orbital methods, as well as ab initio procedure at the HF/3-21G level of theory, were employed to optimize geometries of the transition states for the first step of the hydrolysis of cocaine and model esters, including methyl acetate [59, 60] for which experimental activation energy in aqueous solution is available. However, the geometry optimization of the first transition state for the cocaine benzoyl-ester hydrolysis was successful with only the MNDO, PM3, and SM3 methods. No first-order saddle point corresponding to the expected transition state structure was found on the AM1 and HF/3-21G potential energy surfaces. Thus, it was necessary to further examine this putative transition state with higher levels of theory. Further, the energy barrier, 24.6 kcal/mol [57], predicted by the semiempirical molecular orbital calculations for the first step of the hydrolysis of neutral cocaine in aqueous solution was likely overestimated, as the energy barrier, 23.4 kcal/mol [59, 60], determined by the same kind of calculations for the first step of the methyl acetate hydrolysis was significantly larger than the reported experimental activation energy, 10.45 kcal/mol [61] or 12.2 kcal/mol [62], in aqueous solution.

As discussed below in detail, first-principles electronic structure calculations have provided accurate predictions of the reaction pathways and the corresponding energy barriers, not only for the first step of hydrolysis of cocaine free base at the benzoyl ester group, but also for the entire reaction processes of hydrolysis of cocaine free base at both the benzoyl ester and methyl ester groups.

2.1.1
Geometries of Transition States and Intermediates

In a more sophisticated computational study [63], first-principles reaction coordinate calculations on the hydrolysis of cocaine free base (neutral cocaine) were performed by using Becke's three-parameter hybrid exchange functional [64] and the Lee–Yang–Parr correlation functional (B3LYP) [65] with the 6-31+G(d) basis set. Vibrational frequencies were evaluated at the optimized geometries to confirm all the first-order saddle points and local minima found on the potential energy surfaces, and to evaluate zero-point vibration energies (ZPVE). Intrinsic reaction coordinate (IRC) calculations [66, 67] were also performed to verify the expected connections of the first-order saddle points with local minima found on the potential energy surfaces [63]. The important geometries optimized at the B3LYP/6-31+G(d) level for the base-catalyzed hydrolysis of neutral cocaine and three model esters are depicted in Figs. 2 and 3 [63]. The reaction coordinate calculations indicate that the mechanisms of the base-catalyzed hydrolysis of the cocaine benzoyl ester and methyl ester groups are indeed similar to the usual two-step $B_{AC}2$ route of hydrolysis of alkyl esters [43, 47]. The first step is the formation of a tetrahedral intermediate by the attack of hydroxide oxygen at the carbonyl carbon of cocaine methyl ester or benzoyl ester group. The second step is the decomposition of the tetrahedral intermediate to products through breaking the $C-O$ bond between the carbonyl carbon and ester oxygen [63].

For the cocaine benzoyl ester hydrolysis, the nucleophilic hydroxide ion can approach from the two faces, denoted by Re and Si, of the carbonyl to

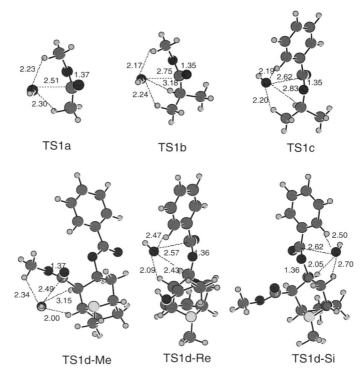

Fig. 2 Geometries of the transition states optimized at the B3LYP/6-31+G(d) level for the first step of the hydrolysis of CH_3COOCH_3, $(CH_3)_2CHCOOCH_3$, $C_6H_5COOCH(CH_3)_2$, the cocaine methyl-ester, and the cocaine benzoyl-ester [63]. Internuclear distances are given in angstrom

form two stereoisomer tetrahedral intermediates (S and R). The two transition state structures, denoted by TS1d-Re and TS1d-Si, optimized at the B3LYP/6-31+G(d) level for the two competing pathways of the first step of the cocaine benzoyl-ester hydrolysis are depicted in Fig. 2, together with those optimized for the first step of the hydrolysis of CH_3COOCH_3 (TS1a), $(CH_3)_2CHCOOCH_3$ (TS1b), $C_6H_5COOCH(CH_3)_2$ (TS1c), and the cocaine methyl-ester (TS1d-Me) [63]. As one can see from Fig. 2, all of the six transition state structures for the first step are very similar to each other as far as the position of the nucleophilic hydroxide relative to the carbonyl. The distances between the hydroxide oxygen and carbonyl carbon are 2.49–2.75 Å.

As the second step of the ester hydrolysis, the decomposition of the tetrahedral intermediate requires a proton transfer from the hydroxide/hydroxyl oxygen to the ester oxygen, while the C–O bond between the carbonyl carbon and ester oxygen gradually breaks. Two competing pathways were examined for the second step [63]: one associated with the direct proton

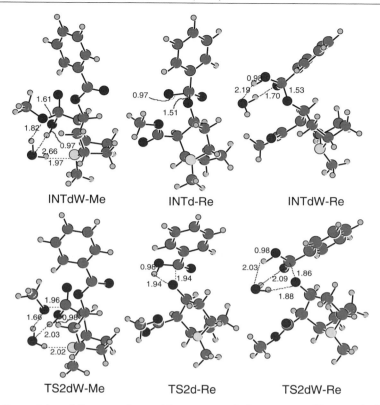

Fig. 3 Geometries of the second transition states and the corresponding tetrahedral intermediates optimized at the B3LYP/6-31+G(d) level for the hydrolysis of the cocaine methyl-ester and benzoyl-ester groups [63]. Internuclear distances are given in angstrom

transfer from the hydroxide/hydroxyl oxygen to the ester oxygen; and the other associated with a water-assisted proton transfer. Figure 3 depicts the optimized geometries of the transition state for the water-assisted proton transfer (TS2dW-Me) during the cocaine methyl-ester hydrolysis and the transition states for the direct proton transfer (TS2d-Re) and water-assisted proton transfer (TS2dW-Re) during the cocaine benzoyl-ester hydrolysis initialized by the hydroxide attack from the Re face. Figure 3 also shows the optimized geometries of the tetrahedral intermediates INTdW-Me, INTd-Re, and INTdW-Re corresponding to transition states TS2dW-Me, TS2d-Re, and TS2dW-Re, respectively. For the water-assisted proton transfer pathway involving transition state TS2dW-Re (or TS2dW-Me), the water molecule hydrogen-bonding with the ester oxygen in the tetrahedral intermediate INTdW-Re (or INTdW-Me) gradually transfers a proton to the ester oxygen through the hydrogen bond, while the hydroxide/hydroxyl proton gradually transfers to the water oxygen.

2.1.2
Energy Barriers for the Formation of the Tetrahedral Intermediates

The geometries optimized at the B3LYP/6-31+G(d) level were employed to carry out the single-point energy calculations at the MP2/6-31+G(d) level [63]. Solvent shifts of the energies were accounted for by performing self-consistent reaction field (SCRF) energy calculations using the geometries optimized at the B3LYP/6-31+G(d) level in gas phase. The energy barrier for reaction in aqueous solution was taken as a sum of the energy change calculated at the MP2/6-31+G(d)//B3LYP/6-31+G(d) level in gas phase and the corresponding solvent shift determined by the SCRF calculations at the HF/6-31+G(d) level.

The solute–solvent interaction can be divided into a long-range electrostatic interaction and short-range non-electrostatic interactions (such as cavitation, dispersion, and Pauli repulsion) [68–72]. The dominant long-range electrostatic interaction was evaluated by using the GAMESS [73] implementation of the surface and volume polarization for electrostatic interactions (SVPE) [74–76]. The SVPE model is also known as the fully polarizable continuum model (FPCM) [47, 48, 63, 77–89] because it fully accounts for both surface and volume polarization effects in the SCRF calculation. The contributions of short-range non-electrostatic interactions to the energy barriers were estimated by using the polarizable continuum model (PCM) [68] implemented in the Gaussian98 program [90] with the default choices of the program for the recommended standard parameters. The total solvent shift [63] was taken as a sum of the long-range electrostatic interaction contribution determined by the SVPE calculation and the total contribution of the short-range non-electrostatic interactions determined by the PCM calculation.

The energy barriers determined for the ester hydrolyses in aqueous solution are summarized in Table 1. The total energy of the individual reactants, $RCOOR' + HO^-$, in gas phase is about $14-26$ kcal/mol higher than the first transition state (TS1). Theoretical studies of the alkaline hydrolysis of alkyl esters revealed that for the ester hydrolysis in gas phase, between the individual reactants and TS1, there is a hydrogen-bonded reactant complex (denoted by HBR) [46] whose energy is lower than TS1. Thus, the energy barrier for the first step of the hydrolysis, i.e., formation of the tetrahedral intermediate, in gas phase is the energy change from HBR to TS1. However, in aqueous solution various SCRF calculations gave the same qualitative result that the individual reactants are more stable than both TS1 and HBR, whereas HBR is still more stable than TS1 [47]. It follows that in aqueous solution the HBR structure is not stable, and that the reaction goes directly from the individual reactants to TS1. This is because the interaction between solvent water and the individual reactants is stronger than that between methyl acetate and hydroxide ion. Hence, the energy barrier for the first step of the hydrolysis in

Table 1 Energy barriers (in kcal/mol) calculated for the base-catalyzed hydrolysis of neutral cocaine and model esters in aqueous solution [63][a]

Reaction	ΔE(Gas)[b]	Solvent shift[c] Electrostatic (SVPE)	Non-electrostatic (PCM)[e]	Total	Energy barrier
CH₃COOCH₃					
CH$_3$COOCH$_3$					
Reactants → TS1a	– 14.31	25.15 [24.98][d]	0.52	25.67	11.4[f]
(CH$_3$)$_2$CHCOOCH$_3$					
Reactants → TS1b	– 14.05	20.97	1.05	22.02	8.0
C$_6$H$_5$COOCH(CH$_3$)$_2$					
Reactants → TS1c	– 19.16	26.69	1.32	28.00	8.8
Cocaine (methyl-ester)					
Reactants → TS1d-Me	– 18.68	24.81	0.84	25.65	7.0
INTdW-Me → TS2dW-Me	2.51	2.51	– 0.23	2.28	4.8
Cocaine (benzoyl-ester)					
Reactants → TS1d-Re	– 26.33	32.40	1.55	33.95	7.6
Reactants → TS1d-Si	– 22.99	30.26	1.25	31.51	8.5
INTd-Re → TS2d-Re	3.52	7.49	1.37	8.86	12.4
INTdW-Re → TS2dW-Re	2.16	1.32	– 0.33	0.99	3.2

[a] All calculations used geometries optimized at the B3LYP/6-31+G(d) level in gas phase
[b] Energy change determined at the MP2/6-31+G(d)//B3LYP/6-31+G(d) level in gas phase. The ZPVE corrections were made for all the values
[c] Unless otherwise indicated, the solvent shifts were determined by performing the SVPE and PCM calculations at the HF/6-31+G(d) level
[d] Values in brackets were determined by carrying out the SVPE calculations at the MP2/6-31+G(d) level
[e] Total contribution of non-electrostatic interactions between solute and solvent
[f] The corresponding experimental activation energies reported for hydrolysis of CH$_3$COOCH$_3$ in aqueous solution were 10.45 kcal/mol [61] and 12.2 kcal/mol [62]

aqueous solution is the energy change from the individual solvated reactants to the solvated first transition state TS1. As shown in Table 1, the extremely large solvent shifts of the energy barriers for the first step of the ester hydrolysis are attributed mainly to the contributions of the long-range electrostatic interactions between the solutes and solvent.

As one can see from Table 1, the solvent shift determined for the first step of the hydrolysis of methyl acetate (the rate-determining step) by the SVPE calculations at the MP2/6-31+G(d) level differs from the shift determined at the HF/6-31+G(d) level by less than 0.2 kcal/mol. The calculated energy barrier, 11.4 kcal/mol, is in good agreement with the experimental determinations of activation energy, 10.45 or 12.2 kcal/mol, reported for the hydrolysis of methyl acetate in aqueous solution [61, 62].

As seen in Table 1, the energy barrier, 7.6 kcal/mol, calculated for the first step of the cocaine benzoyl ester hydrolysis through the hydroxide attack from the Re face of the carbonyl is ~ 1 kcal/mol lower than that through hydroxide attack from the Si face. The energy barrier, 7.0 kcal/mol, calculated for the first step of the cocaine methyl ester hydrolysis is slightly lower than the lowest barrier, 7.6 kcal/mol, for the first step of the cocaine benzoyl-ester hydrolysis. The energy barriers calculated for the first step of the cocaine hydrolysis are all significantly lower than the barrier for the first step of the hydrolysis of methyl acetate. To understand the changes of the calculated energy barriers from methyl acetate hydrolysis to cocaine hydrolysis, the cocaine hydrolysis will be compared with the hydrolysis of other two simplified cocaine models, $(CH_3)_2CHCOOCH_3$ and $C_6H_5COOCH(CH_3)_2$, representing the cocaine methyl-ester and benzoyl-ester, respectively.

Methyl acetate, CH_3COOCH_3, is a minimal model of the cocaine methyl ester in which the two β carbon atoms for the carboxylic acid moiety of the methyl ester are all simplified as hydrogen atoms. $(CH_3)_2CHCOOCH_3$ is a slightly larger model of the cocaine methyl ester in which the two β carbon atoms for the carboxylic acid moiety of the cocaine methyl ester are represented as methyl groups. Correspondingly, transition state structures TS1a and TS1b may be regarded as two simplified models of transition state structure TS1d-Me, as seen in Fig. 2. The energy barrier, 8.0 kcal/mol, calculated for the $(CH_3)_2CHCOOCH_3$ hydrolysis is 3.4 kcal/mol lower than that for the CH_3COOCH_3 hydrolysis but matches cocaine methyl ester hydrolysis very well at only 1.0 kcal/mol higher. It follows that substitution of the two α hydrogen atoms in R with two methyl groups significantly decreases the energy barrier for the first step of the ester hydrolysis, and that further substitution of the β hydrogen for the carboxylic acid moiety slightly decreases the energy barrier. The significant decrease of the energy barrier upon substitution of the two α hydrogen atoms in R with two methyl groups may be attributed mainly to the stronger $C - H \cdots O$ hydrogen bond [91–93] between the hydroxide oxygen and one of the β hydrogen atoms in the first transition state (TS1b or TS1d-Me). The fact that the hydrogen bond with the β hydrogen is stronger than the hydrogen bond with the α hydrogen is caused by the steric effect. In the transition state, the β hydrogen is sterically more favorable than the α hydrogen to form a hydrogen bond with the hydroxide oxygen. Thus, $(CH_3)_2CHCOOCH_3$ is a reasonable model for the cocaine methyl ester. Similarly, $C_6H_5COOCH(CH_3)_2$ models the cocaine benzoyl ester in which the two β carbon atoms for the alcohol moiety of the cocaine benzoyl ester are represented as methyl groups. Correspondingly, transition state structure TS1c may be regarded as a model of transition state structure TS1d-Re. The energy barrier, 8.8 kcal/mol, calculated for the $C_6H_5COOCH(CH_3)_2$ hydrolysis is very close to that of the cocaine benzoyl ester (TS1d-Re) at only 1.2 kcal/mol higher. Thus, $C_6H_5COOCH(CH_3)_2$ is a reasonable model for the cocaine benzoyl-ester.

2.1.3
Energy Barriers for the Decomposition of the Tetrahedral Intermediates

The energy barrier for the second step of the cocaine hydrolysis, i.e., decomposition of the tetrahedral intermediate, is the energy change from the intermediate (INT) to the second transition state (TS2) no matter whether the hydrolysis occurs in gas phase or in aqueous solution. The energy barriers were also determined by using the same computational protocol described above. Since the calculated energy barrier for the first step of the hydrolysis associated with transition state TS1d-Re is lower than that associated with transition state TS1d-Si, the whole reaction pathway, individual reactants → TS1d-Re → INTd-Re → TS2d-Re → individual products, was considered only for the cocaine benzoyl ester hydrolysis involving the direct proton transfer. The energy barrier, 12.4 kcal/mol, calculated for the second step of the hydrolysis associated with transition state TS2d-Re is 4.8 kcal/mol higher than the corresponding first step. For the cocaine benzoyl-ester hydrolysis involving the water-assisted proton transfer, the calculated energy barrier, 3.2 kcal/mol, associated with transition state TS2dW-Re is 4.4 kcal/mol lower than the first step. It follows that the direct participation of the solvent water molecule in the proton transfer process decreases the energy barrier by 9.2 kcal/mol. This is why the energy barrier for the second step of the hydrolysis involving the water-assisted proton transfer is significantly lower, whereas the energy barrier for the second step involving the direct proton transfer is significantly higher than the first step. Thus, the reaction pathway involving the water-assisted proton transfer should dominate the hydrolysis in aqueous solution. Similar results were also reported for the second step of the methyl acetate hydrolysis [47].

For the second step of the cocaine methyl ester hydrolysis involving the water-assisted proton transfer, the calculated energy barrier, 4.8 kcal/mol, associated with transition state TS2dW-Me, is also lower than the corresponding first step. So, with the direct participation of the solvent water molecule in the proton transfer process, the first step of the hydrolysis in aqueous solution should be rate-determining, whether for the cocaine benzoyl ester hydrolysis or for the cocaine methyl ester hydrolysis. This conclusion provides theoretical support for the design of analogs of the first transition state for the cocaine benzoyl ester hydrolysis to elicit anti-cocaine catalytic antibodies [22, 25].

2.2
Hydrolysis of Protonated Cocaine

Under physiological conditions (pH 7.4), cocaine (pK_a 8.6) exists mainly as the protonated amine. The reaction pathways discussed above for the ester hydrolysis of neutral cocaine predict similar rates of reaction for methyl ester

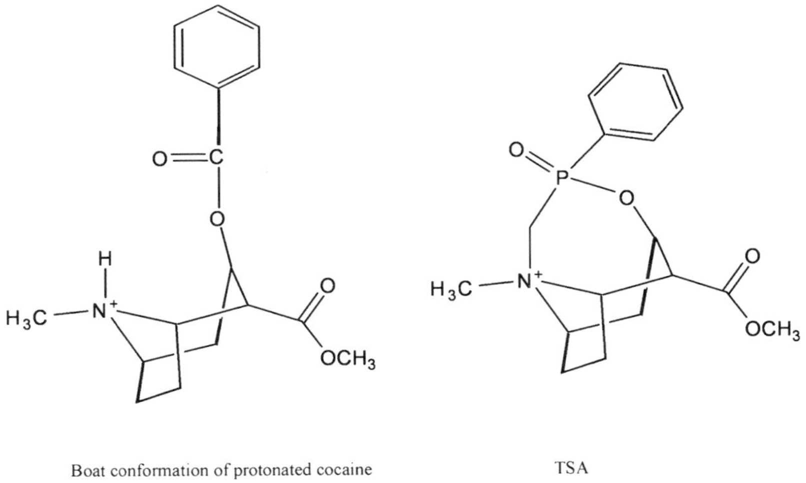

Boat conformation of protonated cocaine   TSA

Fig. 4 Boat conformation of protonated (–)-cocaine and the corresponding TSA structure [89]

hydrolysis and benzoyl ester hydrolysis. However, the methyl ester rapidly hydrolyzes in vivo and in aqueous solution at neutral pH. Experimental kinetic studies [45] suggested that internal participation of the protonated amine in the alkaline hydrolysis of the cocaine methyl ester could account for its lability relative to the benzoyl ester.

Further, for antibody catalysis the methyl ester is too small to be an effective epitope but the participation could be induced if an antibody were able to recruit cocaine from the chair conformation to the less stable boat form (see Fig. 4 for the structure) and reorient the syn-protonated amine and benzoyl ester into proximity. Antibodies can provide significant binding energy and in principle antibody binding could effect conformer selection and promotion of substrate-assisted catalysis. To examine this idea, a detailed computational analysis [89] of the energetics of this reaction was also performed for design of novel TSA structures for the alkaline hydrolysis of boat cocaine in comparison with the hydrolysis of chair cocaine.

2.2.1
Reaction Pathways for Chair Cocaine

The first-principles reaction coordinate calculations at the B3LYP/6-31+G(d) level led to the optimized geometries of the rate-determining transition states (i.e., the transition states for the first reaction step of the ester hydrolysis), as depicted in Fig. 5. For benzoyl ester hydrolysis of protonated cocaine in its chair conformation, the nucleophilic hydroxide ion can also approach from two faces, denoted by Si and Re, of the carbonyl to form two stereoisomer

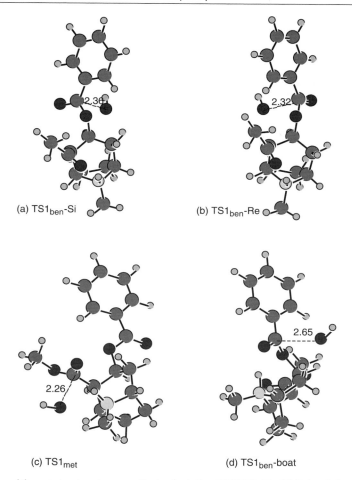

(a) TS1$_{ben}$-Si

(b) TS1$_{ben}$-Re

(c) TS1$_{met}$

(d) TS1$_{ben}$-boat

Fig. 5 Transition state structures optimized at the B3LYP/6-31+G(d) level for the ester hydrolysis of protonated cocaine [89]: **a** and **b** for the benzoyl ester hydrolysis of the chair cocaine; **c** for the methyl ester hydrolysis of the chair cocaine; and **d** for the benzoyl ester hydrolysis of the boat cocaine

tetrahedral intermediates (R and S). The two transition state structures, optimized at the B3LYP/6-31+G(d) level for the two competing pathways of the first step of the benzoyl ester hydrolysis, are denoted by TS1$_{ben}$ – Si and TS1$_{ben}$ – Re. The transition state for the first step of the cocaine methyl ester hydrolysis is denoted by TS1$_{met}$. These three transition state structures optimized for the hydrolysis of protonated cocaine are similar to the corresponding transition state structures optimized for the alkaline hydrolysis of neutral cocaine. A remarkable difference is that the internuclear distances between the hydroxide oxygen and the carbonyl carbon become significantly shorter for the protonated cocaine: 2.36 Å, 2.32 Å, and 2.26 Å in $TS1_{ben}$ – Si, $TS1_{ben}$ – Re, and $TS1_{met}$, respectively, for protonated cocaine hydrolysis compared to the corresponding distances 2.62 Å, 2.57 Å, and 3.15 Å for neutral cocaine. These results verify earlier predictions [57, 58] made using SM3 semiempirical methods, where PM3 geometry optimizations in aqueous solution resulted in internuclear hydroxide oxygen-carbonyl carbon distances of 2.23 Å for $TS1_{ben}$ – Si and 2.13 Å for $TS1_{ben}$ – Re.

The calculated energetic results are summarized in Table 2. It should be pointed out that the ΔG(gas) values listed in Table 2 are simply the Gibbs free energy changes from the separated reactants to the corresponding transition states when the solvent effects are ignored, but these values are not the free energy barriers for the corresponding reactions in the gas phase. This is because previous studies [46–48] have demonstrated that for the reaction of an ester, the ester and hydroxide ion first form a hydrogen-bonded complex in the gas phase (a local minimum on the potential energy surface) before going to the transition state, whereas such a hydrogen-bonded complex does not exist in aqueous solution. So, the free energy barrier for the imaginary reaction in the gas phase should be the free energy change from the hydrogen-bonded complex to the first transition state.

Table 2 Calculated Gibbs free energies (in kcal/mol) of the transition states relative to the corresponding separated reactants for the ester hydrolyses of protonated cocaine in solution [89] [a]

Transition state	ΔG(gas) [b]	Solvent shift [c] Electrostatic (SVPE)	Non-electrostatic (PCM) [d]	ΔG(solution) [e] Without non-electrostatic	With non-electrostatic
$TS1_{ben}$ – Si	– 82.7	101.8	0.5	19.1	19.6
$TS1_{ben}$ – Re	– 85.3	102.1	0.1	16.8	16.9
$TS1_{met}$	– 100.4	112.9	– 1.4	12.5	11.1
$TS1_{ben}$-boat	– 81.7	98.3	0.2	16.6	16.8

ΔG is given in kcal/mol, at $T = 298.15$ K and $P = 1$ atm [a] All calculations used geometries optimized at the B3LYP/6-31+G(d) level in gas phase. The reactants are hydroxide ion and the protonated cocaine in its chair or boat conformation

[b] Calculated at the MP2/6-31+G(d)//B3LYP/6-31+G(d) level in gas phase, including zero-point vibration and thermal corrections

[c] The electrostatic part of the solvent shift was determined by performing the SVPE calculations at the HF/6-31+G(d) level, whereas the non-electrostatic contribution was determined by the PCM calculation

[d] Total contribution of short-range non-electrostatic solute-solvent interactions

[e] Gibbs free energy barrier in aqueous solution calculated as the ΔG(gas) value plus the electrostatic solvent shift determined by the SVPE calculation, without or with the non-electrostatic contributions determined by the PCM calculation

As seen in Table 2, the solvent effects are crucial for calculating realistic free energy barriers and, not surprisingly, the calculated solvent shifts are dominated by the solute–solvent electrostatic interactions. The estimated short-range non-electrostatic contributions to the free energy barriers are negligible compared to the electrostatic contributions to the solvent shifts.

The calculated free energy barriers (at $T = 298.15$ K and $P = 1$ atm) associated with transition states $TS1_{ben}$ – Si and $TS1_{ben}$ – Re for the benzoyl ester hydrolysis of protonated cocaine are 19.1 and 16.8 kcal/mol, respectively. Thus, the reaction pathway for hydroxide oxygen attacking from the Re face of the carbonyl should be dominant, which is consistent with the conclusion obtained from the energy barriers calculated for neutral cocaine hydrolysis. This is not surprising because the proton attached to the tropane N atom does not participate in the benzoyl ester hydrolysis of chair cocaine. So, the effects of the cocaine protonation on the energy barriers for the benzoyl ester hydrolysis of chair cocaine should be insignificant. Similar computations [63] on neutral cocaine hydrolysis predicted the energy barriers (i.e., the free energy barriers at $T = 0$ K) to be 8.5 and 7.6 kcal/mol, corresponding to the transition states $TS1_{ben}$ – Si and $TS1_{ben}$ – Re, respectively. The corresponding free energy barriers calculated at $T = 0$ K for the benzoyl ester hydrolysis of protonated cocaine are 9.9 and 7.0 kcal/mol. These two values become 19.1 and 16.8 kcal/mol, respectively, at $T = 298.15$ K and $P = 1$ atm. The differences between the calculated free energy barriers at $T = 0$ K and the corresponding free energy barriers at $T = 298.15$ K are primarily attributed to entropic effects, particularly the translational entropy changes from the separated reactants to the transition states.

However, the free energy barrier calculated for the methyl ester hydrolysis of protonated cocaine (2.5 kcal/mol at $T = 0$ K and 12.5 kcal/mol at $T = 298.15$ K and $P = 1$ atm) is significantly lower than that for the dominant pathway of the benzoyl ester hydrolysis (7.0 kcal/mol at $T = 0$ K and 16.8 kcal/mol at $T = 298.15$ K and $P = 1$ atm). It is also significantly lower than that for the methyl ester hydrolysis of neutral cocaine (7.0 kcal/mol at $T = 0$ K) [63]. The significant decrease of the free energy barrier, ~ 4 kcal/mol, can be attributed to the intramolecular acid catalysis of alkaline hydrolysis of the cocaine methyl ester. This catalysis results from the interplay between two opposing factors. First, the carbonyl oxygen of the methyl ester moiety hydrogen-bonds to the tropane N through the proton at the N atom in the transition state ($TS1_{met}$) and the corresponding reactant (cocaine). The optimized internuclear distance between the carbonyl oxygen of the methyl ester moiety and the hydrogen on the tropane N is 1.801 Å in the reactant and 1.932 Å in the transition state. This NH·····O distance slightly increases in going from the reactant to the transition state, as the hydroxide oxygen gradually approaches the carbonyl carbon to form a tetrahedral intermediate. On the other hand, during the conversion of reactants to transition state $TS1_{met}$, the partial negative charge at the carbonyl oxygen becomes progressively larger. According to the natural population analysis (NPA) [89] at the B3LYP/6-31+G(d) level, the net atomic charge at the carbonyl oxygen of the methyl ester moiety is – 0.649 in the reactant and – 0.703 in the transition state. So, there are two opposite factors affecting the change of the NH······O hydrogen bond strength in going from the reactant to the transition state: one is the increase of the bond distance, and the other is the increase of the negative charge on the oxygen atom when the changes of the charges on the N and H atoms are negligible. The aforementioned decrease ($\sim$ 4 kcal/mol) in the free energy barrier implies that the increase of the negative charge on the oxygen atom is predominant, making the NH······O hydrogen bonding slightly stronger in the transition state. The stronger intramolecular hydrogen bonding should contribute more effectively to TS stabilization, which explains the decrease in the free energy barrier. Furthermore, for ester hydrolysis of neutral cocaine (without a proton at the tropane N atom), the energy barrier calculated for the methyl ester hydrolysis of neutral cocaine is almost the same as that for the dominant pathway of the benzoyl ester hydrolysis [63]. With the tropane N atom being protonated, the proton is involved in the bond formation and breaking process such that the barrier becomes $\sim$ 4 kcal/mol lower for the methyl ester hydrolysis of protonated cocaine. The calculated relative magnitudes of the free energy barriers for the hydrolysis of the protonated cocaine are qualitatively consistent with the recently reported experimental results [45] of the investigations on the hydrolysis kinetics of cocaine under physiological conditions, because the cocaine methyl ester hydrolysis was found to be faster than the cocaine benzoyl ester hydrolysis.

2.2.2
Reaction Pathway for Boat Cocaine

The transition state structure, denoted by $TS1_{ben}$-boat, optimized at the B3LYP/6-31+G(d) level for the benzoyl ester hydrolysis of boat cocaine is also depicted in Fig. 5. Because the carbonyl oxygen of the benzoyl ester moiety also hydrogen-bonds to the tropane N atom through the proton at the N for boat cocaine, one might also expect similar intramolecular acid catalysis of the benzoyl ester hydrolysis of boat cocaine as seen in the methyl ester hydrolysis of chair cocaine discussed above. The free energy barrier (7.2 kcal/mol at $T = 0$ K and 16.6 kcal/mol at $T = 298.15$ K and $P = 1$ atm) calculated for the benzoyl ester hydrolysis of boat cocaine is significantly higher than that for the methyl ester hydrolysis of chair cocaine and, at $T = 298.15$ K and $P = 1$ atm, is only 0.2 kcal/mol lower than that for the dominant pathway of the benzoyl ester hydrolysis of chair cocaine. This is because the optimized distance between the carbonyl oxygen of the benzoyl ester moiety and the hydrogen on the tropane N significantly increases from 1.632 Å in the reactant (boat cocaine) to 2.027 Å in the transition state $TS1_{ben}$-boat while the net negative charge (NPA charge) at the carbonyl oxygen of the benzoyl ester moiety increases slightly from − 0.657 in the reactant to − 0.670 in the transition state. Note that the atomic charges determined by NPA or any other theoretical approach may only be used to qualitatively assess the change of the charge, as the absolute charges calculated are closely dependent on the theoretical approach used in the calculation. Qualitatively, the calculated increase of the negative charge from the reactant to the transition state for the benzoyl ester hydrolysis of the boat cocaine is smaller than that calculated for the methyl ester hydrolysis of chair cocaine, implying that the factor of the charge increase for the benzoyl ester hydrolysis of the boat cocaine is less significant than that for the methyl ester hydrolysis of chair cocaine. Overall, the effects of the two opposite factors (i.e., the increase of the NH······O distance and increase of the negative charge on the O atom) on the free energy barrier nearly cancel out for the methyl ester hydrolysis of chair cocaine. Alternatively, one can describe the intramolecular catalysis of boat cocaine as a process that overcomes the unfavorable steric interactions produced by crowding when the more favorable chair conformation converts to the boat conformation. Intramolecular catalysis of this more crowded species has roughly the same free energy of activation as does regular intermolecular catalysis of the benzoyl ester.

Of greater interest is that the free energy barrier for the benzoyl ester hydrolysis of boat cocaine is not higher than for the benzoyl ester hydrolysis of chair cocaine [89]. This result implies that a TSA structure (Fig. 4) for intramolecular hydrolysis might yield catalysts that would recruit a functional group from the substrate. Thus these theoretical calculations answer a global question: Is the benzoyl ester hydrolysis of boat cocaine even plausible? The similar free energy barriers calculated for the benzoyl ester hydrolysis of the chair and boat cocaine structures support this concept. Intramolecular hydrogen bonding could be useful in generating antibody-based catalysts that recruit cocaine to the boat conformation, and an analog that elicited antibodies to approximate the protonated tropane N and the benzoyl O more closely than the natural boat conformer might increase the contribution from hydrogen bonding.

2.2.3
Development of Anti-cocaine Catalytic Antibodies

The first anti-cocaine catalytic antibody was reported by Landry and associates in 1993 [22]. The transition-state analog used to elicit the first anti-cocaine catalytic antibody was a stable structure of the transition state for the benzoyl ester hydrolysis of chair cocaine [22, 25]. Based on the computational analysis discussed above, the new TSA structure depicted in Fig. 4 was synthesized and 85 cocaine esterases out of 450 anti-analog antibodies were elicited [94] – a performance markedly superior to that of a previously employed simple phosphonate ester as a stable analog of the transition state structure for the benzoyl ester hydrolysis of chair cocaine [22, 25, 95]. In turn, the encouraging experimental results [94] support the thrust of the computational studies [89]. (As noted in [89] the computational studies described in [89], published in 2005, were actually completed far before the experimental studies described in 2004 in [94], but publication had been considerably delayed.)

3
Mechanism for BChE-Catalyzed Hydrolysis of Cocaine

3.1
3D Structure of BChE

To uncover the detailed reaction pathway for BChE-catalyzed hydrolysis of a substrate, one first needs to know the 3D structure of the enzyme. The first X-ray crystal structure of BChE was reported in later 2003 [96]. However, some computational studies on BChE–cocaine binding and on the fundamental reaction pathway were reported in literature prior to the report of the first X-ray crystal structure of BChE. The computational studies [97–100] reported prior to the report of the X-ray crystal structure were based on a homology model of BChE constructed from the solved X-ray crystal structure of *Torpedo californica* acetylcholinesterase (AChE). A detailed comparison [101] of the X-ray crystal structure with the homology model of BChE reveals that the overall structure of the homology model is very close to that of the X-ray crystal structure; the only significant difference can be seen at the acyl binding pocket. The similarity between the X-ray crystal structure and homology model of BChE, along with further computational modeling using the X-ray crystal structure, confirms the fundamental structural and mechanistic insights [100] obtained from the computational studies based on the homology model.

3.2
Fundamental Reaction Pathways

3.2.1
Similarity between Structures of Cocaine and Butyrylcholine

Reaction coordinate calculations for a chemical reaction begin with a concept of the orientation of the reactants. Different starting structures for the enzyme–substrate complex can lead to completely different reactions. For such enzymatic reactions, one needs to know the structure of the prereactive

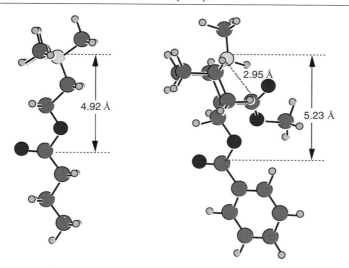

Fig. 6 Geometries of BCh and (–)-cocaine optimized at the B3LYP/6-31+G(d) level [100]

enzyme–substrate complex, which may also be called the "near attack conformation" (NAC) defined by Bruice et al. [102] and discussed more carefully
by Shurki et al. [103]. The initial insights into the enzyme–substrate binding
came from a comparison of the optimized geometry of butyrylcholine (BCh)
with those of (–)-cocaine and (+)-cocaine [100].

The geometries of BCh and (–)-cocaine optimized at the B3LYP/6-31+G(d)
level are depicted in Fig. 6. Note that cocaine mainly exists in its protonated
form under physiological condition because its pK_a is 8.6 [45], and thus both
BCh and cocaine have positively charged quaternary ammonium groups.
In human BChE, W82 is thought to be the key factor in the stabilization
of positively charged substrates in the BChE-substrate complexes, although
this interaction should be more properly classified as a cation-π interaction [104]. While the positively charged quaternary ammonium is positioned
to effectively bind with W82 in the prereactive BChE-substrate complex, the
carbonyl carbon of the substrate must be positioned proximal to S198 O^γ for
nucleophilic attack. Thus the distance between the carbonyl carbon and the
quaternary ammonium is critical and according to the optimized geometries
depicted in Fig. 6, this distance is 4.92 Å for the excellent substrate BCh. The
optimized C to N distance for the substrate cocaine benzoyl ester (5.23 Å)
is similar to that of BCh. The C to N distance for the cocaine methyl ester
(2.95 Å) is remarkably shorter. This helps to explain why (–)-cocaine and (+)-
cocaine bind with BChE in such a way as to hydrolyze at the benzoyl ester,
instead of at the methyl ester. In contrast, for the non-enzymatic hydrolysis
of cocaine under physiological conditions (pH 7.4, 37 °C) the methyl ester
hydrolyzes faster than the benzoyl [45].

3.2.2
BChE–Substrate Complexes

Based on the structural similarity discussed above, the relative positions of the positively charged quaternary ammonium and the carbonyl group of the benzoyl ester moiety in the prereactive BChE–cocaine complexes could be similar to those in the corresponding BChE–BCh complex. The main structural difference between the BChE-(–)-cocaine and BChE-(+)-cocaine complexes exists only in the relative position of the methyl ester group (Fig. 1). Hence, the initial structures of both the BChE-(–)-cocaine and BChE-(+)-cocaine complexes used in molecular modeling and simulations were generated from a 3D model of human BChE with substrate butyrylcholine (BCh), constructed by Harel et al. [105, 106] by replacement of substrate. (–)-Cocaine and (+)-cocaine were positioned similarly to BCh: the carbonyl group of the benzoyl ester was superimposed on the carbonyl group of BCh, and the nitrogen at the positively charged tropane nucleus was superimposed on the nitrogen of the positively charged quaternary ammonium.

The energy-minimized geometries of the BChE-(–)-cocaine and BChE-(+)-cocaine complexes are depicted in Fig. 7a and b, showing the interactions of the carbonyl group of the benzoyl ester with the hydroxyl oxygen (O^γ) of S198 and with the oxyanion hole formed by the peptidic NH functions of G116, G117, and A199. In the minimized structures of BChE binding with (–)-cocaine and (+)-cocaine, the internuclear distances between the carbonyl carbon of the benzoyl ester and S198 O^γ are 3.19 and 3.18 Å, respectively. During the MD simulations from 100 ps to 500 ps, the time-average values of the distance between the carbonyl carbon and S198 O^γ are ~ 3.51 and ~ 3.53 Å for (–)-cocaine and (+)-cocaine, respectively [100]. These C to O^γ distances are all comparable to the distances between the carbonyl carbon and the hydroxide oxygen (2.99–3.56 Å) [46] in the optimized geometries of the prereactive complexes of carboxylic acid esters with hydroxide ion [46–48]. The distances between the carbonyl oxygen of the benzoyl ester and the NH hydrogen of G116, G117, and A199 are 1.98, 2.59, and 1.92 Å, respectively, in the minimized structures of BChE with (–)-cocaine. The respective O to H distances in the minimized structures of BChE with (+)-cocaine are 1.82, 2.42, and 2.39 Å. The respective time-average values of the O to H distances are ~ 3.67, ~ 2.21, and ~ 2.46 Å for the MD simulation on BChE with (–)-cocaine, and ~ 3.80, ~ 2.39, and ~ 3.01 Å for the MD simulation with (+)-cocaine [100]. In addition, the MD trajectories also reveal that in both the BChE-(–)-cocaine and BChE-(+)-cocaine complexes, the cocaine nitrogen atom stays at nearly the same position as the BCh nitrogen atom in the structure of the BChE model constructed by Harel et al. [106]. These limited results suggest that both (–)-cocaine and (+)-cocaine may bind with human BChE so as to allow S198 O^γ to approach the carbonyl carbon of the benzoyl ester.

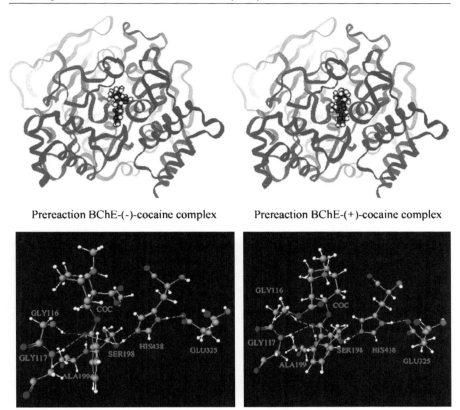

Prereaction BChE-(-)-cocaine complex   Prereaction BChE-(+)-cocaine complex

Fig. 7 Minimized structures of prereactive BChE-(–)-cocaine and BChE-(+)-cocaine complexes [100]: **a** ribbon; **b** residues forming the catalytic triad and the three-pronged oxyanion hole

The energy-minimized structures of the prereactive BChE-(–)-cocaine and BChE-(+)-cocaine complexes depicted in Fig. 7 are similar to the prereactice enzyme–substrate structure proposed for BChE binding with other positively charged substrates, i.e., butyrylthiocholine and succinyldithiocholine [104, 107]; they are all positioned horizontally at the bottom of the substrate-binding gorge of BChE. To better understand BChE binding with cocaine, (–)-cocaine and (+)-cocaine were also docked to the BChE active site in order to model the non-prereactive enzyme–substrate complexes. In the MD-simulated non-prereactive BChE-(–)-cocaine and BChE-(+)-cocaine complexes, (–)-cocaine and (+)-cocaine are positioned vertically in the substrate-binding gorge between D70 and W82. The MD trajectories for the non-prereactive BChE-(–)-cocaine and BChE-(+)-cocaine complexes are also very stable. In addition, the simulated non-prereactive BChE-(–)-cocaine and BChE-(+)-cocaine complexes are very close to the simulated

Michaelis–Menten complexes reported by Sun et al. [97]. All these suggest that the binding of BChE with (–)-cocaine and (+)-cocaine is similar to those proposed with butyrylthiocholine and succinyldithiocholine. Both the non-prereactive and prereactive enzyme–substrate complexes could exist before going to the chemical reaction steps.

To better compare the modeled non-prereactive complex with the pre-reactive complex for (–)-cocaine and for (+)-cocaine, the protein backbone atoms in the non-prereactive complex were superimposed with the corresponding atoms in the prereactive complex [100]. It turns out that the overall protein structures in the non-prereactive and prereactive complexes are very close to each other, while the orientations of the substrate are nearly vertical to each other. So, for both (–)-cocaine and (+)-cocaine, the substrate needs to rotate about 90° [100] during the change from the non-prereactive complex to the prereactive complex; more specifically, (–)-cocaine needs to rotate slightly more than (+)-cocaine. The energy barrier of the change from the non-prereactive complex to the prereactive complex for (–)-cocaine is expected to differ from that for (+)-cocaine. This is because the relative positions of the C-2 methyl ester group of substrate are different and, therefore, some amino acid residues hindering the rotation of one substrate might not hinder the rotation of another. Specific residues possibly hindering substrate rotation will be discussed below.

It is apparent that a detailed mechanistic understanding of the difference between the catalytic activity of BChE for (+)-cocaine and for (–)-cocaine could lead to important insights for the rational design of esterases with a high catalytic activity for hydrolysis of natural (–)-cocaine.

3.2.3
Possible Reaction Pathways

For both (–)-cocaine and (+)-cocaine, the relative positions of the nitrogen and the benzoyl carbonyl in the simulated prereactive BChE-substrate complex are essentially the same as those reported for BCh in BChE [100]. One may expect that BChE-catalyzed hydrolysis of (–)-cocaine and (+)-cocaine follow a reaction pathway similar to that for BChE-catalyzed hydrolysis of BCh. A remarkable difference between (–)-cocaine and (+)-cocaine is associated with the relative positions of the C-2 methyl ester group. The C-2 methyl ester group of (–)-cocaine remains on the same side of the carbonyl of the benzoyl ester as the attacking hydroxyl oxygen (S198 O^{γ}), whereas the C-2 methyl ester of (+)-cocaine remains on the opposite side. This difference could cause a difference in hydrogen bonding, electrostatic, and van der Waals interactions during the catalytic process, and result in a significant difference in free energies of activation. Nevertheless, the basic BChE mechanism for both enantiomers may resemble the common catalytic mechanism for ester hydrolysis in other serine hydrolases [100, 108], including the thoroughly investigated AChE [109–112]. Thus, based on the modeling and simulation of the prereactive complexes and the knowledge about ester hydrolysis in other serine hydrolases, a possible reaction pathway for BChE-catalyzed hydrolysis of cocaine can be hypothesized. Scheme 1 depicts

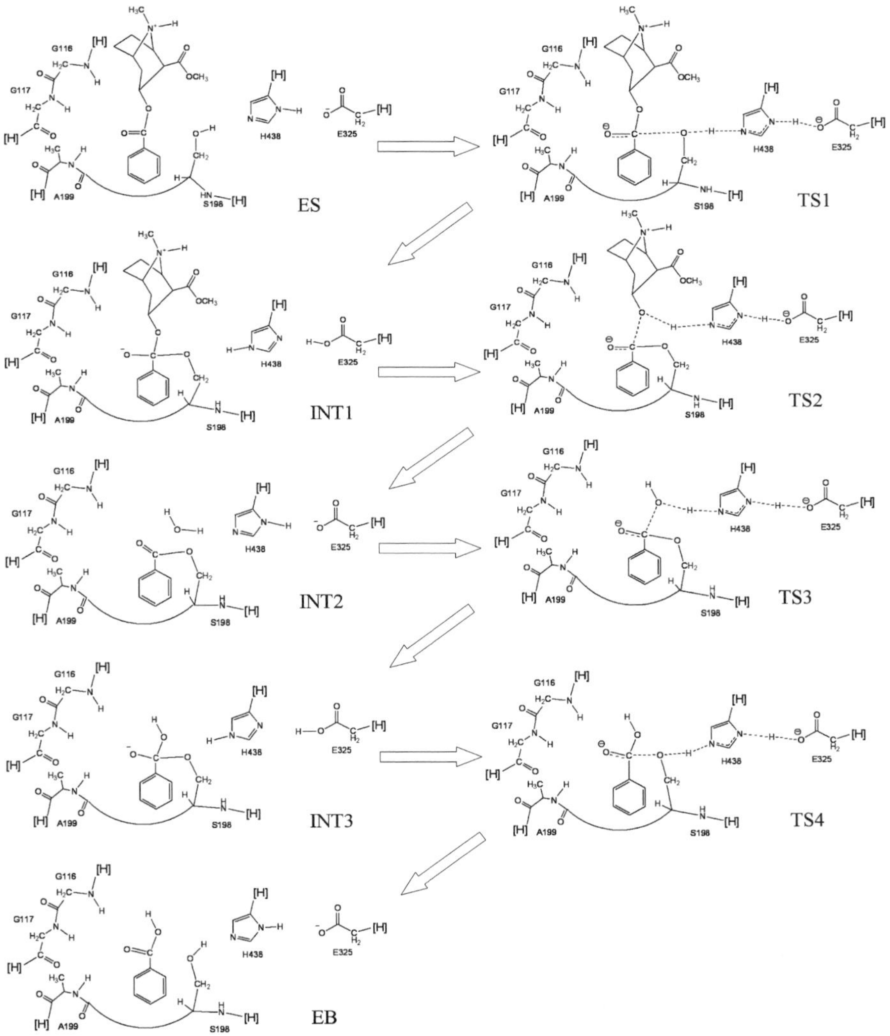

Scheme 1 Schematic representation of BChE-catalyzed hydrolysis of (–)-cocaine. Only the QM-treated high-layer part of the reaction system in the ONIOM (QM/MM) calculations [113] are drawn. Notation [H] refers to a non-hydrogen atom in the MM-treated low-layer part of the protein, and the cut covalent bond with this atom is saturated by a hydrogen atom. The transition covalent bonds existing in all of the transition states are indicated with *dashed lines*

(–)-cocaine and important groups from the catalytic triad (S198, E325, and H438) and three-pronged oxyanion hole (G116, G117, and A199). The proposed hydrolysis of cocaine consists of both the acylation and deacylation stages demonstrated for ester hydrolysis by other serine hydrolases. A significant difference might exist in the number of potential hydrogen bonds involving the carbonyl oxygen in the oxyanion hole. The three-pronged oxyanion hole formed by peptidic NH groups of G116, G117, and A199 in BChE (or by peptidic NH groups of G118, G119, and A201 in AChE) contrasts with the two-pronged oxyanion hole of many other serine hydrolases. Schematic representation of the pathway for (+)-cocaine hydrolysis should be similar to Scheme 1, differing only in the relative position of the C-2 methyl ester group in the acylation.

As depicted in Scheme 1 [113], the acylation is initialized by S198 O^γ attack at the carbonyl carbon of the cocaine benzoyl ester to form the first tetrahedral intermediate (INT1) through the first transition state (TS1). During the formation of INT1, the C – O bond between the carbonyl carbon and S198 O^γ gradually forms, while the proton at S198 O^γ gradually transfers to the imidazole N atom of H438, which acts as a general base. The second step of the acylation is the decomposition of INT1 to the metabolite ecgonine methyl ester and acyl-BChE (INT2a) through the second transition state (TS2). During the change from INT1 to INT2a, the proton gradually transfers to the benzoyl ester oxygen, while the C – O bond between the carbonyl carbon and the ester oxygen gradually breaks. Also, during the first step of acylation, the carbonyl oxygen may potentially form up to three hydrogen bonds with the NH groups of G116, G117, and A199.

In the MD-simulated prereactive enzyme–substrate complex [100], only one or two of the three NH groups weakly hydrogen-bond to the carbonyl oxygen of (+)-cocaine or (–)-cocaine during the simulations. No hydrogen bonding was noted between the carbonyl oxygen and the NH group of G116. These potential hydrogen bonds are expected to increase in strength from ES to TS1 and to INT1 due to the expected increase of net negative charge on the carbonyl oxygen. By the same logic, these potential hydrogen bonds are expected to progressively weaken from INT1 to TS2 and to INT2. The deacylation is initialized by water (oxygen) attack at the carbonyl carbon with participation of H438 as a general base, and is the reverse of acylation with respect to bond breaking/formation and potential hydrogen bonds.

3.2.4
Reaction Coordinates and Energy Barriers

To examine the above mechanistic hypotheses, detailed first-principles reaction coordinate calculations [100] were performed on the fundamental reaction pathways for BChE-catalyzed hydrolysis of (–)-cocaine with a BChE active site model. The BChE active site model used in the reaction coordinate calculations includes only the six amino acids indicated in Scheme 1 with the following simplifications: S198 is represented by methanol, H438 is represented by imidazole, E325 is represented by acetate (CH_3COO^-), and G116, G117, and A199 are all represented by ammonia molecules. This BChE active site model consists of 34 atoms. So, a total of 78 atoms were included in the ab initio calculations [100] on the model of (–)-cocaine hydrolysis. Reaction coordinate calculations on this model system are expected to provide a qualitative picture for the formation and breaking of covalent bonds at the reaction center and to estimate the intrinsic energy barriers and Gibbs free energy barriers for the enzymatic reaction.

The reaction coordinate calculations [100] confirmed the mechanistic hypothesis depicted in Scheme 1, i.e., the entire chemical reaction process consists of four individual steps (ES → TS1→ INT1 → TS2 → INT2 → TS3 → INT3 → TS4 → EB). The calculated energy barriers (ΔE_a) and Gibbs free energy barriers (ΔG_a) are summarized in Table 3.

Similar computational studies [46–48] on the reaction pathways for various non-enzymatic ester hydrolyses demonstrate that electron correlation effects are important only for final energy evaluations, but are not important for the geometry optimizations. The geometry optimizations at a lower HF level (using a smaller basis set) followed by single-point energy calculations at the MP2/6-31+G(d) level are adequate for predicting energy barriers in excellent agreement with the corresponding experimental data. The calculations at higher levels, e.g., replacing MP2 with QCISD(T) or using larger basis set, do not change the calculated energy barriers substantially [46, 47]. As seen in Table 3, the results calculated at the MP2/6-31+G(d)//HF/3-21G level show that for both the acylation and deacylation stages, the energy bar-

Table 3 Energy barriers (ΔE_a) and Gibbs free energy barriers (ΔG_a), in kcal/mol, calculated for BChE-catalyzed hydrolysis of (–)-cocaine at 298 K and 1 atm [100] [a]

Method	ΔE_a				ΔG_a			
	Step 1	Step 2	Step 3	Step 4	Step 1	Step 2	Step 3	Step 4
MP2/6-31+G(d)	4.0	3.1	16.6	6.5	5.6	3.6	19.0	6.6
			(17.0)				(18.2)	
B3LYP/6-31+G(d)	6.2		16.9		7.8		19.2	
			(18.5)				(19.8)	
B3LYP/6-31+G(d,p)	5.5		16.2		7.1		18.5	
			(17.5)				(18.7)	
B3LYP/6-31++G(d,P)	5.6		16.2		7.2		18.5	
			(17.5)				(18.7)	

[a] Values in parentheses were calculated using the geometries optimized at the B3LYP/6-31G(d) level. Other values were calculated using the geometries optimized at the HF/3-21G level

rier and free energy barrier predicted for the first step (i.e., the first or third step of the entire chemical reaction process) is always higher than that for the corresponding second step (i.e., the second or fourth step of the entire chemical reaction process). Hence, the B3LYP energy calculations using the 6-31+G(d) and larger basis sets were also performed on these two critical reaction steps. The energy barriers calculated at the B3LYP/6-31+G(d) level are close to the corresponding values at the MP2/6-31+G(d) level. Increasing the basis set from 6-31+G(d) to 6-31++G(d,p), the changes of the calculated barriers are smaller than 0.7 kcal/mol.

Since the third reaction step (i.e., the first step of deacylation) is associated with the highest energy barrier and the highest Gibbs free energy barrier, the geometries of TS3 and INT2 were also optimized at the B3LYP/6-31G(d) level to evaluate the barrier for this highest-barrier reaction step using the B3LYP/6-31G(d) geometries. It has been found that the barriers calculated using the B3LYP/6-31G(d) geometries are close to those from the HF/3-21G geometries, particularly for the Gibbs free energy barriers. These comparisons indicate that the energy barriers and Gibbs free energy barriers predicted at the MP2/6-31+G(d)//B3LYP/6-31G(d) and MP2/6-31+G(d)//HF/3-21G levels are reliable.

Because the third reaction step is predicted to have the highest barrier, this reaction step is expected to be rate determining if the effects of the remaining protein environment on the calculated barriers can be neglected. Further, because the third reaction steps for (−)-cocaine and for (+)-cocaine are identical, their hydrolysis rates in BChE would be expected to be the same if this step of the chemical reaction process were really rate determining for the entire catalytic process. In fact, (−)-cocaine hydrolysis in BChE is about 1000 to 2000 times slower than (+)-cocaine, suggesting that some other factors, such as the change from the non-prereactive complex to the prereactive complex (ES), are important for (−)-cocaine or for both (−)-cocaine and (+)-cocaine. Furthermore, if a chemical reaction step in the acylation, rather than the transformation from non-prereactive BChE-(−)-cocaine complex to the prereactive BChE-(−)-cocaine complex, is rate determining, the enzymatic reaction rate is expected to be pH-dependent. An experimental study [97] revealed that the rate of the BChE-catalyzed hydrolysis of (−)-cocaine is not significantly affected by the pH of the reaction solution, whereas the rate of the BChE-catalyzed hydrolysis of (+)-cocaine is clearly pH-dependent. So, it is likely that the rate-determining step is the change from the non-prereactive complex to the prereactive complex for BChE-catalyzed hydrolysis of (−)-cocaine [100].

3.3
MD Simulations of Cocaine Binding with BChE Mutants

The enzyme–substrate binding and fundamental reaction pathways discussed above provide a rational base for the design of more active BChE mutants for catalytic hydrolysis of (–)-cocaine. Now that the change from the non-prereactive complex to the prereactive complex is probably the rate-determining step of the BChE-catalyzed hydrolysis of (–)-cocaine, useful BChE mutants could be designed to specifically accelerate the change from the non-prereactive BChE-(–)-cocaine complex to the prereactive BChE-(–)-cocaine complex. A detailed analysis of the MD-simulated structures of wild-type BChE binding with (–)-cocaine and (+)-cocaine revealed that Y332 is a key residue hindering the structural change from the non-prereactive BChE-(–)-cocaine complex to the prereactive BChE-(–)-cocaine complex [100, 101]. A number of possible mutants of BChE were proposed for wet experimental tests [97–101, 114]. The earliest design of BChE mutants was only based on the modeled or simulated structure of the non-prereactive BChE-(–)-cocaine complex with wild-type BChE; the possible dynamics of the proposed BChE mutants were not examined. Some of the proposed mutants indeed have an improved catalytic efficiency against (–)-cocaine [97–99, 101, 114].

In order to more reliably predict the BChE mutants with a possibly higher catalytic efficiency against (–)-cocaine, MD simulations were also performed on the structures of (–)-cocaine binding with a number of hypothetical BChE mutants in their non-prereactive and prereactive complexes [101]. Table 4 summarizes the average values of some important geometric parameters in the simulated complexes.

In the simulated non-prereactive complex, the average distance between the carbonyl carbon of cocaine benzoyl ester and S198 O^γ is 7.6 Å for A328W/Y332A BChE and 7.1 Å for A328W/Y332G BChE, as seen in Table 4. In the simulated prereactive complex, the average values of this important internuclear distance become 3.87 and 3.96 Å for A328W/Y332A and A328W/Y332G BChEs, respectively. Compared to the simulated wild-type BChE-(–)-cocaine prereactive complex, the average distances between the carbonyl carbon of the cocaine benzoyl ester and S198 O^γ in the prereactive complex of (–)-cocaine with A328W/Y332A and A328W/Y332G BChEs are all slightly longer, whereas the average distances between the carbonyl oxygen of the cocaine benzoyl ester and the NH of G116, G117, and A199 residues are all shorter. This suggests that (–)-cocaine more strongly binds with A328W/Y332A and A328W/Y332G BChEs in the prereactive complexes. More importantly, the (–)-cocaine rotation in the active site of A328W/Y332A and A328W/Y332G BChEs from the non-prereactive complex to the prereactive complex did not cause considerable changes of the positions of A332 (or G332), W328, and F329 residues, compared to the (–)-cocaine rotation in the active site of wild-type BChE. These results suggest that A328W/Y332A and A328W/Y332G BChEs should be associated with lower energy barriers than the wild-type for the (–)-cocaine rotation from the non-prereactive complex to the prereactive complex. Further, (–)-cocaine binding with A328W/Y332G BChE is very similar to the binding with A328W/Y332A BChE, but the pos-

Table 4 Time-averaged values of some key geometric parameters (Å and degree) in the simulated non-prereactive and prereactive BChE-cocaine complexes [101]

BChE-cocaine binding[a]	Average values of the geometric parameters[c]						RMSD[d]	
	$<D1>_{non}$	$<D1>$	$<D2>$	$<D3>$	$<D4>$	$<\Theta>$	nonpre	pre
Wild-type	5.60	3.27	5.77	2.71	3.37	67	1.14	1.27
Wild-type with (+)-cocaine[b]	7.64	3.69	2.88	2.30	2.83	61	1.15	1.13
A328W/Y332A	7.11	3.87	3.30	2.14	3.01	51	1.58	1.65
A328W/Y332G	7.06	3.96	2.28	2.52	2.42	60	1.20	1.35
A328W/Y332A/Y419S	5.18	5.84	5.64	4.56	6.97	164	2.66	2.62

[a] Refers to (−)-cocaine binding with wild-type human BChE or (−)-cocaine binding with a mutant BChE, unless indicated otherwise

[b] Refers to (+)-cocaine binding with wild-type human BChE

[c] $<D1>_{non}$ and $<D1>$ represent the average distances between the S198 O^{γ} atom and the carbonyl carbon of the cocaine benzoyl ester in the simulated non-prereactive and prereactive BChE-cocaine complexes, respectively. $<D2>$, $<D3>$, $<D4>$ refer to the average values of the simulated distances from the carbonyl oxygen of the cocaine benzoyl ester to the NH hydrogen atoms of G116, G117, and A199 residues, respectively. $<\Theta>$ is the average value of the dihedral angle formed by the S198 O^{γ} atom and the plane of the carboxylate group of the cocaine benzoyl ester

[d] Root-mean-square deviation (RMSD) of the coordinates of backbone atoms in the simulated structure from those in the X-ray crystal structure of BChE. *nonpre* and *pre* refer to the non-prereactive and prereactive BChE-cocaine complexes, respectively

ition change of F329 residue caused by the (−)-cocaine rotation was significant only in A328W/Y332A BChE, thus suggesting that the energy barrier for the (−)-cocaine rotation in A328W/Y332G BChE should be slightly lower than that in A328W/Y332A BChE.

Concerning (−)-cocaine binding with A328W/Y332A/Y419S BChE, Y419 stays deep inside the protein and does not directly contact with the cocaine molecule. The Y419S mutation was considered because this mutation was initially expected to further increase the free space of the active site pocket so that the (−)-cocaine rotation could be easier. However, as seen in Table 4, the average distance between the carbonyl carbon of cocaine benzoyl ester and S198 O^{γ} atom in the simulated prereactive complex was as long as 5.84 Å. The average distances between the carbonyl oxygen of the cocaine benzoyl ester and the NH hydrogen atoms of G116, G117, and A199 residues are between 4.56 and 6.97 Å; no any hydrogen bond between them. In addition to the internuclear distances, another interesting geometric parameter is the dihedral angle, Θ, formed by S198 O^{γ} and the plane of the carboxylate group of the cocaine benzoyl ester. As seen in Table 4, the Θ values in the prereactive complexes of cocaine with wild-type BChE and all of the BChE mutants, other than A328W/Y332A/Y419S BChE, all slightly deviate from the ideal value of 90° for the nucleophilic attack of S198 O^γ at the carbonyl carbon of cocaine. The Θ value in the prereactive complex of (–)-cocaine with A328W/Y332A/Y419S BChE is 164°, which is considerably different from the ideal value of 90°.

The above discussion suggests that the energy barriers for the (–)-cocaine rotation in A328W/Y332A and A328W/Y332G BChEs from the non-prereactive complex to the prereactive complex, the rate-determining step for the BChE-catalyzed hydrolysis of (–)-cocaine, should be lower than that in wild-type BChE. Thus, the MD simulations predict that both A328W/Y332A and A328W/Y332G BChEs should have a higher catalytic efficiency than wild-type BChE for (–)-cocaine hydrolysis. Further, the MD simulations also suggest that the energy barrier for the (–)-cocaine rotation in A328W/Y332G BChE should be slightly lower than that in A328W/Y332A BChE and, therefore, the catalytic efficiency of A328W/Y332G BChE for the (–)-cocaine hydrolysis should be slightly higher than that of A328W/Y332A BChE. In addition, the MD simulations predict that A328W/Y332A/Y419S BChE should have no catalytic activity, or have a considerably lower catalytic efficiency than the wild-type, for (–)-cocaine hydrolysis because (–)-cocaine binds with the mutant BChE in a way that is not suitable for the catalysis. Following the computational predictions, the wet experimental studies (including site-directed mutagenesis, protein expression, and enzyme activity assay against (–)-cocaine) were carried out [101]. The experimental kinetic data qualitatively confirms the theoretical predictions based on the MD simulations. In particular, the catalytic efficiency of A328W/Y332G BChE is indeed slightly higher than that of A328W/Y332A BChE against (–)-cocaine, and A328W/Y332A/Y419S BChE is indeed inactive against (–)-cocaine [101].

3.4
Evolution of Hydrogen Bonding during the Reaction Process

3.4.1
Theoretical Issue for MD Simulation of a Transition State

To examine the evolution of hydrogen bonding during the reaction process through MD simulations, one needs to perform MD simulations on all of the transition states, in addition to the routine MD simulations on the reactants, intermediates, and products. One must address a critical issue [115–118] before performing any MD simulation on a transition state. In principle, MD simulation using a classical force field (molecular mechanics) can only simulate a stable structure corresponding to a local minimum on the potential energy surface, whereas a transition state during a reaction process is always associated with a first-order saddle point on the potential energy surface. Hence, MD simulation using a classical force field cannot directly simulate a transition state without any restraint on the geometry of the transition state.

Nevertheless, if one can technically remove the freedom of imaginary vibration in the transition state structure, then the number of vibrational freedoms (normal vibration modes) for a non-linear molecule will decrease from $3N - 6$ to $3N - 7$ (or less). The transition state structure is associated with a local minimum on the potential energy surface within a subspace of the reduced vibrational freedoms, although it is associated with a first-order saddle point on the potential energy surface with all of the $3N - 6$ vibrational freedoms. Theoretically, the vibrational freedom associated with the imaginary vibrational frequency in the transition state structure can be removed by appropriately freezing the reaction coordinate. The reaction coordinate corresponding to the imaginary vibration of the transition state is generally characterized by a combination of some key geometric parameters. These key geometric parameters are bond lengths of the forming and breaking covalent bonds for BChE-catalyzed hydrolysis of cocaine. Thus, one just needs to maintain the bond lengths of the forming and breaking covalent bonds during the MD simulation on a transition state [115–117].

Technically, one can maintain the bond lengths of the forming and breaking covalent bonds by simply fixing all atoms within the reaction center, by using some constraints on the forming and breaking covalent bonds, or by redefining the forming and breaking covalent bonds. It should be pointed out that the purpose of performing such type of MD simulation on a transition state is to examine the dynamic change of the protein environment surrounding the reaction center and the interaction between the reaction center and the protein environment. The detailed MD procedure for transition state simulation has been described in the latest literature [115–117].

3.4.2
Structures from MD Simulations and QM/MM Optimizations

All transition states and intermediates need to be simulated, allowing the protein structure to have a sufficiently long time to adapt to the fixed reaction center geometries obtained from the earliest ab initio reaction coordinate calculations [100]. The MD trajectories actually became stable quickly [115], so were the internuclear distances involved in the potential $N - H \cdots O$ hydrogen bonds, i.e., the distances from the carbonyl oxygen (denoted by O31) of cocaine benzoyl ester to the NH hydrogen atoms of G116, G117, and A199. The numerical results concerning the MD trajectories are summarized in Table 5. As seen in Table 5, the RMSD values are all smaller than 2.0 Å for all of the MD trajectories, demonstrating that the backbone of BChE did not dramatically change in going from the prereactive BChE–cocaine complex (ES) to the transition states, intermediates, and product.

As seen in Table 5, in the QM/MM-optimized geometries, the distances optimized for TS3 and INT3 (associated with the rate-determining step of the chemical reaction process) at the B3LYP/6-31+G(d):Amber level

Table 5 Summary of the MD-simulated and optimized key distances (Å) and the root-mean-square deviation (RMSD) of the simulated structures from the initial structure [115]

Structure	Method	Distance[a]					RMSD[b]
		D1	D2	D3	D4	D5	
ES	MD[c]	3.79	2.14	4.47	3.11	4.89	1.28
	QM/MM(a)[d]	4.05	2.16	4.60	3.41	5.22	
TS1	MD	4.59	2.91	1.92	2.00	3.61	1.59
	QM/MM(a)	4.12	2.24	2.04	2.34	4.36	
INT1	MD	3.92	1.94	2.36	2.08	4.01	1.57
	QM/MM(a)	4.09	2.03	1.93	2.06	4.04	
TS2	MD	3.49	1.98	2.22	2.26	4.13	1.53
	QM/MM(a)	3.55	1.80	1.90	2.36	4.54	
INT2	MD	3.12	2.15	2.50	3.94	3.37	1.76
	QM/MM(a)	2.60	1.89	2.81	5.15	3.41	
TS3	MD	3.76	1.98	2.68	3.86	1.96	1.61
	QM/MM(a)	3.76	2.14	1.87	4.04	2.07	
	QM/MM(b)[d]	4.10	2.27	1.85	3.99	2.07	
INT3	MD	3.01	2.03	1.88	4.05	5.68	1.70
	QM/MM(a)	2.04	1.80	1.64	4.27	6.07	
	QM/MM(b)	2.23	1.73	1.69	4.18	6.00	
TS4	MD	3.73	2.04	1.87	3.04	4.26	1.53
	QM/MM(a)	3.24	1.91	1.69	2.96	4.84	
EB	MD	4.14	3.22	4.23	3.64	2.06	1.48
	QM/MM(a)	3.37	2.14	2.12	4.87	2.87	
BChE[e]	MD				3.64	4.35	1.49
AChE-ACh[f]	MD	2.32	1.92	3.53	4.17	5.40	1.37

[a] D1, D2, and D3 represent the distances between the carbonyl oxygen of cocaine benzoyl ester and the NH hydrogen of G116, G117, and A199, respectively. D4 and D5 respectively refer to the distances between the NH hydrogen of G116 and the oxygen atoms $O^{\varepsilon 1}$ and $O^{\varepsilon 2}$ of E197 side chain

[b] Root-mean-square deviation of the coordinates of backbone atoms in the simulated structure from those in the initial structure

[c] Average distances from the stable trajectory of MD simulation

[d] Distances in the geometry optimized by performing the QM/MM calculation at the HF/3-21G:Amber level (a) or at the B3LYP/6-31+G(d):Amber level (b). The bond lengths for all of the transition bonds were fixed in the geometry optimizations

[e] Results for the pure protein without cocaine or any other ligand in the active site

[f] Results for prereactive complex between mouse acetylcholinesterase (AChE) and acetylcholine (ACh). For this system, D1, D2, and D3 represent the distances between the carbonyl oxygen of ACh and the NH hydrogen of G121, G122, and A204, respectively. D4 and D5 respectively refer to the distances between the NH hydrogen of G121 and the oxygen atoms $O^{\varepsilon 1}$ and $O^{\varepsilon 2}$ of E202 side chain

are all reasonably close to the corresponding distances optimized at the HF/3-21G:Amber level, showing that the HF/3-21G level is adequate for the treatment of high-layer atoms in this study. This is consistent with the conclusion based on the earliest ab initio QM calculations [100] on the active site model. The earliest ab initio QM calculations [100] demonstrated that the HF/3-21G level is adequate for the geometry optimizations and that the energy barriers calculated at the B3LYP/6-31+G(d)//HF/3-21G level are all close to those calculated at the higher levels using the geometries optimized at the B3LYP/6-31G(d) level. Further, the QM/MM-optimized distances are also close to the corresponding distances in the MM-optimized structures, suggesting that it is reasonable using the classical force field for this kind of enzymatic reaction system.

It should be noted that the significance of the QM/MM calculations [115] described here is limited, because the geometries of the transition states were not fully optimized and, thus, the energy barriers cannot be evaluated explicitly based on these QM/MM calculations. More meaningful QM/MM reaction coordinate calculations [113] will be discussed later in this chapter.

The computational results summarized in Table 5 qualitatively confirm the existence of the oxyanion hole consisting of G116, G117, and A199 during BChE-catalyzed hydrolysis of cocaine, because G116, G117, and A199 all had hydrogen bonding or close interaction with O31 of cocaine in at least one transition state or intermediate. However, the N – H$\cdots$O hydrogen bonds are mostly between cocaine O31 (i.e., the carbonyl oxygen at the benzoyl ester group) and the NH hydrogen atoms of G117 and A199. The NH hydrogen of G116 was not hydrogen-bonded to cocaine O31, except very briefly in INT2 and INT3. Only G117 had an N – H$\cdots$O hydrogen bond with cocaine O31 in the MD-simulated prereactive BChE–cocaine complex ES. The simulated average H$\cdots$O distance in this N – H$\cdots$O hydrogen bond is 2.14 Å. After further energy minimization with Amber7 and the QM/MM geometry optimization, this H$\cdots$O distance became 2.02 and 2.16 Å, respectively, as seen in Table 5. Changing from the prereactive complex (ES) to the transition states and intermediates, another N – H$\cdots$O hydrogen bond formed between cocaine O31 and A199 while the N – H$\cdots$O hydrogen bond with G117 was fully or partially maintained. For example, the N – H$\cdots$O hydrogen bond between cocaine O31 and G117 existed partially in TS1 and fully in INT1. Thus, compared to ES, the transition states and intermediates were stabilized further by the hydrogen bonding of cocaine O31 with A199.

3.4.3
Hydrogen Bonding Energies

To better represent the overall strength of hydrogen bonding of cocaine O31 with the oxyanion hole in each MD-simulated structure, the total number of hydrogen bonds with cocaine O31 was estimated [115] by using a cut-

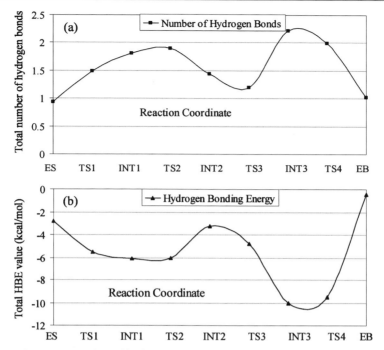

Fig. 8 Total average number of hydrogen bonds (**a**) and the corresponding total hydrogen bonding energy (**b**) between the carbonyl oxygen of cocaine benzoyl ester and the oxyanion hole in the simulated the prereactive BChE–cocaine complex (ES), transition states, intermediates, and product for BChE-catalyzed hydrolysis of (–)-cocaine [115]

off value of 2.5 Å for the H···O distances in order to roughly determine whether any hydrogen bond existed in any snapshot of MD simulation. Thus, the total number of hydrogen bonds with cocaine O31 can be obtained for each snapshot. The simulated total number of hydrogen bonds with cocaine O31 was estimated as the average over all of the snapshots taken in stable range of the MD trajectory. Figure 8a shows the simulated total numbers of hydrogen bonds estimated in this way for all of the structures involved in the reaction process: ES → TS1 → INT1 → TS2 → INT2 → TS3 → INT3 → TS4 → EB. Here, TS refer to the transition states, INT represent the intermediates, and EB refers to the BChE–benzoate acid complex. The results depicted in Fig. 8a clearly suggest that the total average number of hydrogen bonds between cocaine O31 and the oxyanion hole gradually increased from ES to TS1, INT1, and TS2, whereas the total average number of hydrogen bonds gradually decreased from TS2 to INT2 and TS3 before a remarkable increase from TS3 to INT3. In going from INT3 to TS4 and EB, the total average number of hydrogen bonds gradually decreased again.

As seen in Table 5, the H···O distances optimized at various levels of theory are consistent with the change of the simulated total average number of hydrogen bonds during the reaction process. For example, the optimized distances revealed only one N – H···O hydrogen bond with G117 in ES, a stronger N – H···O hydrogen bond with A199 and a weaker N – H···O hydrogen bond with G117 in TS1, and stronger N – H···O hydrogen bonds with both G117 and A199 in INT1. This supports the conclusion of the gradual increase of the hydrogen bonding for ES → TS1 → INT1.

With each of the simulated H···O distances, the hydrogen bonding energy (HBE) were also estimated by using the general HBE equation implemented in AutoDock 3.0 program suite [119]. Specifically, for each hydrogen bond with cocaine O31, a HBE value was evaluated with each snapshot of the MD-simulated structure. The final HBE of the MD-simulated hydrogen bond was considered to be the average HBE value of all snapshots taken from the stable MD trajectory. The total hydrogen bonding energy between cocaine O31 and the oxyanion hole in each MD-simulated state is depicted in Fig. 8b.

Comparing Fig. 8b with Fig. 8a, one can clearly see that the calculated total hydrogen bonding energy is mostly proportional to the simulated total number of hydrogen bonds, but the exception exists in some cases, particularly for the change from INT2 to TS3. The exception is due to the reason that different hydrogen bonds may have quite different average H···O distances and, therefore, have quite different hydrogen bonding energies. Thus, the total hydrogen bonding energy should be a better indicator of the overall hydrogen bonding between cocaine O31 and the oxyanion hole. In going from ES to TS1, the simulated total number of hydrogen bonds increases from 0.93 to 1.49 and, correspondingly, the calculated total HBE value changes from – 2.9 kcal/mol to – 5.5 kcal/mol, suggesting that the hydrogen bonding decreases the energy barrier for the first reaction step by ∼ 2.6 kcal/mol. In going from INT1 to TS2, the simulated total number of hydrogen bonds changes from 1.81 to 1.90 and the calculated total HBE value changes from – 6.1 kcal/mol to – 6.0 kcal/mol, indicating that the hydrogen bonding effects on the energy barrier for the second reaction step are negligible. In going from INT2 to TS3, the simulated total number of hydrogen bonds decreases from 1.44 to 1.20, whereas the calculated total HBE value changes from – 3.2 kcal/mol to – 4.8 kcal/mol, showing that the average hydrogen bonding strength per bond in TS3 is stronger than that in INT2. These energetic data suggest that the hydrogen bonding effects might decrease the energy barrier for the third reaction step by ∼ 1.6 kcal/mol. In going from INT3 to TS4, the simulated total number of hydrogen bonds decreases from 2.22 to 1.99 and the calculated total HBE value changes from – 10.1 kcal/mol to – 9.5 kcal/mol, suggesting that the hydrogen bonding may slightly increase the energy barrier for the fourth reaction step by ∼ 0.6 kcal/mol.

3.4.4
Comparison with AChE-Catalyzed Hydrolysis of Acetylcholine

It is interesting to compare the mechanism of BChE-catalyzed hydrolysis of cocaine with that of AChE-catalyzed hydrolysis of neurotransmitter acetylcholine (ACh), because AChE and BChE have very similar active sites including the same type of catalytic triad and the same type of oxyanion hole. In terms of mouse AChE, the catalytic triad consists of S203, H447, and E334 and the oxyanion hole consists of G121, G122, and A204. The only significant difference is that the cavity of BChE active site is larger so that it can accommodate a larger substrate like cocaine. McCammon et al. [120–122] reported MD simulations and QM/MM calculations on the mouse AChE–ACh system. Their MD simulations [121, 122] were performed on the prereactive AChE–ACh complex, whereas their QM/MM calculations [120] were carried out on the initial step of the acylation stage of the AChE-catalyzed ACh hydrolysis. Their QM/MM results [120] indicate that in the AChE–ACh Michaelis complex, two hydrogen bonds exist between the carbonyl oxygen of ACh and the peptidic NH groups of G121 and G122. In going from the AChE–ACh Michaelis–Menten complex to the (first) transition state and (first) intermediate, the distance between the carbonyl oxygen of ACh and NH group of A204 becomes shorter, and the third hydrogen bond is formed both in the transition state and in the tetrahedral intermediate [120]. Based on the structural similarity of these two closely related cholinesterases, one might easily assume that the $N - H \cdots O$ hydrogen bonding between the carbonyl oxygen of the substrate and the three residues of oxyanion hole in the BChE–cocaine and AChE–ACh systems should be very close to each other during the catalytic hydrolysis processes. However, a remarkable difference exists on the role of the first residue, i.e., G121 in AChE or G116 in BChE, of the oxyanion hole. The MD simulations and QM/MM calculations [115] on BChE-catalyzed cocaine hydrolysis revealed that G116 never hydrogen-bonded to O31 of cocaine during the acylation stage and was rarely involved in hydrogen bonding with O31 of cocaine during the deacylation stage. The simulated structures ES, TS1, INT1, and TS2 [115] are associated with the acylation stage, in which all of the cocaine atoms were included in the simulations and calculations. The simulated structures INT2, TS3, INT3, and TS4 are all associated with the deacylation stage, in which the free product ecgonine methyl ester had left the active site and cocaine only had benzoate atoms included in the simulations and calculations.

To make sure that this remarkable difference between the structures simulated for these two hydrolysis processes was not an artifact of the possibly different computational strategies used for the two different reaction processes, the prereactive complex between ACh and mouse AChE (using X-ray crystal structure 1MAH in the Protein Data Bank [123]) was also simulated by using the same MD approach as used for the BChE–cocaine system. As seen in Table 5, the MD simulation of the prereactive AChE–ACh complex also revealed an $N-H\cdots O$ hydrogen bond between the carbonyl oxygen of ACh and the NH hydrogen of G121, which is qualitatively consistent with the computational results reported by McCammon et al. [120–122]. This consistency supports the remarkable difference between the two catalytic hydrolysis processes.

To further examine the possible effect of the bulky ligand (cocaine) on the hydrogen bond between G116 and E197, an additional MD simulation [115] on BChE was carried out in a water bath without cocaine or any other ligand. As seen in Table 5, the simulated BChE structure did not show a hydrogen bond between G116 and E197, showing that the ligand has a role in the hydrogen bonding between the NH hydrogen of G116 and the carboxyl group of E197 side chain.

3.5
Effects of Protein Environment on the Energy Barriers

The key to the rational design of high-activity mutants of BChE against (–)-cocaine is to further understand the protein environmental effects on the reaction pathway, particularly the transition states involved and the corresponding energy barriers. This is because, to increase the catalytic activity of BChE for (–)-cocaine, one needs to design necessary mutation(s) to modify the protein environment such that the modified protein environment can more favorably stabilize the transition states and, therefore, lower the energy barriers, particularly for the rate-determining step(s). Understanding the protein environmental effects on the reaction pathway and energy barriers should help to rationally design BChE mutants with a lower energy barrier and, therefore, a higher catalytic activity for (–)-cocaine. However, ab initio reaction coordinate calculations [100] with an active site model can only account for breaking and formation of covalent bonds during the catalytic reaction process; the more complex protein environmental effects on the reaction pathway and energy barriers cannot be accounted for very well. Recently, extensive hybrid quantum mechanical/molecular mechanical (QM/MM) calculations [113] were performed on the entire BChE-(–)-cocaine and BChE-(+)-cocaine systems to optimize the geometries of the transition states and the corresponding prereactive enzyme–substrate complexes and of intermediates involved in the BChE-catalyzed hydrolysis of (–)- and (+)-cocaine, and to predict the corresponding energy barriers. The calculated results reveal remarkable effects of the protein environment on the energy barriers and provide useful insights into rational design of BChE mutants with lower energy barriers for the catalytic hydrolysis of (–)-cocaine.

3.5.1
QM/MM-Optimized Geometries of Transition States

The geometry optimizations using the QM/MM method have led to the converged geometries of the transition states (TS1, TS2, TS3, and TS4) and the corresponding prereactive BChE–cocaine complex (ES) and intermediates (INT1, INT2, and INT3) for the hydrolyses of (–)- and (+)-cocaine catalyzed by human BChE (see Scheme 1 for the atoms treated quantum mechanically in the QM/MM calculations) [113]. The QM/MM-optimized geometries [113] of the transition states, intermediates, and prereactive BChE–cocaine complex and their connections on the potential energy surface are consistent with the assumed enzymatic reaction pathway involving four reaction steps depicted in Scheme 1. For example, in the first step associated with TS1, the hydroxyl oxygen of S198 gradually attacks the carbonyl carbon of cocaine benzoyl ester, while the proton of the hydroxyl group gradually transfers to a nitrogen atom of H438 side chain, and the H438 side chain gradually transfers another proton to an oxygen atom of E325 side chain. This reaction step and the third reaction step (associated with TS3) both belong to the standard general base-catalysis mechanism, whereas both the second and fourth steps (associated with TS2 and TS4) follow the standard specific acid-catalysis mechanism. The results obtained from the QM/MM calculations qualitatively confirm the fundamental reaction pathway proposed for BChE-catalyzed hydrolysis of cocaine based on the earliest reaction coordinate calculations with a simplified active site model [100]. It follows that the protein environment neglected in the earliest reaction coordinate calculations do not change the fundamental reaction pathway for this enzymatic reaction, as far as the breaking and formation of covalent bonds are concerned.

However, the protein environment significantly affects the hydrogen bonding between the carbonyl oxygen of cocaine and the oxyanion hole (G116, G117, and A199) during the enzymatic reaction process [113]. Such type of hydrogen bonding with the oxyanion hole is crucial for the transition state stabilization, particularly for the first and third reaction steps, because the carbonyl oxygen atom in TS1 and TS3 possesses more negative charge than that in ES and INT2. The key internuclear distances concerning the hydrogen bonding in the QM/MM-optimized geometries of the transition states are all qualitatively consistent with those summarized in Table 5.

3.5.2
Energy Barriers

Table 6 summarizes the energy barriers predicted for BChE-catalyzed hydrolysis of (–)- and (+)-cocaine by performing the QM/MM calculations at the MP2/6-31+G(d):Amber level for all of the reaction steps, along with the corresponding energy barriers calculated for the (–)-cocaine hydrolysis with

Table 6 Energy barriers (ΔE_a, in kcal/mol) calculated for BChE-catalyzed hydrolysis of (–)- and (+)-cocaine [113]

Method and substrate		ΔE_a			
		Step 1	Step 2	Step 3	Step 4
Neglecting protein environment[a]	(–)-Cocaine	4.0	3.1	16.6 (17.0)	6.5
Including protein environment[b]	(–)-Cocaine	13.0	0.1	14.2	7.2
	(+)-Cocaine	12.1	0.4	14.2	7.2

[a] Calculated for a simplified model system [100] at the MP2/6-31+G(d)//HF/3-21G level. The value in parenthesis was calculated at the MP2/6-31+G(d)//B3LYP/6-31G(d) level
[b] Calculated for the real enzymatic reaction system by using the QM/MM method at the MP2/6-31+G(d):Amber level with the geometries optimized at the HF/3-21G:Amber level

a simplified active site model of BChE (neglecting the protein environment) for comparison. A comparison between the two sets of energy barriers listed in Table 6 reveals that the protein environmental effects dramatically change the energy barrier calculated for the first reaction step of the (–)-cocaine hydrolysis. The energy barriers calculated for the other steps are relatively less sensitive to the inclusion of the protein environment. The protein environmental effects increase the energy barrier for the first step of the (–)-cocaine hydrolysis by ~ 9 kcal/mol, decrease the energy barriers for the second and third steps by ~ 2–3 kcal/mol, and slightly increase the energy barrier for the fourth step. As a result, the second reaction step becomes almost barrierless and the energy barrier calculated for the fourth step is still much lower than that calculated for the third step. Based on the QM/MM results listed in Table 6, the third reaction step has the highest energy barrier, 14.2 kcal/mol; the energy barrier of 13.0 kcal/mol calculated for the first step of the (–)-cocaine hydrolysis is close to the barrier calculated for the third step. The energy barrier of 12.1 kcal/mol calculated for the first step of the (+)-cocaine hydrolysis is slightly lower than that the first step of the (–)-cocaine hydrolysis.

Note that the third and fourth reaction steps of BChE-catalyzed hydrolysis of (+)-cocaine are the same as the corresponding third and fourth reaction steps of BChE-catalyzed hydrolysis of (–)-cocaine. The highest energy barrier being associated with the third step means that (–)- and (+)-cocaine should be hydrolyzed by BChE at the same rate if the chemical reaction process is the rate-determining stage for both (–)- and (+)-cocaine. Further, if the chemical reaction process is the rate-determining stage, the catalytic rate constant k_{cat} should be dependent on the pH of the reaction solution, because H438 in the catalytic triad can be protonated and lose its catalytic role at low pH. However, BChE-catalyzed hydrolysis of (+)-cocaine ($k_{cat} = 1.07 \times 10^2$ s^{-1} and $K_M = 8.5$ μM) was observed to be three orders-of-magnitude faster than BChE-catalyzed hydrolysis of (–)-cocaine ($k_{cat} = 6.5 \times 10^{-2}$ s^{-1} and $K_M = 9.0\,\mu$M) [97] and only the k_{cat} value for (+)-cocaine was pH-dependent. So, the experimental data [97] clearly indicate that the rate-determining stage should be the chemical reaction process for (+)-cocaine, whereas the formation of the prereactive BChE-substrate binding complex (ES) should be the rate determining stage for (–)-cocaine. The calculated energy barriers further demonstrate that the third reaction step is the rate-determining step for (+)-cocaine.

The highest energy barrier calculated for BChE-catalyzed cocaine hydrolysis is ~ 3.7 kcal/mol higher than that (10.5 kcal/mol) calculated at the similar level, i.e., MP2/6-31+G(d) QM/MM, by McCammon et al. [120] for the first step of the AChE-catalyzed hydrolysis of ACh (the first step was recognized as the rate-determining step for the enzymatic reaction). The difference in the energy barrier between the two enzymatic reactions can be attributed to the aforementioned difference in the number of N – H$\cdots$O hydrogen bonds of the substrate with the oxyanion hole of the enzyme during the reaction processes.

The difference between the QM/MM-calculated energy barriers for the rate-determining steps of the two enzymatic reaction systems is consistent with the experimental observation that the k_{cat} value (1.6×10^4 s^{-1}) [124] for AChE-catalyzed hydrolysis of ACh was about 150-fold larger than that ($k_{cat} = 1.07 \times 10^2$ s^{-1}) [97] for BChE-catalyzed hydrolysis of (+)-cocaine. Based on the widely used classical transition-state theory (CTST) [125], the experimental k_{cat} difference of $\sim$ 150-fold suggests an energy barrier difference of ~ 3.0 kcal/mol when $T = 298.15$ K, which is in good agreement with the calculated barrier difference of ~ 3.7 kcal/mol [113].

3.5.3
Insights into Rational Design of BChE Mutants

It has been known that the formation of the prereactive BChE-(–)-cocaine complex (ES) is the rate-determining step of BChE-catalyzed hydrolysis of (–)-cocaine. Hence, the earliest rational design of BChE mutants has been focused on how to accelerate the ES formation process; for example, A328W/Y332A mutant of BChE was found to have a $\sim$ 9.4-fold improved catalytic efficiency (k_{cat}/K_M) against (–)-cocaine [98, 99]. However, it was not clear whether the energy barrier for the first step of BChE-catalyzed hydrolysis of (–)-cocaine is higher than that for the third step or not. If the energy barrier for the first step were significantly higher than that for the third step, it would mean that the catalytic efficiency of BChE against (–)-cocaine should still be significantly lower than that against (+)-cocaine, even if the chemical reaction process became rate determining. In that case, the site-directed mutagenesis designed to only speedup the ES formation process can be expected to make a limited improvement of the catalytic efficiency against (–)-cocaine. If the energy barrier for the third reaction step were the highest within the chemical reaction process, the catalytic efficiency of BChE against (–)-cocaine would be the same as that against (+)-cocaine when the chemical reaction process became rate determining. The energy barriers determined by the QM/MM calculations on BChE-catalyzed hydrolyses of (–)- and (+)-cocaine further demonstrate that the third reaction step indeed has the highest energy barrier (14.2 kcal/mol) within the chemical reaction processes, but the energy barrier of 13.0 kcal/mol calculated for the first step of (–)-cocaine hydrolysis is close to that for the third step. Further, the energy barrier for the first step is rather sensitive to the change of the protein environment because the protein environmental effects dramatically increase the energy barrier calculated for the first step, although the energy barriers for the subsequent steps look less sensitive to the change of the protein environment. These computational results suggest that it would be possible to design a BChE mutant that has a catalytic efficiency against (–)-cocaine comparable to wild-type BChE against (+)-cocaine, if the designed mutation could considerably speedup the ES formation process without significantly changing the energy barrier for any step of the chemical reaction process. However, a mutation designed to speedup the ES formation process could also change the energy barriers for the chemical reaction steps, especially for the first reaction step because the energy barrier calculated for this step is so sensitive to the protein environmental effects. So, future rational design of the high activity mutants of BChE against (–)-cocaine should pay attention to whether the mutation could also increase or decrease the energy barrier(s) for the first and/or third step of the chemical reaction process.

These computational insights help to understand available experimental data better. It has been found that the catalytic rate constant k_{cat} of A328W/Y332A BChE is pH-dependent for both (–)- and (+)-cocaine hydrolyses and that the A328W/Y332A mutation does not change the catalytic efficiency against (+)-cocaine [99]. The experimental kinetic data show that the chemical reaction process becomes the rate-determining for both (–)- and (+)-cocaine hydrolyses catalyzed by A328W/Y332A mutant of BChE, but the energy barriers for the rate-determining step of the two reactions must be different. Taking these experimental data and QM/MM results into account together, it is very likely that the rate-determining step of (+)-cocaine hydrolysis catalyzed by A328W/Y332A mutant of BChE is still the third reaction step. However, the rate-determining step of (–)-cocaine hydrolysis catalyzed by A328W/Y332A mutant of BChE becomes the first reaction step and has a significantly higher energy barrier than the third step. A closer look [113] at the detailed TS1 structure optimized for hydrolysis of (–)-cocaine catalyzed by wild-type BChE reveals that the TS1 structure is likely stabilized by a cation-π interaction between the protonated tropane nitrogen of (–)-cocaine and the benzene ring of Y332 side chain. The QM/MM-optimized distance between the protonated tropane nitrogen of (–)-cocaine and the center of the benzene ring of Y332 side chain was ~ 4.9 Å. Such a cation-π interaction will disappear when Y332 changes to Ala. Hence, while the Y332A mutation can help to remove the hindrance for the ES formation, the Y332A mutation may also destabilize the TS1 structure for (–)-cocaine hydrolysis. The Y332A mutation having no significant effect on the third reaction step may be explained by the fact that the tropane group of (–)-cocaine has left the active site after the second reaction step. Thus, there is no cation-π interaction in the TS3 structure whether Y332 changes to Ala or not. This mechanistic understanding suggests that starting from A328W/Y332A mutant of BChE, the rational design of further mutation(s) to improve the catalytic activity against (–)-cocaine should primarily aim to decrease the energy barrier for the first reaction step without significantly affecting the ES formation and other chemical reaction steps.

3.6
Computational Design of BChE Mutants
Based on Transition State Simulations

Generally speaking, for rational design of a mutant enzyme with a higher catalytic activity for a given substrate, one needs to design a mutation that can accelerate the rate-determining step of the entire catalytic reaction process while the other steps are not slowed down by the mutation. As discussed above, extensive computational modeling and experimental data indicate that the formation of the prereactive BChE-(–)-cocaine complex (ES) is the rate-determining step of (–)-cocaine hydrolysis catalyzed by wild-type BChE [97–101], whereas the rate-determining step of the corresponding (+)-cocaine hydrolysis is the chemical reaction process consisting of the acylation and deacylation stages, or four individual reaction steps [113]. This mechanistic understanding is consistent with the experimental observation [99] that the catalytic rate constant of wild-type BChE against (+)-cocaine is pH-dependent, whereas that of the same enzyme against (–)-cocaine is independent of the pH. The pH-dependence of the rate constant for (+)-cocaine hydrolysis is clearly associated with the protonation of H438 residue in the catalytic triad (S198, H438, and E325). For the first and third steps of the reaction process, when H438 is protonated, the catalytic triad cannot function and, therefore, the enzyme becomes inactive. The lower the pH of the reaction solution, the higher the concentration of the protonated H438, and the lower the concentration of the active enzyme. Hence, the rate constant was found to decrease with decreasing the pH of the reaction solution for the enzymatic hydrolysis of (+)-cocaine [99].

Based on the above mechanistic understanding, the earlier efforts for rational design of the BChE mutants were focused on how to improve the ES formation process. Several BChE mutants [97–99, 101, 114], including A328W, A328W/Y332A, A328W/Y332G, and F227A/S287G/A328W/Y332M, have been found to have a significantly higher catalytic efficiency (k_{cat}/K_M) against (–)-cocaine; these mutants of BChE have a $\sim$ nine- to 34-fold improved catalytic efficiency against (–)-cocaine. The aforementioned analysis of experimental and computational data clearly shows that the rate-determining step of (–)-cocaine hydrolysis catalyzed by the A328W/Y332A mutant should be the first step of the chemical reaction process. Further, recently reported computational modeling [97–101] also suggests that the formation of the prereactive BChE-(–)-cocaine complex (ES) is hindered mainly by the bulky side chain of Y332 residue in wild-type BChE, but the hindering can be removed by the Y332A or Y332G mutation [101, 117]. Therefore, starting from the A328W/Y332A or A328W/Y332G mutant, the rational design of further mutation(s) to improve the catalytic efficiency of BChE against (–)-cocaine can aim to decrease the energy barrier for the first reaction step without significantly affecting the ES formation and other chemical reaction steps [117].

For rational design of high-activity mutants of BChE, a unique computational strategy [117] has been developed to virtually screen various possible BChE mutants based on MD simulations of the rate-determining transition state (i.e., TS1).

In the design of a high-activity mutant of BChE against (–)-cocaine, one would like to predict some possible mutations that can lower the energy of the transition state for the first chemical reaction step (TS1) and, therefore, lower the energy barrier for this critical reaction step. Apparently, a mutant associated with the stronger hydrogen bonding between the carbonyl oxygen of (–)-cocaine benzoyl ester and the oxyanion hole of the BChE mutant in the TS1 structure may potentially have a more stable TS1 structure and, therefore, a higher catalytic efficiency for (–)-cocaine hydrolysis. Hence, the hydrogen bonding with the oxyanion hole in the TS1 structure is a crucial factor affecting the transition state stabilization and the catalytic activity. The possible effects of some mutations on the hydrogen bonding were examined by performing MD simulations on the TS1 structures for (–)-cocaine hydrolysis catalyzed by wild-type BChE and its various mutants [117]. The MD simulation in water was performed for 1 ns or longer to make sure a stable MD trajectory was obtained for each simulated TS1 structure with wild-type or mutant BChE. Figure 9 depicts plots of four important H$\cdots$O distances in the MD-simulated TS1 structure versus the simulation time for (–)-cocaine hydrolysis catalyzed by A199S/S287G/A328W/Y332G BChE, along with the root-mean-square deviation (RMSD) of the simulated positions of backbone atoms from those in the corresponding initial structure. The H$\cdots$O distances in the simulated TS1 structures for wild-type BChE and its three mutants are summarized in Table 7. The H$\cdots$O distances between the carbonyl oxygen of (–)-cocaine and the peptidic NH hydrogen atoms of G116, G117, and A199 (or S199) of BChE are denoted by D1, D2, and D3, respectively, in Table 7 and Fig. 9. D4 in Table 7 and Fig. 9 refers to the H$\cdots$O distance between the carbonyl oxygen of (–)-cocaine and the hydroxyl hydrogen of S199 side chain in the simulated TS1 structure corresponding to the A199S/S287G/A328W/Y332G mutant.

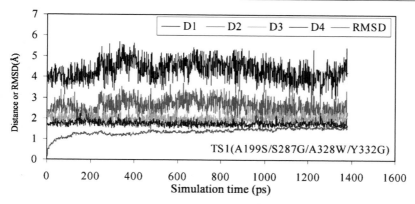

Fig. 9 Plots of the key internuclear distances (in Å) versus the time in the simulated TS1 structure for (−)-cocaine hydrolysis catalyzed by A199S/S287G/A328W/Y332G BChE [117]. Traces D1, D2, and D3 refer to the distances between the carbonyl oxygen of (−)-cocaine and the NH hydrogen of G116, G117, and S199, respectively. Trace D4 is the internuclear distance between the carbonyl oxygen of (−)-cocaine and the hydroxyl hydrogen of the S199 side chain in A199S/S287G/A328W/Y332G BChE. RMSD represents the root-mean-square deviation (in Å) of the simulated positions of the protein backbone atoms from those in the initial structure

As seen in Table 7, the simulated H···O distance D1 is always too long for the peptidic NH of G116 to form a N – H···O hydrogen bond with the carbonyl oxygen of (−)-cocaine in all of the simulated TS1 structures. In the simulated TS1 structure for wild-type BChE, the carbonyl oxygen of (−)-cocaine formed a firm N – H···O hydrogen bond with the peptidic NH hydrogen atom of A199 residue; the simulated H···O distance (D3) was 1.61 to 2.35 Å, with an average D3 value of 1.92 Å. Meanwhile, the carbonyl oxygen of (−)-cocaine also had a partial N – H···O hydrogen bond with the peptidic NH hydrogen atom of G117 residue; the simulated H···O distance (D2) was 1.97 to 4.14 Å (the average D2 value was 2.91 Å). The average D2 and D3 values became 2.35 and 1.95 Å, respectively, in the simulated TS1 structure for the A328W/Y332A mutant. These distances suggest a slightly weaker N – H···O hydrogen bond with A199, but a stronger N – H···O hydrogen bond with G117, in the simulated TS1 structure for the A328W/Y332A mutant that the corresponding N – H···O hydrogen bonds for the wild-type. The average D2 and D3 values (2.25 and 1.97 Å, respectively) in the simulated TS1 structure for the A328W/Y332G mutant are close to the corresponding distances for the A328W/Y332A mutant. The overall strength of the hydrogen bonding between the carbonyl oxygen of (−)-cocaine and the oxyanion hole of the enzyme is not expected to change considerably when wild-type BChE is replaced by the A328W/Y332A or A328W/Y332G mutant.

However, the story for the simulated TS1 structure for (−)-cocaine catalyzed by the A199S/S287G/A328W/Y332G mutant was remarkably different.

Table 7 Summary of the MD-simulated key distances (in Å) and the calculated total hydrogen-bonding energies (HBE, in kcal/mol) between the oxyanion hole and the carbonyl oxygen of (–)-cocaine benzoyl ester in the first transition state (TS1) [117]

Transition state	Distance[a]					Total HBE[b]
		D1	D2	D3	D4	
TS1 structure for (–)-cocaine hydrolysis catalyzed by wild-type BChE	Average	4.59	2.91	1.92		– 5.5 (– 4.6)
	Maximum	5.73	4.14	2.35		
	Minimum	3.35	1.97	1.61		
	Fluctuation	0.35	0.35	0.12		
TS1 structure for (–)-cocaine hydrolysis catalyzed by A328W/Y332A mutant of BChE	Average	3.62	2.35	1.95		– 6.2 (– 4.9)
	Maximum	4.35	3.37	3.02		
	Minimum	2.92	1.78	1.61		
	Fluctuation	0.23	0.27	0.17		
TS1 structure for (–)-cocaine hydrolysis catalyzed by A328W/Y332G mutant of BChE	Average	3.60	2.25	1.97		– 6.4 (– 5.0)
	Maximum	4.24	3.17	2.76		
	Minimum	2.89	1.77	1.62		
	Fluctuation	0.23	0.24	0.17		
TS1 structure for (–)-cocaine hydrolysis catalyzed by A199S/S287G/A328W/Y332G mutant of BChE	Average	4.39	2.60	2.01	1.76	– 14.0 (– 12.0)
	Maximum	5.72	4.42	2.68	2.50	
	Minimum	2.87	1.76	1.62	1.48	
	Fluctuation	0.48	0.36	0.17	0.12	

[a] D1, D2, and D3 represent the internuclear distances between the carbonyl oxygen of cocaine benzoyl ester and the NH hydrogen of residues #116 (i.e. G116), #117 (i.e. G117), and #199 (i.e. A199 or S199) of BChE, respectively. D4 is the internuclear distance between the carbonyl oxygen of cocaine benzoyl ester and the hydroxyl hydrogen of S199 side chain in the A199S/S287G/A328W/Y332G mutant
[b] Total HBE value is the average of the HBE values calculated by using the instantaneous distances in all of the snapshots. The value in parenthesis is the total HBE value calculated by using the MD-simulated average distances

As one can see from Table 7 and Fig. 9, when residue #199 becomes a serine (i.e., S199), the hydroxyl group on the side chain of S199 can also hydrogen-bond to the carbonyl oxygen of (–)-cocaine to form an O – H⋯O hydrogen bond, in addition to the two N – H⋯O hydrogen bonds with the peptidic NH of G117 and S199. The simulated average H⋯O distances with the peptidic NH hydrogen of G117, peptidic NH hydrogen of S199, and hydroxyl hydrogen of S199 are 2.60, 2.01, and 1.76 Å, respectively. Due to the additional O – H⋯O hydrogen bond, the overall strength of the hydrogen bonding with the modified oxyanion hole of A199S/S287G/A328W/Y332G BChE should be significantly stronger than that of wild-type, A328W/Y332A, and A328W/Y332G BChEs.

To better represent the overall strength of hydrogen bonding between the carbonyl oxygen of (–)-cocaine and the oxyanion hole in a MD-simulated TS1 structure, the hydrogen bonding energy (HBE) associated with each simulated H$\cdots$O distance was estimated [117]. For each hydrogen bond with the carbonyl oxygen of (–)-cocaine, a HBE value can be evaluated with each snapshot of the MD-simulated structure. The final HBE of the MD-simulated hydrogen bond is considered to be the average HBE value of all snapshots taken from the stable MD trajectory. The estimated total HBE value for the hydrogen bonds between the carbonyl oxygen of (–)-cocaine and the oxyanion hole in each simulated TS1 structure is also shown in Table 7.

The HBE for each hydrogen bond was also estimated by using the MD-simulated average H$\cdots$O distance. As seen in Table 7, the total hydrogen-bonding energies (i.e., – 4.6, – 4.9, – 5.0, and – 12.0 kcal/mol for the wild-type, A328W/Y332A, A328W/Y332G, and A199S/S287G/A328W/Y332G BChEs, respectively) estimated in this way are systematically higher (i.e., less negative) than the corresponding total hydrogen-bonding energies (i.e., – 5.5, – 6.2, – 6.4, and – 14.0 kcal/mol) estimated in the aforementioned way. However, the two sets of total HBE values are qualitatively consistent with each other in terms of the relative hydrogen-bonding strengths in the four simulated TS1 structures. In particular, the two sets of total HBE values consistently reveal that the overall strength of the hydrogen bonding between the carbonyl oxygen of (–)-cocaine and the oxyanion hole in the simulated TS1 structure for A199S/S287G/A328W/Y332G BChE is significantly stronger than that for wild-type, A328W/Y332A, and A328W/Y332G BChEs.

The computational results discussed above suggest that the transition state for the first chemical reaction step (TS1) of (–)-cocaine hydrolysis catalyzed by the A199S/S287G/A328W/Y332G mutant should be significantly more stable than that by the A328W/Y332A or A328W/Y332G mutant, due to the significant increase of the overall hydrogen bonding between the carbonyl oxygen of (–)-cocaine and the oxyanion hole of the enzyme in the TS1 structure. The aforementioned analysis of the literature also indicates that the first chemical reaction step associated with TS1 should be the rate-determining step of (–)-cocaine hydrolysis catalyzed by a BChE mutant including Y332A or Y332G mutation, although the formation of the prereactive enzyme–substrate complex (ES) is the rate-determining step for (–)-cocaine hydrolysis catalyzed by wild-type BChE. This suggests a clear correlation between the TS1 stabilization and the catalytic efficiency of A328W/Y332A, A328W/Y332G, and A199S/S287G/A328W/Y332G BChEs for (–)-cocaine hydrolysis: the more stable the TS1 structure, the lower the energy barrier, and the higher the catalytic efficiency. Thus, the MD simulations predict that A199S/S287G/A328W/Y332G BChE should have a higher catalytic efficiency than A328W/Y332A or A328W/Y332G BChE for (–)-cocaine hydrolysis.

The computational predictions based on the transition state simulations were followed by wet experiments [117]. The wet experiments have revealed that A199S/S287G/A328W/Y332G BChE has a $\sim$ 456-fold improved catalytic efficiency against (–)-cocaine compared to the wild-type, or A199S/S287G/A328W/Y332G BChE has a k_{cat}/K_M value of $\sim (4.15 \pm 0.37) \times 10^8$ M min^{-1} for (–)-cocaine hydrolysis [117]. The catalytic efficiency of A199S/S287G/A328W/Y332G BChE against (–)-cocaine is significantly higher than that of AME-359 (i.e., F227A/S287G/A328W/Y332M BChE, $k_{cat}/K_M = 3.1 \times 10^7$ M min^{-1}), whose catalytic efficiency against (–)-cocaine is the highest within all of the previously reported BChE mutants) [114]. AME-359 has a $\sim$ 34-fold improved catalytic efficiency against (–)-cocaine compared to wild-type BChE. By using the designed A199S/S287G/A328W/Y332G BChE as an exogenous enzyme in humans, when the concentration of this mutant is kept the same as that of the wild-type BChE in plasma, the halflife of (–)-cocaine in plasma should be reduced from $\sim$ 45–90 min to only $\sim$ 6–12 s, considerably shorter than the time required for cocaine to cross the blood–brain barrier to reach the CNS. Hence, the outcome of the rational design and discovery study could eventually result in a valuable, efficient anti-cocaine medication.

4
Concluding Remarks

Reaction mechanisms of cocaine hydrolysis have been studied extensively through the combined use of a variety of state-of-the-art techniques of molecular modeling. The computational techniques used include homology modeling, molecular docking, molecular dynamics simulations, first-principles electronic structure calculations, and hybrid quantum mechanical/molecular mechanical calculations. These state-of-the-art computational studies have led to detailed mechanistic insights into the reaction pathways and energy profiles for non-enzymatic hydrolysis of cocaine in water and for cocaine hydrolysis catalyzed by human butyrylcholinesterase (BChE). These detailed mechanistic insights provide a solid basis for rational design of possible anti-cocaine medication. In particular, the information about the transition states and their stabilization for non-enzymatic hydrolysis of cocaine in water has been very useful in rational design of stable analogs of the transition states for cocaine hydrolysis to elicit new anti-cocaine catalytic antibodies. By using the computational insights into the catalytic mechanisms for BChE-catalyzed hydrolysis of (–)- and (+)-cocaine, a novel computational design strategy based on the simulation of the rate-determining transition state has been developed to design high-activity mutants of BChE for hydrolysis of (–)-cocaine. This has led to the exciting discovery of BChE mutants with considerably improved catalytic efficiency against (–)-cocaine. One of the discovered new BChE mutants (i.e., A199S/S287G/A328W/Y332G) has a $\sim$ 456-fold improved catalytic efficiency against (–)-cocaine. The encouraging outcome of the computational design and discovery effort demonstrates that the unique computational design approach based on transition-state simulation holds promise for rational enzyme redesign and drug discovery.

Concerning future directions of the mechanism-based design of anti-cocaine therapeutic, the catalytic efficiency of BChE could be improved further by continuing use of the novel computational design strategy followed by wet experimental tests. The catalytic efficiency of the discovered anti-cocaine catalytic antibodies could also be improved through computational design of possible mutants of the catalytic antibodies. For this purpose, one first needs to understand the detailed catalytic mechanisms for (–)-cocaine hydrolysis catalyzed by the catalytic antibodies and then see how to stabilize the rate-determining transition state for the hydrolysis process. The similar computational protocols that have been used for BChE and its mutants can be employed to study the catalytic mechanism of cocaine hydrolysis catalyzed by the catalytic antibodies and to design their possible high-activity mutants. In addition to high-activity mutants of BChE and catalytic antibodies, one may also consider the alternative options of developing high-activity mutants of other enzymes for accelerating (–)-cocaine metabolism. The similar computational protocols used for BChE and its mutants can also be employed to study the catalytic mechanisms of cocaine hydrolysis catalyzed by other enzymes and to design their possible high-activity mutants.

Finally, the similar computational strategy discussed in this chapter in combination with appropriate wet experiments may also be useful in rational redesign of many other metabolic enzymes for the therapeutic treatments of metabolic diseases.

Acknowledgements The financial support from the National Institute on Drug Abuse (NIDA) of the National Institutes of Health (NIH) (grant R01 DA013930) is gratefully acknowledged.

References

1. Gawin FH, Ellinwood EHN Jr (1988) Eng J Med 318:1173
2. Landry DW (1997) Sci Am 276:28
3. Singh S (2000) Chem Rev 100:925
4. Sparenborg S, Vocci F, Zukin S (1997) Drug Alcohol Depend 48:149
5. Gorelick DA (1997) Drug Alcohol Depend 48:159
6. Gorelick DA, Gardner EL, Xi ZX (2004) Drugs 64:1547
7. Baird TJ, Deng SX, Landry DW, Winger G, Woods JH (2000) J Pharmacol Exp Ther 295:1127
8. Carrera MRA, Ashley JA, Wirsching P, Koob GF, Janda KD (2001) Proc Natl Acad Sci USA 98:1988
9. Deng SX, de Prada P, Landry DW (2002) J Immunol Methods 269:299
10. Kantak KM (2003) Expert Opin Pharmacother 4:213

11. Carrera MRA, Kaufmann GF, Mee JM, Meijler MM, Koob GF, Janda KD (2004) Proc Natl Acad Sci USA 101:10416
12. Dickerson TJ, Kaufmann GF, Janda KD (2005) Expert Opin Biol Ther 5:773
13. Meijler MM, Kaufmann GF, Qi LW, Mee JM, Coyle AR, Moss JA, Wirsching P, Matsushita M, Janda KD (2005) J Am Chem Soc 127:2477
14. Rogers CJ, Mee JM, Kaufmann GF, Dickerson TJ, Janda KD (2005) J Am Chem Soc 127:10016
15. Carrera MRA, Ashley JA, Parsons LH, Wirsching P, Koob GF, Janda KD (1995) Nature 378:727
16. Fox BS (1997) Drug Alcohol Depend 100:153
17. Carrera MRA, Ashley JA, Zhou B, Wirsching P, Koob GF, Janda KD (2000) Proc Natl Acad Sci USA 97:6202
18. Carrera MRA, Ashley JA, Wirsching P, Koob GF, Janda KD (2001) Proc Natl Acad Sci USA 98:1988
19. Carrera MRA, Trigo JM, Roberts AJ, Janda KD (2005) Pharmacol Biochem Behav 81:709
20. Fox BS, Kantak KM, Edwards MA, Black KM, Bollinger BK, Botka AJ, French TL, Thompson TL, Schad VC, Greenstein JL, Gefter ML, Exley MA, Swain PA, Briner TJ (1996) Nat Med 2:1129
21. Landry DW, Yang GXQ (1997) J Addict Dis 16:1
22. Landry DW, Zhao K, Yang GXQ, Glickman M, Georgiadis TM (1993) Science 259:1899
23. Matsushita M, Hoffman TZ, Ashley JA, Zhou B, Wirsching P, Janda KD (2001) Bioorg Med Chem Lett 11:87
24. Cashman JR, Berkman CE, Underiner GE (2000) J Pharm Exp Ther 293,952:961
25. Yang G, Chun J, Arakawa-Uramoto H, Wang X, Gawinowicz MA, Zhao K, Landry DW (1996) J Am Chem Soc 118:5881
26. Kamendulis LM, Brzezinski MR, Pindel EV, Bosron WF, Dean RA (1996) J Pharmacol Exp Ther 279:713
27. Poet TS, McQueen CA, Halpert JR (1996) Drug Metab Dispos 24:74
28. Pan WJ, Hedaya MA (1999) J Pharm Sci 88:468
29. Sukbuntherng J, Martin DK, Pak Y, Mayersohn M (1996) J Pharm Sci 85:567
30. Browne SP, Slaughter EA, Couch RA, Rudnic EM, McLean AM (1998) Biopharm Drug Dispos 19:309
31. Carmona GN, Baum I, Schindler CW, Goldberg SR, Jufer R (1996) Life Sci 59:939
32. Lynch TJ, Mattes CE, Singh A, Bradley RM, Brady RO, Dretchen KL (1997) Toxicol Appl Pharmacol 145:363
33. Mattes CE, Lynch TJ, Singh A, Bradley RM, Kellaris PA, Brady RO, Dretchen KL (1997) Toxicol Appl Pharmacol 145:372
34. Mattes CE, Belendiuk GW, Lynch TJ, Brady RO, Dretchen KL (1998) Addict Biol 3:171
35. Lockridge O, Blong RM, Masson P, Froment M-T, Millard CB, Broomfield CA (1997) Biochemistry 36:786
36. Gately SJ (1991) Biochem Pharmacol 41:1249
37. Gately SJ, MacGregor RR, Fowler JS, Wolf AP, Dewey SL, Schlyer DJ (1990) J Neurochem 54:720
38. Darvesh S, Hopkins DA, Geula C (2003) Nature Rev Neurosci 4:131
39. Giacobini E (ed) (2003) Butyrylcholinesterase: its function and inhibitors. Dunitz Martin, Great Britain
40. Lerner RA, Benkovic SJ, Schultz PG (1991) Science 252:659
41. Jones RAY (1979) Physical and Mechanistic organic chemistry. Cambridge University Press, Cambridge

42. McMurry J (1988) Organic chemistry, 2nd edn. Cole, California
43. Lowry TH, Richardson KS (1987) Mechanism and Theory in organic chemistry, 3rd edn. Harper and Row, New York
44. Williams A (1987) In: Page MI, Williams A (eds) Enzyme mechanisms. Burlington, London, p 123
45. Li P, Zhao K, Deng S, Landry DW (1999) Helvetica Chim Acta 82:85
46. Zhan CG, Landry DW, Ornstein RL (2000) J Am Chem Soc 122:1522
47. Zhan CG, Landry DW, Ornstein RL (2000) J Am Chem Soc 122:2621
48. Zhan CG, Landry DW, Ornstein RL (2000) J Phys Chem A 104:7672
49. Bender ML, Thomas RJ (1961) J Am Chem Soc 83:4189
50. Bender ML, Matsui H, Thomas RJ, Tobey SW (1961) J Am Chem Soc 83:4193
51. Bender ML, Heck HA (1967) J Am Chem Soc 89:1211
52. Bender ML, Ginger RD, Unik JP (1958) J Am Chem Soc 80:1044
53. O'Leary MH, Marlier JF (1979) J Am Chem Soc 101:3300
54. Guthrie JP (1991) J Am Chem Soc 113:3941
55. Hengge A (1992) J Am Chem Soc 114:6575
56. Marlier JF (1993) J Am Chem Soc 115:5953
57. Sherer EC, Turner GM, Lively TN, Landry DW, Shields GC (1996) J Mol Model 2:62
58. Sherer EC, Yang G, Turner GM, Shields GC, Landry DW (1997) J Phys Chem A 101:8526
59. Sherer EC, Turner GM, Shields GC (1995) Int J Quantum Chem Quantum Biol Symp 22:83
60. Turner GM, Sherer EC, Shields GC (1995) Int J Quantum Chem Quantum Biol Symp 22:103
61. Fairclough RA, Hinshelwood CN (1937) J Chem Soc 538
62. Rylander PN, Tarbell DS (1950) J Am Chem Soc 72:3021
63. Zhan CG, Landry DW (2001) J Phys Chem A 105:1296
64. Becke AD (1993) J Chem Phys 98:5648
65. Lee C, Yang W, Parr RG (1988) Phys Rev B 37:785
66. Gonzalez C, Schlegel HB (1989) J Chem Phys 90:2154
67. Gonzalez C, Schlegel HB (1990) J Phys Chem 94:5523
68. Tomasi J, Persico M (1994) Chem Rev 94:2027
69. Cramer CJ, Truhlar DG (1996) In: Tapia O, Bertran J (eds) Solvent effects and chemical reactions. Kluwer, Dordrecht, p 1
70. Cramer CJ, Truhlar DG (1999) Chem Rev 99:2161
71. Chipman DM (1997) J Chem Phys 106:10194
72. Chipman DM (1999) J Chem Phys 110:8012
73. Schmidt MW, Baldridge KK, Boatz JA, Elbert ST, Gordon MS, Jensen JH, Koseki S, Matsunaga N, Nguyen KA, Su SJ, Windus TL, Dupuis M, Montgomery JA (1993) J Comput Chem 14:1347
74. Zhan CG, Bentley J, Chipman DM (1998) J Chem Phys 108:177
75. Zhan CG, Chipman DM (1998) J Chem Phys 109:10543
76. Zhan CG, Chipman DM (1999) J Chem Phys 110:1611
77. Zheng F, Zhan CG, Ornstein RL (2001) J Chem Soc Perkin Trans 2 2355
78. Zheng F, Zhan CG, Ornstein RL (2002) J Phys Chem B 106:717
79. Zhan CG, Dixon DA, Sabri MI, Kim MS, Spencer PS (2002) J Am Chem Soc 124:2744
80. Dixon DA, Feller D, Zhan CG, Francisco SF (2002) J Phys Chem A 106:3191
81. Zhan CG, Norberto de Souza O; Rittenhouse R, Ornstein RL (1999) J Am Chem Soc 121:7279
82. Zhan CG, Zheng F (2001) J Am Chem Soc 123:2835

83. Dixon DA, Feller D, Zhan CG, Francisco SF (2003) Int J Mass Spectrom 227:421
84. Zhan CG, Dixon DA, Spencer PS (2003) J Phys Chem B 107:2853
85. Zhan CG, Dixon DA, Spencer PS (2004) J Phys Chem B 108:6098
86. Chen X, Zhan CG (2004) J Phys Chem A 108:3789
87. Chen X, Zhan CG (2004) J Phys Chem A 108:6407
88. Xiong Y, Zhan CG (2004) J Org Chem 69:8451
89. Zhan CG, Deng SX, Skiba JG, Hayes BA, Tschampel SM, Shields GC, Landry DW (2005) J Comput Chem 26:980
90. Frisch MJ, Trucks GW, Schlegel HB, Scuseria GE, Robb MA, Cheeseman JR, Zakrzewski VG, Montgomery JA, Stratmann RE, Burant JC, Dapprich S, Millam JM, Daniels AD, Kudin KN, Strain MC, Farkas O, Tomasi J, Barone V, Cossi M, Cammi R, Mennucci B, Pomelli C, Adamo C, Clifford S, Ochterski J, Petersson GA, Ayala PY, Cui Q, Morokuma K, Malick DK, Rabuck AD, Raghavachari K, Foresman JB, Cioslowski J, Ortiz JV, Stefanov BB, Liu G, Liashenko A, Piskorz P, Komaromi I, Gomperts R, Martin RL, Fox DJ, Keith T, Al-Laham MA, Peng CY, Nanayakkara A, Gonzalez C, Challacombe M, Gill PMW, Johnson B, Chen W, Wong MW, Andres JL, Gonzalez AC, Head-Gordon M, Replogle ES, Pople JA (1998) Gaussian 98. Gaussian, Pittsburgh, PA
91. Gu Y, Kar T, Scheiner S (1999) J Am Chem Soc 121:9411
92. Meadows ES, De Wall SL, Barbour LJ, Fronczek FR, Kim MS, Gokel GW (2000) J Am Chem Soc 122:3325
93. Vargas R, Garza J, Dixon DA, Hay BP (2000) J Am Chem Soc 122:4750
94. Deng S, Bharat N, de Prada P, WLandry DW (2004) Org Biomol Chem 2:288
95. Mets B, Winger G, Cabrera C, Seo S, Jamdar S, Yang G, Zhao K, Briscoe RJ, Almonte R, Woods JH, Landry DW (1998) Proc Natl Acad Sci USA 95:10176
96. Nicolet Y, Lockridge O, Masson P, Fontecilla-Camps JC, Nachon F (2003) J Biol Chem 278:41141
97. Sun H, Yazal JE, Lockridge O, Schopfer LM, Brimijoin S, Pang YP (2001) J Biol Chem 276:9330
98. Sun H, Shen ML, Pang YP, Lockridge O, Brimijoin S (2002) J Pharmacol Exp Ther 302:710
99. Sun H, Pang YP, Lockridge O, Brimijoin S (2002) Mol Pharmacol 62:220
100. Zhan CG, Zheng F, Landry DW (2003) J Am Chem Soc 125:2462
101. Hamza A, Cho H, Tai HH, Zhan CG (2005) J Phys Chem B 109:4776
102. Bruice TC, Lightstone FC (1999) Acc Chem Res 32:127
103. Shurki A, Štrajbl M, Villá J, Warshel A (2002) J Am Chem Soc 124:4097
104. Masson P, Legrand P, Bartels CF, Froment M-T, Schopfer LM, Lockridge O (1997) Biochemistry 36:2266
105. Ekholm M, Konschin H (1999) J Mol Struct (THEOCHEM) 467:161
106. Harel M, Sussman JL, Krejci E, Bon S, Chanal P, Massoulie J, Silman I (1992) Proc Natl Acad Sci USA 89:10827
107. Masson P, Xie W, Froment MT, Levitsky V, Fortier PL, Albaret C, Lockridge O (1999) Biochim Biophys Acta 1433:281
108. Hu C-H, Brinck T, Hult K (1998) Int J Quantum Chem 69:89
109. Wlodek ST, Clark TW, Scott L, McCammon JA (1997) J Am Chem Soc 119:9513
110. Wlodek ST, Antosiewicz J, Briggs JM (1997) J Am Chem Soc 119:8159
111. Zhou HX, Wlodek ST, McCammon JA (1998) Proc Natl Acad Sci USA 95:9280
112. Malany S, Sawai M, Sikorski RS, Seravalli J, Quinn DM, Radic Z, Taylor P, Kronman C, Velan B, Shafferman A (2000) J Am Chem Soc 122:2981
113. Zhan CG, Gao D (2005) Biophys J 89:3863

114. Gao Y, Atanasova E, Sui N, Pancook JD, Watkins JD, Brimijoin S (2005) Mol Pharmacol 67:204
115. Gao D, Zhan CG (2006) Proteins 62:99
116. Gao D, Zhan CG (2005) J Phys Chem B 109:23070
117. Pan Y, Gao D, Yang W, Cho H, Yang GF, Tai HH, Zhan CG (2005) Proc Natl Acad Sci USA 102:16656
118. Gao D, Cho H, Yang W, Pan Y, Yang GF, Tai HH, Zhan CG (2006) Angew Chem Int Ed 45:653
119. Morris GM, Goodsell DS, Halliday RS, Huey R, Hart WE, Belew RK, Olson AJ (1998) J Comput Chem 19:1639
120. Zhang Y, Kua J, McCammon JA (2002) J Am Chem Soc 124:10572
121. Tai K, Shen T, Börjesson U, Philippopoulos M, McCammon JA (2001) Biophys J 81:715
122. Kua J, Zhang YK, Eslami AC, Butler JR, McCammon JA (2003) Protein Science 12:2675
123. Bernstein FC, Koetzle TF, Williams GJ, Meyer EF, Brice MD, Rodgers JR, Kennard O, Shimanouchi T, Tasumi M (1977) J Mol Biol 112:535
124. Rosenberry TL (1975) Proc Nat Acad Sci USA 72:3834
125. Alvarez-Idaboy JR, Galano A, Bravo-Pérez G, Ruíz ME (2001) J Am Chem Soc 123:8387

Top Heterocycl Chem (2006) 4: 161–249
DOI 10.1007/7081_021
© Springer-Verlag Berlin Heidelberg 2006
Published online: 6 May 2006

QSAR Studies on Thiazolidines:
A Biologically Privileged Scaffold

Yenamandra S. Prabhakar (✉) · V. Raja Solomon · Manish K. Gupta ·
S. B. Katti (✉)

Medicinal and Process Chemistry Division, Central Drug Research Institute,
Lucknow-226 001, India
yenpra@yahoo.com, setu_katti@yahoo.com

Abstract A large number of drugs and biologically relevant molecules contain heterocyclic systems. Often the presence of hetero atoms or groupings imparts preferential specificities in their biological responses. Amongst the heterocyclic systems, thiazolidine is a biologically important scaffold known to be associated with several biological activities. Some of the prominent biological responses attributed to this skeleton are antiviral, antibacterial, antifungal, antihistaminic, hypoglycemic, anti-inflammatory activities. This diversity in the biological response profiles of thiazolidine has attracted the attention of many researchers to explore this skeleton to its multiple potential against several activities. Many of these synthetic and biological explorations have been subsequently analyzed in detailed quantitative structure-activity relationship (QSAR) studies to correlate the respective structural features and physicochemical properties with the activities to identify the important structural components in deciding their activity behavior. In this, drugs or any biologically active molecules may be viewed as structural frames consisting of strategically positioned functional groups that will interact effectively with the complementary groups/sites of the receptor. With this in focus, the present article reviews the QSAR studies of diverse biological activities of the thiazolidines published during the past decade.

Keywords Thiazolidines · QSAR · Molecular modeling

Abbreviations

AF	*Aspergillus fumigatus*
AIDS	acquired immune deficiency syndrome
ANN	artificial neural network
AR	aldose reductase
ATPase	adenosin tri-phosphatase
BPN	back propagation neural networks
CA	*Candida albicans*
cAMP	cyclic-adenosine mono phosphate
CN	*Cryptococcus neoformans*
CODESSA	comprehensive descriptors for structural and statistical analysis
CoMFA	comparative molecular field analysis
CoMSIA	comparative molecular similarity analysis
COX	cyclooxygenase
CPE	carrageenan mice paw edema
CP-MLR	combinatorial protocol in multiple linear regression
E/L	enzyme ligand interaction
FA	factor analysis
G/PLS	genetic partial least squares
GAGs	glycosaminoglycans
GFA	genetic function approximation
HBTU	2-(1H-benzotrizo-1-yl)-1,1,3,3-tetramethyluraniumhexafluorophospate
HIV	human immunodeficiency virus
MEDV	molecular electronegativity distance vector
MFA	molecular field analysis
MLR	multiple linear regression
MMPs	metalloproteinases
MOE	molecular operating environment
MOPAC	molecular orbital partial atomic charge
MSA	molecular shape analysis
NNRTIs	non-nucleoside reverse transcriptase inhibitors
NO	nitric oxide
NPY5R	neuropeptide Y5 receptor
NRTIs	nucleoside reverse transcriptase inhibitors

NSAIDs   non-steroidal anti-inflammatory drugs
OA   octopamine agonists
PCR   principle components regression
PLS   partial least square
PPARs   peroxisome proliferator-activated receptors
QSAR   quantitative structure-activity relationship
RSM   receptor surface model
RT   reverse transcriptase
TM   *Tricophyton mentagrophyte*
TxA2   Thromboxane A2
TZDs   thiazolidinediones
VSMP   variable selection and modeling method based on prediction

1
Introduction

The quantitative structure activity relationship (QSAR) study envisages establishment of a mathematical/functional relationship between the chosen activity and selected properties of the congeneric series of compounds. In the early years of QSAR, most of the explorations revolved around the numerical expressions and presentation of the correlations and/or patterns from regression, discriminant and clustering approaches. With advancements in computer hardware performance and software, the graphical expression and visual presentation of modeling results have taken long-strides along with the mathematical/functional relationships. These strategies have been effectively put to use particularly in the area of rational drug design. This is evident from a large number of original research publications and review articles published in the literature [1–3]. Against this backdrop the present article highlights the QSAR and molecular modeling explorations carried out on thiazolidine heterocyclic systems. In chemical parlance the term thiazolidine refers to a five-membered ring system containing sulfur and nitrogen; the numbering system is as shown in Fig. 1.

In small-molecule heterocyclic compounds, thiazolidine is a recognized scaffold for potential drugs and drug candidates. Anticonvulsant, sedative, antidepressant, anti-inflammatory, antihypertensive, antihistaminic and antiarthritic activities are a few among many other biological responses shown by this scaffold (Table 1) [4].

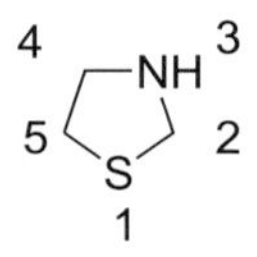

Fig. 1 Structure and numbering of Thiazolidine

Table 1 Diverse biological responses of thiazolidine derivatives [4]

No.	Names[a]	Chemical structure	Use
1	Ralitoline[b]		Anticonvulsant
2	Timiridine esilate[b]		Antidepressant
3	Spiclomazine[b]		Psychotropic
4	Piprozolin		Choleretic
5	Rentiapril[c]		Antihypertensive (ACE inhibitor)
6	Dexetozoline[b]		Antihypertensive
7	Etozolin[b]		Antihypertensive, Diuretic
8	Ozolinone[c]		Diuretic

[a] Names not protected under a registered trademark
[b] WHO recommended non-proprietary names
[c] WHO unprotected non-proprietary names

Table 1 (continued)

No.	Names[a]	Chemical structure	Use
9	Tizolemide[b]		Diuretic
10	Timofibrate[b]		Antihyperlipidemic
11	Metibride[b]		Antihyperlipidemic
12	Carbolidine		Mucolytic
13	Telmesteine[b]		Mucolytic
14	Letosteine[c]		Mucolytic
15	Alonacic[b]		Mucolytic
16	Neosteine[c]		Mucolytic

[a] Names not protected under a registered trademark
[b] WHO recommended non-proprietary names
[c] WHO unprotected non-proprietary names

Table 1 (continued)

No.	Names[a]	Chemical structure	Use
17	Guaisteine[b]		Antitussive
18	Thiadrine		Antitussive, Anti-asthmatic
19	Mezolidon		Antiulcer agent
20	Thiazolidomycin		Antibiotic for several streptomyces species
21	Ferrithiocin		Promotes antibacterial activity of cephalosporins
22	Fezatione		Antifungal, Antitrichophytic
23	Antazonite[c]		Parasticide, Anthelmintic
24	Cyamiazole		Acaricide, Ectoparasiticde

[a] Names not protected under a registered trademark
[b] WHO recommended non-proprietary names
[c] WHO unprotected non-proprietary names

Table 1 (continued)

No.	Names[a]	Chemical structure	Use
25	Nitrodan[c]		Anthelmintic (veterinary)
26	Libecilide[b]		Anti-allergic (inhibits penicillin induced antibodies)
27	Pioglitazone[b]		Oral hypoglycemic agents
28	Troglitazone[b]		Oral hypoglycemic agents
29	Draglitazone[b]		Oral hypoglycemic agents
30	Englitazone[b]		Oral hypoglycemic agents
31	Ciglitazone[b]		Oral hypoglycemic agents
32	Timonacic[b]		Hepatoprotectant Antineoplastic

[a] Names not protected under a registered trademark
[b] WHO recommended non-proprietary names
[c] WHO unprotected non-proprietary names

Table 1 (continued)

No.	Names[a]	Chemical structure	Use
33	Tidiacic[b]		Treatment of liver disorders
34	Epatrestat[b]		Aldose reductase inhibitor
35	Pidotimod[b]		Immunomodulator
36	Benpenolisin[b]		Diagnostic aid (penicillin sensitive)

[a] Names not protected under a registered trademark
[b] WHO recommended non-proprietary names
[c] WHO unprotected non-proprietary names

Furthermore, the avenues of simple chemical transformations in the generation of a diverse variety of pendants on the thiazolidine scaffold has fascinated medicinal chemists and led to exploration of the wealth of biological information involved in these processes. Therefore, a conscious effort is made in the chemistry part of the present review to cull out select modifications of the thiazolidine skeleton namely thiazolidine-4-carboxylic acids and 4-thiazolidinones from the vast array of thiazolidines.

2
Chemistry

2.1
Thiazolidine

The parent compound thiazolidine can be obtained in high yield as hydrochloride by the reaction of cysteamine hydrochloride with aqueous formaldehyde at ambient temperature (Scheme 1) [5].

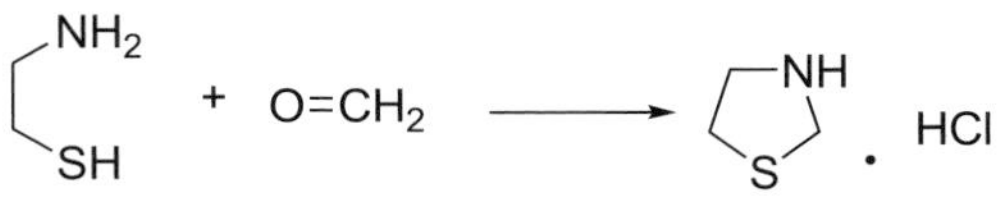

Scheme 1 General synthesis of thiazolidine

By logical extension of the scheme one can generate substituted thiazolidine and its homologs by the reaction of appropriate carbonyl compounds with suitable α-aminothiols (Scheme 2) [6].

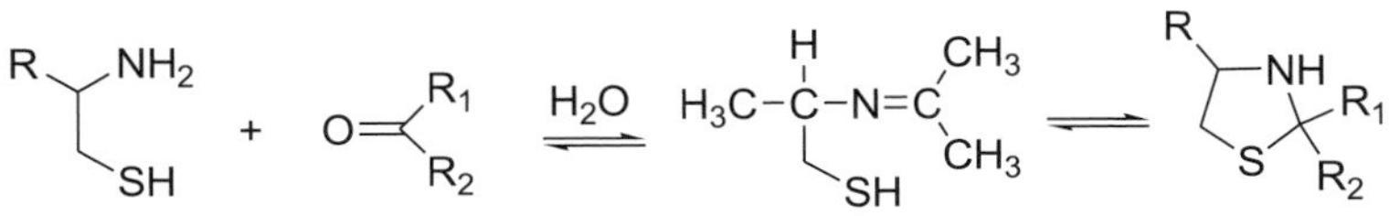

Scheme 2

Furthermore, many thiazolidines are in equilibrium with its acyclic thiol form alternately in aqueous media, with the β-mercaptoalkylamine and the carbonyl compound from which they were formed. The position of the equilibrium depends on the nature of the ring substituents (Scheme 2) [7]. Thiazolidines with unsubstituted ring nitrogen (N3) are readily hydrolyzed by boiling in aqueous acidic or basic solution and the hydrolysis is completed in the presence of a compound which reacts with either of the cleavage products. The presence of substituents on the nitrogen atom (N3) stabilizes the thiazolidine ring. Under strong acidic or basic conditions or at high temperature, 1,3-thiazolidines undergo epimerization followed by decomposition to the aldehyde and aminothiol [8]. Simple oxidative transformations of the thiazolidine skeleton lead to a number of scaffolds namely, N-oxidies, 2- or 3-thiazolines, and thiazoles. Oxidation can also occur at the sulfur atom (S1), to afford sulfoxide or sulfone derivatives [7]. Thus, thiazolidine is amenable to chemo selective reactions leading to different structural entities (Fig. 2).

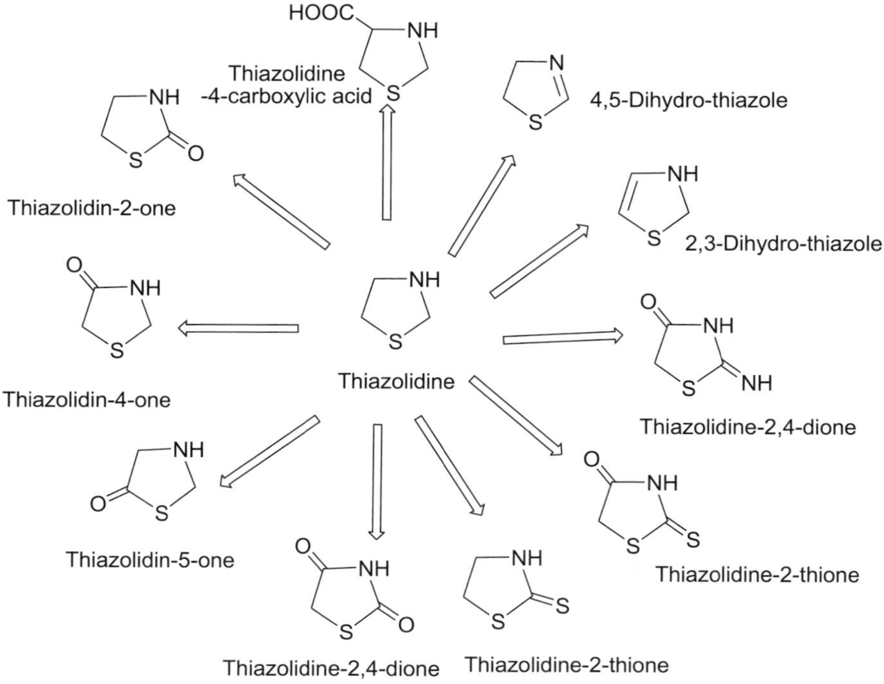

Fig. 2 Some representative chemical transformations of thiazolidine

2.2
Thiazolidine-4-carboxylic Acids

Discovery of thiazolidine-4-carboxylic acids was made accidentally by Birch and Harris [9] during a study of the effect of formaldehyde on the titration curves of amino acids. Later, Schubert [10] explained the formation of thiazolidine-4-carboxylic acid by condensation of cysteine and formaldehyde followed by intramolecular cyclization. Accordingly, a large number of thiazolidine-4-carboxylic acids can be synthesized by condensation of aldehydes and ketones with cysteine and/or pencillamine as shown in Scheme 3.

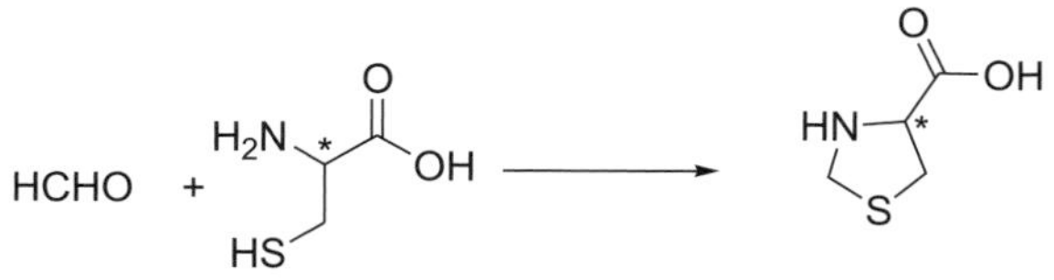

Scheme 3 General synthesis of thiazolidine-4-carboxylic acid

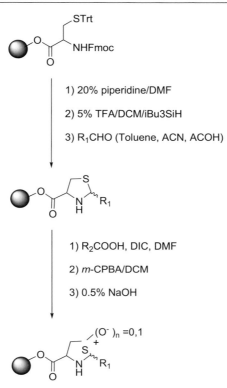

Scheme 4 A solid-phase synthetic protocol of thiazolidine derivatives [13]

Although cyclization can take place in many ways, according to Kallen [11] the most preferred pathway involves imine formation followed by intramolecular cyclization. During the course of cyclization a new chiral center at C-2 is created thereby giving rise to a diastereomeric mixture namely 2R, 4R and 2S, 4R. An interesting situation arises when the reactant aldehyde is also chiral. The stereochemistry at the newly formed center is controlled by the stereochemistry of the aldehyde [12]. In view of the biological importance of thiazolidine, Patek et al. reported a solid-phase synthesis protocol (Scheme 4) [13]. This enables the synthesis of compound libraries for quick lead optimization.

2.3
Properties of Thiazolidine-4-carboxylic Acids

The unsubstituted thiazolidine-4-carboxylic acid, i.e. thioproline, is remarkably stable towards both alkali as well as acid while the 2-substituted thiazolidine-4-caroxylic acids are unstable under similar conditions [14].

However, the N3-substituted thiazolidines are stable in solution because the substituent group prevents the ring opening. The parent compounds are generally solids and exhibit poor solubility in different solvents whereas N3-substitution increases solubility in most of the organic solvents. The ring heteroatoms (S1 and N3), C-2 and the C4-pendant carboxylic group are amenable to further chemical transformation in generating libraries of compounds.

2.4
Thiazolidin-4-ones

Several protocols for the synthesis of thiazolidin-4-ones are available in the literature [15–23]. Essentially they are three component reactions involving an amine, a carbonyl compound, and a mercapto acid. The process can be either a one-pot three-component condensation or a two-step process as shown in Scheme 5. The reaction has been suggested to proceed via imine formation followed by the attack of the sulfur nucleophile on the imine carbon. The last step involves intramolecular cyclization with the elimination of water to give thiazolidin-4-ones. This step appears to be critical for obtaining high yields. Therefore, variations have been affected in this step to facilitate removal of water. Most commonly used protocols employ azeotropic distillation, molecular sieves, and use of other desiccants like anhydrous zinc chloride [24], sodium sulfate [25], or magnesium sulfate [26]. These protocols require prolonged heating at 70–80 °C for 17–20 h and give moderate to good yields.

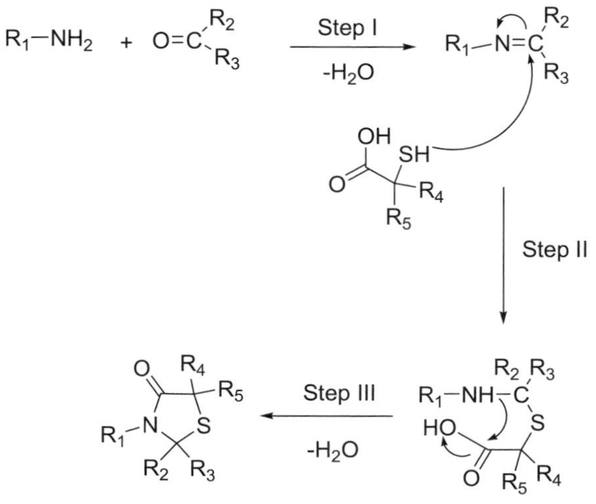

Scheme 5 A synthesis of thiazolidine-4-one derivatives

More recently, an improved protocol has been reported wherein N,N''-dicyclohexylcarbodiimide (DCC) or 2-(1H-benzotrizo-1-yl)-1,1,3,3-tetramethyluraniumhexafluorophospate (HBTU) (Scheme 6) is used as a dehydrating agent to accelerate the intramolecular cyclization resulting in faster reaction and improved yields [27, 28]. The DCC/HBTU-mediated protocol has the advantage of mild reaction conditions, a very short reaction time, and product formation in almost quantitative yields. More importantly, yields of the thiazolidinones are independent of the nature of the reactants. This modification is compatible with a solid-phase combinatorial approach to generate a library of compounds.

Recently, Holmes et al. reported the solution and polymer-supported synthesis of 4-thiazolidinones and 4-metathiaznones derived from amino acids (Scheme 7) [29]. A three-component condensation of an amino ester or resin-bound amino acid (glycine, alanine, β-alanine, phenylalanine, and valine),

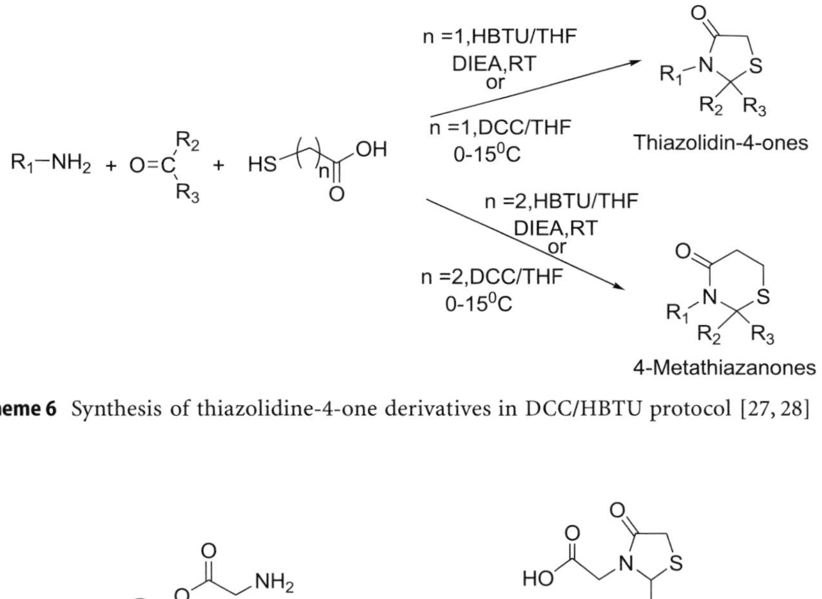

Scheme 6 Synthesis of thiazolidine-4-one derivatives in DCC/HBTU protocol [27, 28]

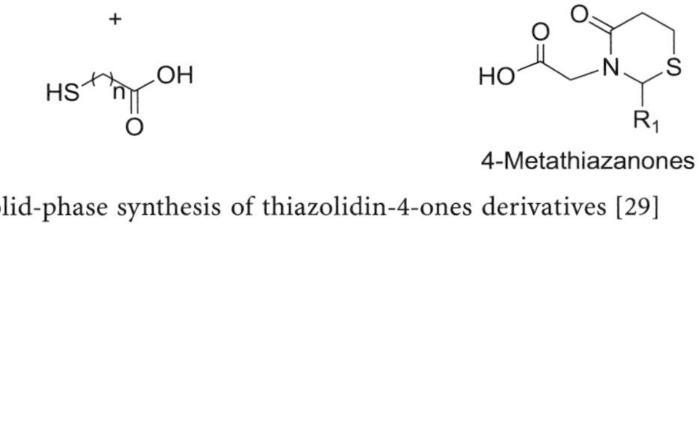

Scheme 7 A solid-phase synthesis of thiazolidin-4-ones derivatives [29]

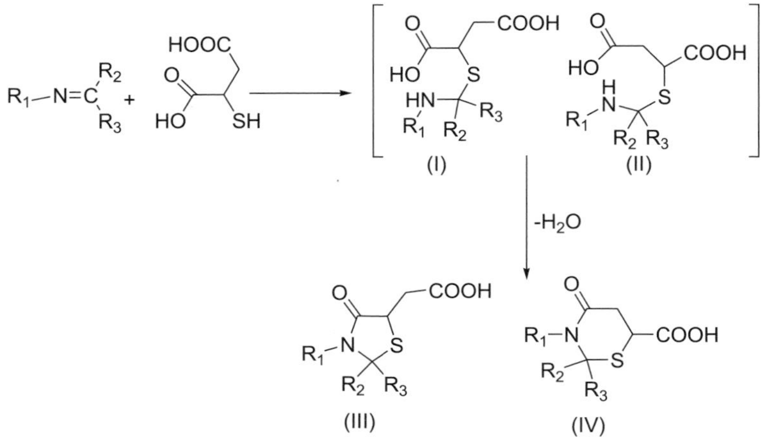

Scheme 8

an aldehyde (benzaldehyde, *o*-tolualdehyde, *m*-tolualdehyde, *p*-tolualdehyde, and 3-pyridine carboxaldehyde), and an *α*-mercapto carboxylic acid led to the formation of five- and six-membered heterocycles. In amino acids, the carboxylic acid function served as an anchor group for attachment to the site of support. The condensation of this support bound amine with several alde-hydes and mercapto acetic acids in a one-pot reaction for 2 h at 70 °C with removal of water has afforded the desired products. For successful completion of the reaction, they employed 15 to 25 equiv. of reagent solution relative to resin loading.

A special situation arises when mercaptosuccinic acid is used as a mer-capto component. Both the carboxyl functions are prone to nucleophilic at-tack by the imine nitrogen and give rise to five- (I) as well as six-membered (II) cyclic structures. Subsequently, Poop et al. showed by X-ray crystal struc-ture analysis that a five-membered (III) cyclic structure is preferred over the six-membered (IV) cyclic structures (Scheme 8) [30].

2.5
Properties of Thiazolidine-4-ones

The N3-unsubstituted thiazolidinones are usually solids and often melt with decomposition, but substitution at this position lowers the melt-ing point. The thiazolidinones that do not carry N3-aryl or higher alkyl substitution are partially soluble in water and can be recrystallized from water. Thiazolidinones undergo a limited number of chemical transform-ations as shown in Scheme 9. The methylene carbon atom at the 5-position of thiazolidinone possesses nucleophilic activity and can attack an elec-trophilic center. Most frequently the reaction occurs in the presence of

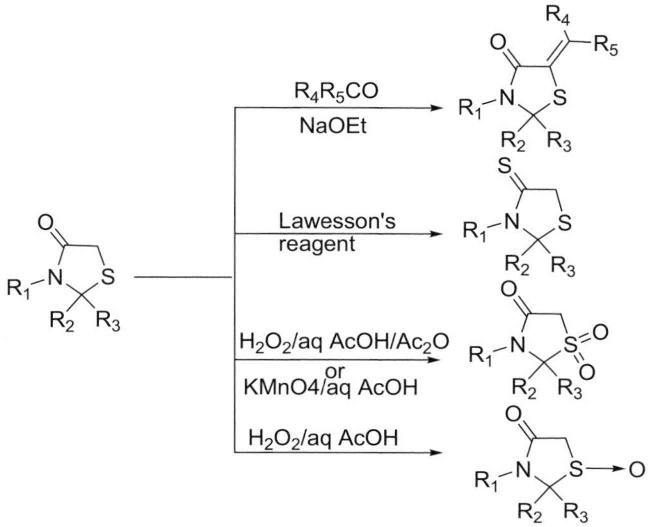

Scheme 9 Different chemical transformations of thiazolidin-4-ones

a base and the anion of the 4-thiazolidinone attacking species as shown in Scheme 9 [31, 32].

The carbonyl group of 4-thiazolidinone is highly unreactive. But in a few cases it undergoes certain reactions for example 4-thiazolidinone on treatment with Lawesson's reagent gives corresponding 4-thione derivatives in almost quantitative yields [33]. Oxidation of 2,3-disubstituted-4-thiazolidinone with hydrogen peroxide in acetic anhydride and acetic acid [34] or by potassium permanganate in aqueous acetic acid solution [35] results in the corresponding sulfones. On the other hand, the oxidation with hydrogen peroxide in aqueous acetic acid solution gives the corresponding sulfoxide [36]. A more detailed account of the chemistry and biological activities of thiazolidine-4-carboxylic acids can be found in earlier reviews [37–40].

The diverse biological activities exhibited by the thiazolidines can be broadly classified into those affecting the host biology including the metabolic disorders and those affecting the parasite function in the host. The former activity profile relates mostly to the inflammatory, ulceration, allergy, and blood coagulation processes and to the metabolic disorders such as obesity, diabetes, and associated pathological manifestations. The latter category of activities is mostly concerned with the human immunodeficiency virus (HIV) and fungi. Apart from these some QSAR reports on the insecticidal activity of thiazolidines have also been included to widen the scope of this article.

3
QSAR and Modeling Studies

QSAR and modeling studies comprise of two basic components. The first one corresponds to the measurements (parameterization) of the chosen object in numeric or boolean form and the other one addresses the development of relational expressions between or among the measurements made on the objects. The reviewed studies have used different physicochemical, quantum chemical, topological, and topographical descriptors from various sources such as Hansch and Leo's monograph [41], Molecular Orbital Partial Atomic Charge (MOPAC) [42], ChemDraw's property/descriptor database [43], Comprehensive Descriptors for Structural and Statistical Analysis (CODESSA) [44, 45], Molecular Electronegativity Distance Vector (MEDV-13) [46], and DRAGON [47] etc for the parameterization of the chemical structure. The 2D-QSAR results presented in the review have been brought out using different statistical procedures and programs namely, simple Multiple Linear Regression (MLR), Principle Components in multiple linear Regression (PCR), Partial Least Square (PLS) [48], Genetic Function Approximation (GFA) [49], Genetic Partial Least Squares (G/PLS) [50], Combinatorial Protocol in Multiple Linear Regression (CP-MLR) [51], Variable Selection and Modeling method based on Prediction (VSMP) [52], Artificial Neural Network (ANN) [53–55], discriminant and cluster analysis.

The 3D-QSAR results explained in this article have come from the application of Comparative Molecular Field Analysis (CoMFA) [56], Comparative Molecular Similarity Analysis (CoMSIA) [57], Molecular Shape Analysis (MSA) [58, 59], Molecular Field Analysis (MFA) [60], and steric and electrostatic field interactions in Receptor Surface Model (RSM) [61, 62] techniques to the datasets. Apart from these, results of some interesting pharmacophore identification [63, 64], virtual screening [65, 66], docking [67, 68] and X-ray crystallographic studies have been discussed to give the complete picture of "virtual to wet chemistry". All these 3D studies have been carried out using various commercial software packages namely Insight II [69], SYBYL [70], MOE [71]. All these packages use "built-in" statistical techniques for the assessment of significance of the developed models. They are similar to one or more statistical procedures of 2D-QSAR. Most of the 2D and 3D results presented in this review have been validated through various techniques such as leave one out, leave many out, and external test sets. However, to maintain uniformity and conciseness, only limited statistical measures of the goodness-of-fit of regression equations have been presented. In all regression equations, n is the number of compounds in the dataset, r is the correlation coefficient, Q^2 is cross-validated r^2 from the leave-one/many out procedure, s is the standard error of the estimate, and F is the F-ratio between the variances of calculated and observed activities. The values given in the parentheses of the regression equation are the standard errors (without arithmetic sign) or 95% confidence limits (with ± arithmetic sign) of the regression coefficients. A mention is made about the nature of each QSAR study/methodology along with the results. The contributing or modeling descriptors for each case have been identified and discussed along with the modeling equations. In the succeeding sections of this article QSAR studies on the diverse biological responses elicited by the various compounds anchored via the thiazolidine skeleton system have been described. Hereafter, "thiazolidine" has been used to broadly address all nitrogen-sulfur containing five-membered heterocyclic systems with different oxidation states. However, care has been taken to specify the name of the chemical congener(s) under each investigation.

3.1
Anti-inflammatory Agents

3.1.1
Cyclooxygenase Inhibitors

Thiazolyl and benzothiazolyl derivatives are known for their anti-inflammatory, analgesic, and antipyretic activities [72–77], for example Meloxicam (Fig. 3) is a known non-steroidal anti-inflammatory agent with a thiazolyl

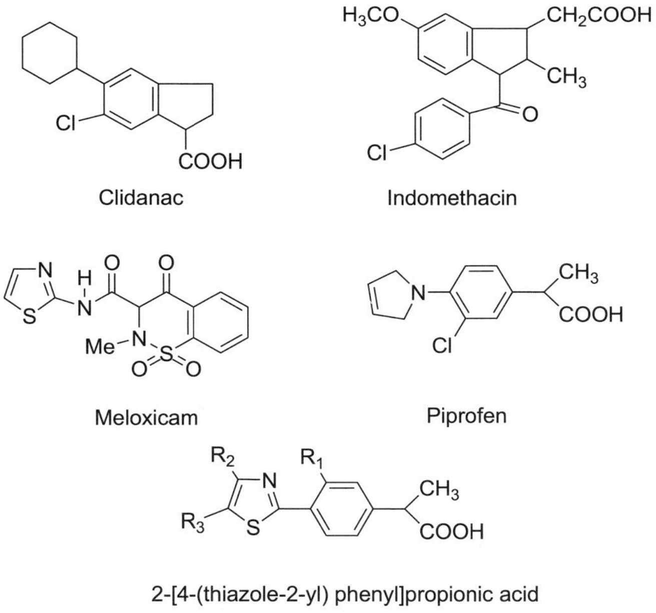

2-[4-(thiazole-2-yl) phenyl]propionic acid

Fig. 3 General structure of 2-[4-(thiazole-2-yl) phenyl]propionic acid derived cyclooxygenase (COX) inhibitors and some of its seed structures

moiety as part of its scaffold [78]. These non-steroidal anti-inflammatory agents are known to act through the inhibition of cyclooxygenase (COX) involved in the arachidonic acid pathway. In this contest, the development of 2-[4-(thiazole-2-yl) phenyl]propionic acid derivatives (Fig. 3, Table 2) as anti-inflammatory agents has its roots in the sub-structural units of indomethacin, clidanac, and piprofen (Fig. 3) [79, 80].

Naito and co-workers investigated the COX inhibitory activity of these 2-[4-(thiazole-2-yl) phenyl]propionic acid derivatives (Fig. 3, Table 2) in terms of different physicochemical and structural descriptors of the substituent positions (R_1, R_2 and R_3) [80]. The following equations show the correlation of the activity with these descriptors in different groups of compound.

$$pI_{50} = 1.08(0.50)\pi(R_1) + 1.18(0.97)\sigma_1(R_1) + 5.64$$
$$n = 11, \quad r = 0.89, \quad s = 0.28, \quad F = 15.81 \tag{1}$$

$$pI_{50} = -4.10(2.75)\sigma_R(R_2) - 0.72(0.28)\Delta L(R_2)$$
$$+ 0.43(0.39)\pi(R_2) + 5.81$$
$$n = 15, \quad r = 0.91, \quad s = 0.46, \quad F = 17.90 \tag{2}$$

$$pI_{50} = 1.03(0.42)\pi(R_1) - 4.48(1.64)\sigma_R(R_2)$$
$$- 0.86(0.13)\Delta L(R_2) + 0.44(0.27)\pi(R_{2,3})$$
$$- 0.40(0.26)\Delta L(R_3) - 1.48(0.52)I(\text{iso}) + 6.11$$
$$n = 45, \quad r = 0.95, \quad s = 0.38, \quad F = 53.97 \,. \tag{3}$$

Equation 1 correlated the activity of these compounds in terms of their R_1 substituent (Table 2; compounds 1–11). This has suggested that higher hydrophobicity and electron-withdrawing properties of the R_1 substituent would lead to better activity [80]. Equation 2 has presented the requirements of the R_2 substituent (Table 2, compounds 1, 12–25) in terms of the STERIMOL length parameter (ΔL) [81]. It has suggested that electron releasing and hydrophobic substituents with a small length would be conducive at the R_2 position for better inhibitory activity. Equation 3 is a collective model of all the compounds together (Table 2) and represents the influence of R_1, R_2 and R_3 substituents on the activity. In this, $\pi(R_{2,3})$ and $I(\text{iso})$ represent the sum of the hydrophobic effects of R_2 and R_3 substituents and an indicator variable defined to account for the *iso*-propyl group at the R_2 position of the compounds, respectively. The negative regression coefficient of $I(\text{iso})$ in this equation has led to the suggestion that the steric restriction at the receptor site for the compounds carrying α-branched alkyl groups are R_2 substituents. The larger negative coefficient of the ΔL term of R_2 (compared to that of R_3) has further confirmed this effect. In summary, this study has indicated that the activity prefers hydrophobic substituents for R_1, R_2, and R_3 positions of these compounds (Fig. 3). It has also suggested that electron-withdrawing

Table 2 COX inhibitory activity of 2-[4-(thiazole-2-yl)phenyl]propionic acid derivatives (Fig. 3) [80]

No.	R_1	R_2	R_3	pI_{50} [b] obsd	cald [c]
1	H	H	H	5.56	6.04
2	F	H	H	6.60	6.51
3	Cl	H	H	6.70	6.81
4	Br	H	H	7.00	6.78
5	CF_3	H	H	6.52	6.58
6	Me	H	H	5.70	6.16
7	OMe	H	H	5.28	5.60
8	SMe	H	H	6.52	6.32
9	OH	H	H	6.22	6.18
10	NO_2	H	H	5.54	5.61
11	NH_2	H	H	6.00	6.02
12	H	Me	H	6.14	6.17
13	H	Et	H	5.52	5.28
14	H	i-Pr	H	4.79	4.16
15	H	Bu	H	3.98	4.13
16	H	acetyl	H	3.98	3.33
17	H	Ph	H	4.44	4.14
18	H	vinyl	H	5.68	5.45
19	H	allyl	H	4.89	4.63
20	H	CH_2OH	H	4.80	4.55
21	H	CH_2OMe	H	4.52	4.29
22	H	$CONH_2$	H	3.12	3.75
23	H	CONHMe	H	2.99	3.08
24	H	$CONMe_2$	H	3.04	3.13
25	H	CF_3	H	5.09	5.01
26	H	H	Me	5.26	5.86
27	H	H	Et	5.49	5.68
28	H	Me	Me	6.00	5.92
29	H	Et	Me	4.99	4.98
30 [a]	H	CH_2CH_2	CH_2	6.15	4.82
31 [a]	H	CH_2CH_2	CH_2CH_2	6.15	4.72
32	F	Me	H	7.00	6.64
33	F	Et	H	5.92	5.75
34	F	i-Pr	H	4.47	4.63
35	F	Bu	H	4.29	4.60
36	F	H	Me	6.52	4.33
37	F	H	Et	6.05	6.15

[a] Compounds are not included in regression analysis study
[b] Cyclooxysenase inhibitory activity
[c] Calculated activity by Eq. 3

Table 2 (continued)

No.	R_1	R_2	R_3	pI_{50} [b] obsd	cald [c]
38	F	Me	Me	6.70	6.39
39	F	Et	Me	5.30	5.45
40	Cl	Me	H	7.00	6.94
41	Cl	Et	H	6.15	6.05
42	Cl	i-Pr	H	4.47	4.93
43	Cl	Bu	H	4.21	4.90
44	Cl	H	Me	7.22	6.63
45	Cl	H	Et	6.70	6.45
46	Cl	Me	Me	6.10	6.69
47	Cl	Et	Me	5.92	5.75

[a] Compounds are not included in regression analysis study
[b] Cyclooxysenase inhibitory activity
[c] Calculated activity by Eq. 3

groups are favorable for R_1 and electron-rich groups are favorable for R_2. In addition to these, steric restrictions have been found to operate in the case of R_2 and R_3 groups.

Both COX-1 and COX-2 enzymes in the arachidonic acid pathway are known to play a role in the inflammatory process. However, the selective inhibition of COX-2 leads to the reduction of the inflammation without any risk of ulceration and bleeding of the stomach or intestinal tract, a side effect commonly associated with most of the non-selective anti-inflammatory drugs. This has made COX-2 an attractive target for many researchers to investigate the structural requirements of its inhibitors. In pursuit of this goal Liu et al. carried out a QSAR study on the COX-2 inhibitory activity of indomethacin (Fig. 3), SC-58125, NS-398 (Fig. 4) together with some thiazolone and oxazolone derivatives (Fig. 4, Table 3) with the MEDV-13 (molecular electronegativity distance vector based on 13 atomic types) descriptors [82]. These descriptors represent the electrotopological state of the molecules [83, 84]. In this study good predictive models (Eq. 4) have been obtained for the COX-2 inhibitory activity of these compounds in terms of the MEDV-13 descriptors.

$$pIC_{50} = 0.089(0.013)x_1 + 0.126(0.029)x_7 + 0.175(0.028)x_{29}$$
$$+ 0.706(0.178)x_{52} + 5.944$$
$$n = 21, \quad r = 0.925, \quad Q^2 = 0.783, \quad s = 0.283, \quad F = 23.62 . \tag{4}$$

In this equation x_1, x_7, x_{29}, and x_{52} are part of MEDV-13 descriptors of the compounds. However, no interpretation has been provided, in terms of prop-

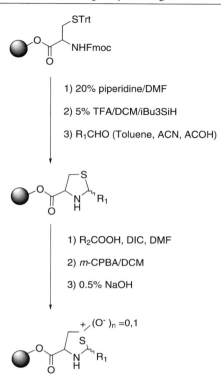

Fig. 4 Structures of SC-58125, NS-398, thiazolones and oxazolones associated with cyclooxygenase-2 (COX-2) inhibitory activity

erties of participating descriptors, to diagnose the inhibitory behavior of the compounds.

In a different study, Vigorita et al. used the docking experiments to analyze and understand the configuration preferences of the optical isomers of 3,3′-(1,2-ethanediyl)-bis[2-(3,4-dimethoxyphenyl)-4-thiazolidinones] (Fig. 5) in relation to the available structures of COX-1 and COX-2 enzymes [85].

These compounds have two equivalent chiral centers (C-2 and C-2′) (Fig. 5). They have been pharmacologically evaluated as a racemic mixture (2R, 2′R/2S, 2′S) and as meso isomers (2R, 2′S). In the docking experiments the differences between the interaction modes of *RR*, *SS*, and *RS* isomers have been analyzed using the ligand internal energies and enzyme/ligand (E/L) interaction energies of the respective isomers with the available structures of COX-1 and COX-2 enzymes (Fig. 6, Table 4) [85].

This study has suggested the features of COX catalytic sites vis-à-vis the interaction modes of *RR*, *SS*, and *RS* enantiomers of 3,3′-(1,2-ethanediyl)-bis[2-(3,4-dimethoxyphenyl)-4-thiazolidinones] with the active sites of the enzyme [85]. A comparison of the E/L interaction energy of these enan-

Table 3 Molecular descriptors and COX-2 inhibitory activities of thiazolone and oxazolone derivatives (Fig. 4) [84]

No.	X	R_1	R_2	R_3	x_1	x_7	x_{29}	x_{52}	pIC$_{50}$[a] obsd	cald[b]
1	NMe	t-Bu	H	NHC($=$NH)NH$_2$	15.73	-8.87	3.22	0	4.09	3.98
2	O	t-Bu	H	NHOEt	15.20	0	1.12	0	4.10	4.78
3	S	t-Bu	H	NHCN	14.71	0	2.65	-0.76	4.26	4.55
4	O	t-Bu	H	OH	14.73	0	-1.51	0	4.47	4.36
5	O	t-Bu	H	NHO-allyl	15.33	0	0.69	0	4.59	4.69
6	S	t-Bu	H	NHC($=$NH)NH$_2$	14.87	0	3.51	-0.78	4.68	4.67
7	S	t-Bu	H	NMeOMe	16.86	-8.20	7.67	0	4.74	4.74
8	S	t-Bu	Me	NHC($=$NH)NH$_2$	16.28	0	4.61	-0.79	4.92	4.73
9	O	t-Bu	H	NHC($=$NH)NH$_2$		0	-1.28	1.18	5.29	5.23
10	S	t-Bu	H	OH	14.73	0	2.21	0	5.33	5.00
11	S	t-Bu	H	NHOEt	14.87	0	6.44	0	5.49	5.70
12	S	t-Bu	H	NHO-allyl	15.35	0	5.90	0	5.57	5.59
13	S	t-Pr	H	NHC($=$NH)NH$_2$	15.48	0	3.39	-0.76	5.74	5.56
14	S	t-Bu	H	Sme	4.89	0	7.83	0	5.74	5.92
15	S	t-Bu	H	NHOMe	15.57	0	6.12	0	5.77	5.64
16	S	t-Bu	H	SH	15.33	0	5.33	0	5.82	5.54
17	S	t-Bu	H	NHOH	14.96	0	5.07	0	5.82	5.49
18	S	i-Pr	H	NHOMe	14.92	0	5.99	0	6.24	6.53
19	Indomethacin				0.13	-2.31	0.13	0	0.13	5.64
20	SC-58125				0	-0.18	5.15	0	5.15	6.66
21	NS-398				0	0.01	0.00	0	0.00	6.10

[a] COX-2 inhibitory activity
[b] Eq. 4

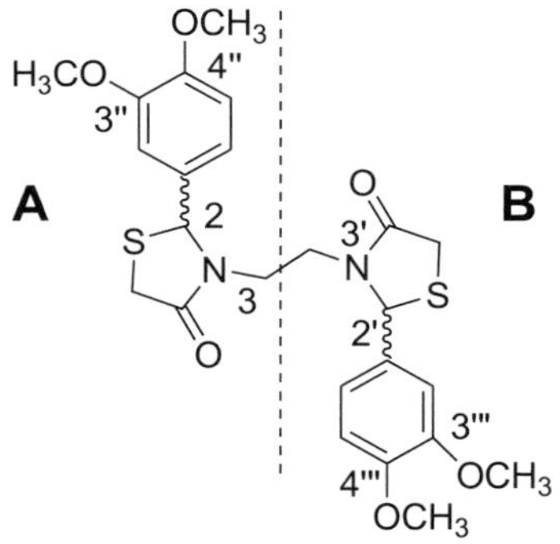

Fig. 5 3,3′-(1,2-ethanediyl)-bis[2-(3,4-dimethoxyphenyl)-4-thiazolidinones]. (Reprinted with permission from [85] Copyright 2003 Elsevier Ltd.)

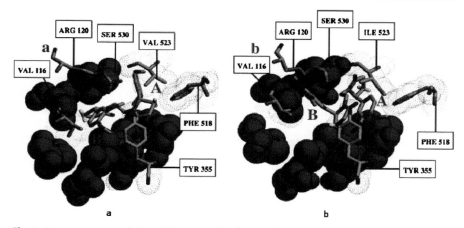

Fig. 6 Entrance view of the *SS* isomer of 3,3′-(1,2-ethanediyl)-bis[2-(3,4-dimethoxyphe-nyl)-4-thiazolidinones] into the COX-2 (**a**) and COX-1 (**b**) binding sites. In this **A** and **B** refer to thiazolidinone moieties as shown in Fig. 5. The steric effect of the non-conserved residue at position 523 (valine in **a** and isolucine in **b**) is reported to be crucial for the distinctive recognition of this compound between COX-1 and COX-2 isoenzymes. (Reprinted with permission from [85]. Copyright 2003 Elsevier Ltd.)

Table 4 Geometrical descriptors[a], E/L (enzyme/ligand) interaction and ligand internal energies [85]

Isomer	Isozyme	N – C – C – N dihedral angle (°)	Ar/Ar distance (Å)	Enzyme/ligand interaction energy (kcal/mol)	Ligand internal energy (kcal/mol)
RR	COX-1	– 163.15	5.706	3.73	41.88
	COX-2	– 77.01	4.478	– 3.31	10.66
SS	COX-1	– 134.85	6.124[b]	14.74	53.01
	COX-2	– 154.59	7.443	– 47.15	12.06
RS	COX-1	– 83.80	7.824	– 37.93	25.62
	COX-2	59.37	4.603	– 46.88	28.93

[a] Global minimum conformations within the enzyme clefts
[b] Folded conformation with the phenyl of moiety "A" in close proximity to the thiazolidinone of moiety "B" (see Fig. 5)

tiomers has indicated that the *SS* isomer has the highest affinity for the COX-2 (Table 4). Further, the internal energy of the *SS* isomer has been reported to be more favorable for binding into the COX-2 catalytic site than into COX-1 (Table 4). It has been shown that the *SS* enantiomer has the capability to deeply penetrate into the catalytic site of the COX-2 (Fig. 6a) by assuming an extended conformation. It has also highlighted the ligand's electrostatic and van der Waals (VdW) interactions with several residues (Arg120, Val116, Ser530, and Tyr385) of the enzyme's binding site. However, in the case of the COX-1, it has been reported that the ligand has to undergo an unfavorable folding (dihedral angle of N – C – C – N is – 134.85°) for binding to the COX-1 site (Fig. 6b). Also, in this situation, one of the thiazolidinone moieties (moiety "A" in Fig. 5) of the ligand has been reported to be in the proximity of a hydrophobic pocket delimited by residues Ile523, Tyr355, and Phe518. All these findings have prompted Vigorita et al. to suggest the favorability of *SS* enantiomers for binding to the COX-2 catalytic site [85].

3.1.2
Nitric Oxide Synthase Inhibitors

Thiazolidines are also known to interfere in the inflammatory process through nitric oxide (NO) and glycosaminoglycans (GAGs) pathways [86, 87]. During the inflammation process, the increased activity of matrix metalloproteinases (MMPs) causes the release of GAGs, followed by the cleavage of collagen and proteoglycans, the primary components of the cartilage. Also, in a feed-back mechanism, the NO, from the NO synthases, determines the inhibition of GAGs and stimulation of chondrocyte production and thereby the levels of MMPs [88–90]. Against this background Panico et al. quantitatively analyzed some 4-alkyl/aryl-2-(N-substituted aminoalkylamido) thiazoles (Fig. 7, Table 5) for their effect on NO production and GAGs release [87].

For these compounds (Table 5) the following equations have been developed to explain their NO and GAGs inhibitory activity.

$$\text{NOinh} = -196.156(94.346)q_{NR1} + 4.742(1.377)pKa - 57.001$$

$$n = 9, \quad r = 0.850, \quad s = 5.383, \quad F = 7.83 \tag{5}$$

$$\text{GAG} = 0.035(0.013)P + 3.381(0.734)pKa - 19.241$$

$$n = 9, \quad r = 0.883, \quad s = 2.508, \quad F = 10.68 . \tag{6}$$

These equations led to the suggestion that the pKa is the major descriptor for deciding the NO and GAG inhibitory activities of the compounds under investigation. As the R_1 substituent in these compounds is primarily responsible

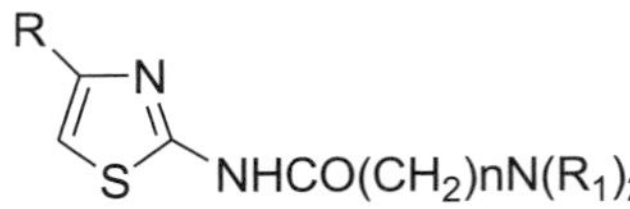

Fig. 7 General structure of 4-alkyl/aryl-2-(N-substituted aminoalkylamido) thiazoles associated with nitric oxide (NO) and glycosaminoglycans (GAGs) release inhibition

Table 5 4-Alkyl/aryl-2-amido thiazoles in modeling the nitric oxide (NO) and glycosaminoglycans (GAGs) release inhibitory activity (Fig. 7) [87]

No.	R	n	$N(R_1)_2$
1	Ph	1	NMe$_2$
2	Ph-OMe	1	Pyrrolidine
3	Ph-OMe	1	Piperidine
4	CH$_2$COOEt	1	Piperidine
5	H	2	Pyrrolidine
6	Ph	2	Pyrrolidine
7	Ph-pOMe	2	NMe$_2$
8	Ph-pOMe	2	Pyrrolidine
9	CH$_2$COOEt	2	Morpholine

for the basicity, the pyrrolidine moiety has been suggested as the most preferred group for this position for improved activity. Apart from the pKa, the charge at NR_1 (qNR_1) and the parachor (P) of the molecules have been found to influence the activities [87].

A different series of 4-alkyl/aryl-2-(N-substituted aminoalkylamido) thiazoles (Fig. 8, Table 6) were analyzed by Litina et al. for their anti-inflammatory activity in terms of the inhibition of carrageenan mice paw edema (%CPE) [90]. In this study the activity (%CPE) was correlated with several physicochemical and indicator parameters of these compounds (Table 6) [91].

$$\log(\%CPE) = 0.078(0.054)c \log D_{7.4} - 0.017(0.013)c \log D_{7.4}^2$$
$$+ 0.029(0.008)SURF + 0.060$$
$$n = 16, \quad r = 0.711, \quad s = 0.107, \quad F = 4.09 \tag{7}$$

(compounds excluded: 11–15 and 22–26)

$$\log(\%CPE) = 0.107(0.031)c \log D_{7.4} - 0.027(0.007)c \log D_{7.4}^2$$
$$+ 0.043(0.005)SURF - 0.681$$
$$n = 13, \quad r = 0.945, \quad s = 0.052, \quad F = 24.95 \tag{8}$$

(compounds excluded: 1, 6, and 13 11–15 and 22–26)

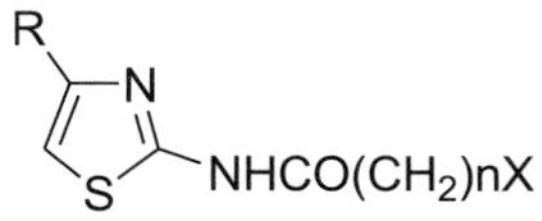

Fig. 8 General structure of 4-alkyl/aryl-2-(N-substituted aminoalkylamido) thiazoles associated with carrageenan induced mice paw edema inhibitory activity

Table 6 Physicochemical properties and carrageenan mice paw edema (%CPE) inhibitory activity of 4-alkyl/aryl-2-amido thiazoles (Fig. 8) [91]

No	R	X	n	RM	c log	$Pc \log D_{7.4}$	SURF	pKa	log %CPE[a] obsd	cald[b]
1	H	NMe_2	1	−0.54	0.31	0.13	53.5	7.11	1.45	1.64
2	H	$cN(CH_2)_4NMe$	1	−0.12	1.10	0.9	56.8	7.17	1.85	1.84
3	Me	NMe_2	1	−0.57	0.81	0.63	50.5	7.11	1.58	1.55
4	Me	NEt_2	1	−0.53	1.71	0.81	47.2	8.24	1.37	1.42
5	Ph	NMe_2	1	−0.27	2.41	2.24	52.3	7.09	1.72	1.68
6	Ph	cNC_4H_8	1	−0.14	3.29	2.40	58.8	8.22	1.69	1.95
7	p-OMePh	NMe_2	1	−0.26	2.43	2.42	49.7	5.55	1.54	1.56
8	p-OMePh	cNC_4H_8	1	−0.16	3.30	3.23	56.8	6.68	1.74	1.83
9	p-OMePh	cNC_5H_{10}	1	−0.03	3.86	3.86	53.6	5.73	1.73	1.64
10	p-Cl-Ph	cNC_5H_{10}	1	−0.3	4.56	4.56	57.1	4.33	1.67	1.70
11	CH_2OCOEt	NMe_2	1	−0.60	0.66	0.29	52.3	7.09	1.65	1.60
12	CH_2OCOEt	NEt_2	1	−0.51	1.56	0.64	49.4	8.22	1.73	1.51
13	CH_2OCOEt	cNC_4H_8	1	−0.45	1.54	0.27	58.8	8.22	1.55	1.88
14	CH_2OCOEt	cNC_5H_{10}	1	−0.54	2.09	1.49	55.7	7.25	1.78	1.82
15	CH_2OCOEt	$cN(CH_2)_4O$	1	−0.59	0.82	0.40	57.4	5.17	1.69	1.83
16	H	cNC_4H_8	2	0.14	1.38	−0.87	60.3	9.65	1.81	1.81
17	Ph	cNC_4H_8	2	−0.26	3.00	0.55	56.8	9.85	1.84	1.82

[a] Log of percent inhibition of carrageenan-induced edema, log %CPE, in mice
[b] Eq. 8

Table 6 (continued)

No	R	X	n	RM	c log	$Pc \log D_{7.4}$	SURF	pKa	log %CPE[a] obsd	cald[b]
18	p-OMePh	NMe_2	2	0.13	2.66	2.45	48.5	7.18	1.73	1.51
19	p-OMePh	cNC_4H_8	2	0.26	3.50	2.7	54.6	8.31	1.75	1.76
20	p-OMePh	cNC_5H_{10}	2	−0.53	4.06	3.78	52.3	7.36	1.60	1.59
21	p-OMePh	$cN(CH_2)_4O$	2	−0.34	2.79	2.7	53.6	6.80	1.74	1.72
22	CH_2OCOEt	NMe_2	2	−0.2	0.89	−0.72	50.7	8.72	1.72	1.42
23	CH_2OCOEt	NEt_2	2	−0.16	1.87	−0.76	48.3	9.85	1.81	1.31
24	CH_2OCOEt	cNC_4H_8	2	−0.03	1.73	−1.13	56.8	9.85	1.54	1.61
25	CH_2OCOEt	cNC_5H_{10}	2	−0.17	2.29	0.40	54.1	8.88	1.63	1.69
26	CH_2OCOEt	$cN(CH_2)_4O$	2	−0.54	1.02	0.13	55.6	6.80	1.77	1.73

[a] Log of percent inhibition of carrageenan-induced edema, log %CPE, in mice
[b] Eq. 8

In deriving these equations several compounds have been excluded as outliers. The parabolic relation in $c \log D_{7.4}$ has suggested 1.98 as the optimum hydrophobicity for easy transportation and rapid delivery to the target biomacromolecule. The positive regression coefficient of the parameter surface tension ($SURF$) has been interpreted as the anti-inflammatory activity of the test compound increases linearly with the increase in its surface tension. On the basis of the above results it has been concluded that the pharmacokinetic properties of these compounds play a key role in their anti-inflammatory activity. These findings have been found to be in good agreement with previously reported results concerning other non-steroidal anti-inflammatory drugs (NSAIDs) [92].

3.1.3
Adenosine Receptor Antagonists

Thiazoles and related analogues have been indicated to influence the inflammatory processes through their antagonistic activity at the adenosine receptors [93–95]. Among the various adenosine receptors, the subtypes A_1 and A_3 have been indicated in inflammation, asthma [96, 97], glaucoma [98], myocardial and cerebral ischemia [99–102]. A systematic template search of quinazolines, isoquinolines and other prototypes has resulted in the induction of the thiazole class as adenosine A_3 receptor antagonists (Fig. 9) [93].

These findings prompted Jung et al. to explore several thiazole and thiadiazole derivatives as potent human adenosine A_3 receptor antagonists (Fig. 9, Table 7) [95]. Using principal component factor analysis (FA) and genetic function approximation (GFA) techniques, Bhattacharya et al. quantitatively analyzed the adenosine A_3 receptor antagonistic activity of these compounds (Fig. 9, Table 7) with the quantum chemical descriptors (Wang–Ford charges from the AM1 method), hydrophobicity ($\log P$) and some indicator parameters [103]. In this study, Eq. 9 has emerged from the factor analysis and

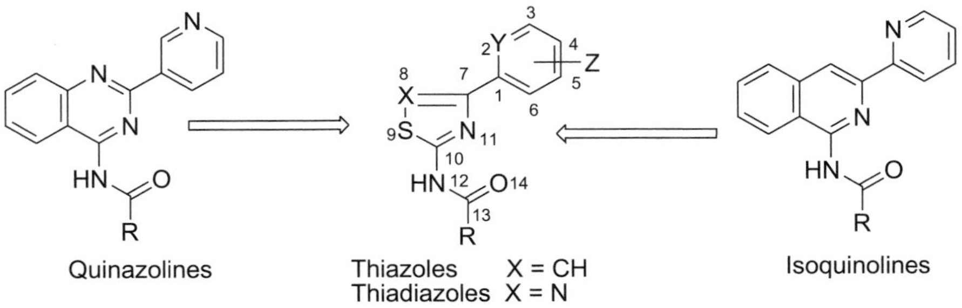

Fig. 9 General structure of thiazole and thiadiazole class of adenosine A_1 and A_3 receptor antagonists and their quinazoline, isoquinoline seed structures

Table 7 Thiazole and thiadiazole analogues in modeling the adenosine A_3 receptor antagonistic activity (Fig. 9) [103–105]

	X	Z	R	pKi [a] obsd	cald[b]	cald[c]	cald[d]
1. (LUF5417)	N	H	4-OMePh	7.09	–	–	7.31
2	CH	H	Me	7.74	8.03	7.89	7.65
3	CH	H	$(Me_3C)O$	5.29	5.22	5.22	5.59
4	CH	H	$NCCH_2$	6.69	7.13	6.50	7.88
5	CH	4-Cl	Me	7.29	7.04	7.48	7.31
6	CH	4-Cl	$PhCH_2$	7.00	6.74	7.07	6.55[e]
7	CH	4-MeO	Me	8.52	8.42	8.24	7.88
8	CH	3-MeO	Me	8.39	8.42	8.39	7.89
9	CH	2MeO	Me	7.09	7.00	7.06	7.47
10	CH	4-MeO	Me	6.28	6.73	6.35	5.93
11	CH	4-MeO	Et	8.62	7.88	8.26	8.09
12	CH	4-MeO	Pr	8.11	8.04	8.28	8.00
13	CH	4-MeO	i-Pr	7.79	7.39	7.80	7.43
14	CH	4-MeO	$NCCH_2$	7.61	8.18	7.57	8.17
15	CH	4-MeO	t-Bu	7.50	7.20	7.67	7.62
16	CH	4-MeO	$(Me_3C)O$	5.49	5.50	5.61	6.06[e]
17	CH	4-MeO	Ph	7.54	7.29	7.62	6.83[e]
18	CH	4-MeO	$PhCH_2$	7.85	7.43	7.85	6.96[e]
19	CH	4-MeO	$PhCH_2CH_2$	7.54	7.08	7.51	7.36
20	CH	4-MeO	p-MeOPhCH$_2$	5.94	7.16	5.97	7.24
21	CH	4-MeO	p-MeOPhCH$_2$CH$_2$	7.54	6.88	7.49	7.89
22	CH	4-MeO	Ph_2CH	6.28	6.97	6.29	6.13
23	CH	4-MeO	Ph_2CHCH_2	6.40	6.41	6.40	6.58
24	CH	4-MeO	2-Furan	7.50	7.68	7.40	6.63
25	CH	4-MeO	Thiophene-2-CH$_2$	7.49	7.24	7.47	6.45[e]
26	CH	4-MeO	2-Thiophene	7.16	7.28	7.38	6.67
27				6.43			6.81
28	N	H	CH_3	8.64	8.92	8.62	8.57[e]
29	N	H	$PhCH_2$	7.10	7.51	6.92	6.99
30	N	4-MeO	Me	9.10	8.88	9.08	8.78
31	N	4-MeO	$PhCH_2$	7.62	7.52	7.59	7.66
32	N	4-MeO	Et	8.95	8.82	9.10	8.60

[a] Adenosine A_3 receptor antagonistic activity
[b] Eq. 10, compounds 1 and 27 are not in the regression [103]
[c] Eq. 14, compounds 1 and 27 are not in the regression [104]
[d] Eq. 15 [105]
[e] Test set compound

Eqs. 10 and 11 have emerged from the GFA technique.

$$pKi = -3.951(2.415)q_2 - 2.623(2.136)q_5 + 2.441(1.489)q_7$$
$$- 0.275(0.261)\log P - 1.498(0.763)I_{OBu_t}$$
$$+ 0.895(0.614)I_{MeEt} + 4.030$$
$$n = 30, \quad r = 0.893, \quad Q^2 = 0.689, \quad s = 0.483, \quad F = 15.0 \qquad (9)$$

$$pKi = -1.940(2.072)q_2 - 11.413(3.878)q_8 - 13.611(5.005)q_9$$
$$- 0.038(0.029)(\log P)^2 - 1.556(0.722)I_{OBu_t} + 3:083$$
$$n = 30, \quad r = 0.902, \quad Q^2 = 0.753, \quad s = 0.452, \quad F = 21.0 \qquad (10)$$

$$pKi = -1.932(2.150)q_2 - 1.413(4.012)q_8 - 13.741(5.189)q_9$$
$$- 0.311(0.260)\log P - 1.518(0.747)I_{OBu_t} + 3.686$$
$$n = 30, \quad r = 0.896, \quad Q^2 = 0.739, \quad s = 0.465, \quad F = 19.6. \qquad (11)$$

The equation from factor analysis (Eq. 9) has explained the activity in terms of $\log P$, I_{MeEt}, (an indicator parameter defined for the presence of the methyl or ethyl substituent at the R position), I_{OBu_t} (an indicator parameter defined for the presence of the *tert*-butyloxy group at the R position) and Wang–Ford charges (q_2, q_5, and q_7) of atoms C2, C5, and C7. This has suggested that less hydrophobic compounds with methyl or ethyl and without the *tert*-butyloxy group at the R position would be favorable for the adenosine A$_3$ receptor antagonistic activity [103]. The GFA technique has resulted in equations with improved statistics (Eqs. 10 and 11). These equations have suggested that less hydrophobic compounds without a *tert*-butyloxy group at the R position would be better for the binding affinity at the receptor. Moreover, the Wang–Ford charges at atoms C2, X8, and S9 have been identified as contributing ones to modulate the binding affinity of these compounds. In this the Wang–Ford charges of X8 and S9 are from part of the thiazole/thiadiazole nucleus and signify their role in deciding the activity of these compounds.

The adenosine A$_3$ receptor antagonistic activity of these compounds have been further analyzed [104] in 3D-QSAR using molecular shape analysis (MSA) and molecular field analysis (MFA) techniques in Cerius2 (version 4.8) software [50]. In this, Jurs atomic charge descriptors were used for the MSA study and H$^+$ point charges and CH$_3$ derived steric fields were used for the MFA study. In this 3D-QSAR study, MSA resulted in Eqs. 12 and 13 and MFA led to Eq. 14.

$$pC = 0.140(0.052)JursPPSA_3 + 61.920(20.616)Jurs\ RNCG$$
$$- 1.024(0.766)JursRPCS - 0.006(0.004)JursSASA$$
$$+ 0.006(0.004)Energy + 1.645(1.022)Fo - 8.735$$
$$n = 30, \quad r = 0.901, \quad Q^2 = 0.705, \quad s = 0.215, \quad F = 16.6 \qquad (12)$$

$$pC = 0.130(0.048)\text{JursPPSA}_3 + 61.291(20.009)\text{Jurs RNCG}$$
$$- 1.056(0.749)\text{JursRPCS} + 0.007(0.004)\text{Energy}$$
$$+ 1.701(0.988)Fo - 0.001(0.000)\text{PMI}_{\text{mag}} - 10.180$$
$$n = 30, \quad r = 0.907, \quad Q^2 = 0.700, \quad s = 0.203, \quad F = 17.8 \tag{13}$$

$$pC = - 0.050\text{H}^+/283 - 0.028\text{H}^+/366 - 0.031\text{H}^+/427$$
$$- 0.019\text{H}^+/448 - 0.054\text{H}^+/466 - 0.052\text{H}^+/534$$
$$- 0.030\text{H}^+/608 - 0.021\text{H}^+/618 - 0.020\text{H}^+/687$$
$$+ 0.033\text{H}^+/757 - 0.010\text{CH}_3/417 - 0.052\text{CH}_3/474$$
$$+ 0.025\text{CH}_3/529 - 0.022\text{CH}_3/756 + 5.59$$
$$n = 30, \quad r = 0.990, \quad Q^2 = 0.716, \quad s = 0.500 . \tag{14}$$

The equations from MSA (Eqs. 12 and 13) have explained the activity in terms of atomic charge weighted positive surface area (JursPPSA$_3$), the charge of the most negative atom divided by the total negative charge (JursRNCG), total molecular solvent-accessible surface area (JursSASA), relative positive charge surface area (JursRPCS), conformational energy of the molecules (Energy), partial moment of inertia, energy of the most stable conformer (PMI$_{\text{mag}}$), and the ratio of common overlap steric volume to the volume of individual molecules (Fo). On the basis of the 3D-QSAR equation from the MFA the authors could explain the 96.1% variance in the activity with cross-validated r^2 as 0.716. In this equation, the numbers suffixed to H$^+$/ and CH$_3$/, respectively, notify the significant polar and steric grid locations involved in modeling the adenosine A$_3$ receptor antagonists. From these studies they have concluded the importance of Jurs descriptors and the interaction energies of various molecular field grid locations in modeling the activity of these analogues [104].

Borghini et al. [105] have also carried out a QSAR study on the adenosine A$_3$ receptor antagonistic activity of these compounds (Fig. 9, Table 7) with the CODESSA descriptors. For this study, 26 compounds were taken in the training set for the model development and six compounds were placed in the test set to verify the developed model. This study resulted in the following correlation (Eq. 15) for the adenosine A$_3$ receptor antagonistic activity of the compounds in terms of electrostatic descriptors namely "HACA-2/TMSA" (ratio of total charge weighted hydrogen-acceptor charged surface area and total molecular surface area), MPCS (minimum partial charge on the S atom), "WNSA-3" (total surface weighted partial negative surface area), and RNH (number of H atoms at the position R).

$$pKi = 17296(260)\text{HACA-2}/\text{TMSA} + 22.73(4.44)\text{RNH}$$
$$+ 46.56(10.83)\text{MPCS} - 0.382(0.009)\text{WNSA-3} - 6.17$$
$$n = 26, \quad r = 0.840, \quad Q^2 = 0.548, \quad s = 0.559, \quad F = 12.23 . \tag{15}$$

Table 8 Thiazole and thiadiazole analogues in modeling the adenosine A_1 receptor antagonistic activity (Fig. 9) [105]

No.	X	Y	R	pKi[a] obsd	cald[b]
1	CH	CH	4-OMePh	7.12	
2	N	CH	4-OMePh	7.49	7.50
3	CH	N	Ph	5.77	5.68
4	CH	N	4-ClPh	6.70	6.32
5	CH	N	4-IPh	5.62	6.20
6	CH	N	4-MePh	5.80	5.94
7	CH	N	4-OMePh	5.49	5.96[c]
8	CH	N	3,4-diClPh	5.80	6.15[c]
9	CH	N	3-ClPh	5.77	5.58
10	CH	N	4-NO$_2$Ph	5.82	6.00
11	CH	N	4-OCH(Me)$_2$Ph	5.24	5.54
12	CH	N	Cyclopentyl	6.04	5.47
13	N	CH	Ph	7.51	7.47
14	N	CH	4-ClPh	7.39	7.90[c]
15	N	CH	4- MePh	7.52	7.39
16	N	CH	4-OHPh	8.14	8.30
17	N	CH	4-OCH2CO$_2$Ph	7.00	7.12
18	N	CH	Cyclohexyl	5.85	6.79
19	N	CH	(*trans*) 4-OMe-cyclohexyl	7.49	6.77
20	N	CH	(*cis*) 4-OMe-cyclohexyl	6.96	6.81
21	N	CH	(*trans*) 4-OH-cyclohexyl	7.70	7.27[c]
22	N	CH	(*cis*) 4-OH-cyclohexyl	7.38	7.15
23	N	CH	NHPh	6.00	5.96
24	CH	N	NHPh	6.03	5.66[c]
25	CH	CH	Ph	7.41	7.52
26	CH	CH	3-ClPh	7.07	7.31
27	CH	CH	4-BrPh	7.48	7.18[c]
28	CH	CH	4-ClPh	7.74	7.52
29	CH	CH	4-NO$_2$Ph	7.66	7.11
30	CH	CH	4- MePh	7.44	7.32
31	CH	CH	4-C(Me)$_3$Ph	5.87	6.51
32	CH	CH	4-CF$_3$Ph	6.78	6.78
33	CH	CH	3,4-diClPh	7.23	7.30
34	CH	CH	2,4-diClPh	7.24	7.17

[a] Adenosine A_1 receptor antagonistic activity
[b] Eq. 16 [105]
[c] Test set compound

This equation also suggested the importance of polarity/partial charges on the heteroatoms of the thiazole moiety for adenosine A_3 receptor antagonistic activity. On this basis, it has been suggested that thiadiazoles capable of accepting hydrogen bonds and carrying a small non-polar alkyl substituent at the position R, would be better for binding to the adenosine A_3 receptor [105].

With the CODESSA descriptors, Borghini et al. [105] also carried out a QSAR study on the adenosine A_1 receptor antagonistic activity of another series of thiazoles/thiadiazoles (Fig. 9) [93, 94]. In this, 27 compounds were included for training the model and seven compounds were used for testing the model (Fig. 9, Table 8). The following three descriptor model in terms of MPCN (minimum partial charge on the N atom), ZXS/ZXR (ZX Shadow/ZX Rectangle), and PP/D^2 (polarity parameter/square distance) has been derived to explain the affinity of these compounds to the adenosine A_1 receptor.

$$pKi = 155.1(17.38)\text{MPCN} + 3.86(0.99)\text{ZXS/ZXR}$$
$$+ 2.12(1.11)\text{PP}/D^2 + 17.28$$
$$n = 27, \quad r = 0.901, \quad Q^2 = 0.742, \quad s = 0.382, \quad F = 33.15 . \tag{16}$$

In Eq. 15, the coefficient of ZXS/ZXR has suggested a preference for an aromatic substituent for the R group. The positive regression coefficients of MPCN and PP/D^2 are in favor of an increase in the partial charge and polarity in the compounds for better activity. This has prompted us to suggest that thiadiazole derivatives with aromatic substituents would enhance the affinity of the compounds for the adenosine A_1 receptor.

3.1.4
Histamine (H_1) Antagonists

Thiazolidines are known to show their action on histamine receptors. The geometrical similarity between 2-aryl-3-[3-(N,N-dimethylamino)propyl]-1,3-thiazolidin-4-ones and different histamine (H_1) antagonists such as bamipine, clemastine, cyproheptadine, triprolidine, promathazine, chlorpheniramine, and carbinoxamine (Fig. 10) [106, 107] prompted Diurno et al. to evaluate these compounds for antihistaminic activity [108]. Singh and coworkers have investigated the QSAR of the antihistaminic (H_1) activity of 2,3-disubstituted thiazolidin-4-ones (Table 9) with the hydrophobicity (π), molar refractivity (MR), Hammett's sigma (σ) constant, dipole moment (μ), and Swain and Lupton's polarity (F) and resonance (R) constants of the substituent groups [109]. In this study the antihistaminic (H_1) activity of these compounds has been found to be best explained by $\sum F$ and π_4 descriptor-

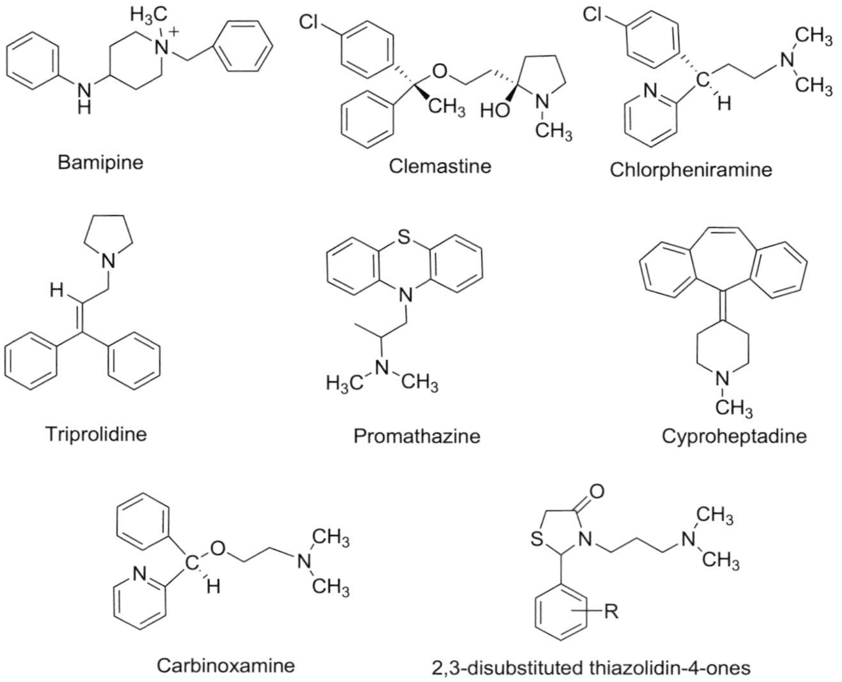

Fig. 10 Selected histamine (H_1) antagonists and 2,3-disubstituted thiazolidin-4-ones embedded with some of their structural features

combinations of R substituent groups (Eq. 17) [109].

$$pA_2 = -2.753(0.715) \sum F + 0.907(0.297)\pi_4 + 7.267$$
$$n = 15, \quad r = 0.832, \quad s = 0.672, \quad F = 13.54 \tag{17}$$

$$pA_2 = -2.573(0.346) \sum F + 1.019(\pm 0.148)\pi_4 + 7.412$$
$$n = 13, \quad r = 0.961, \quad s = 0.320, \quad F = 60.70 . \tag{18}$$

In deriving Eq. 18, compounds 3 and 6 (Table 9) have been excluded as they were found to be outliers in the study. From this, they concluded that the hydrophobic substituent at the 4-position of the phenyl ring and cumulative negative polar effects of all the substituents in the phenyl group are advantageous for antihistaminic activity [109].

Agrawal et al. have studied the anti-histaminic activity of these 2,3-disubstituted thiazolidin-4-one analogues (Table 9) with graph theoretical indices and discovered the following equation for the activity in terms of the rooted Wiener index (Ww), rooted Szeged index (Szw), molecular redundancy index (MRI), and an indicator variable Ip_1 (for the p-alkyl of the aryl

Table 9 Physicochemical properties and anti-histaminic activity of 2,3-disubstituted thiazolidin-4-ones derivatives (Fig. 10) [109]

No.	R	$\sum F$	π_4	pA_2 [a] obsd	cald [b]
1	H	0.00	0.00	7.4	7.4
2	4-F	0.43	0.14	7.1	6.5
3	4-Cl	0.41	0.71	5.6	7.1
4	4-Br	0.44	0.86	7.4	7.2
5	4-Me	− 0.04	0.56	8.3	8.1
6	4-OMe	0.26	− 0.02	5.0	6.7
7	4-NO$_2$	0.67	− 0.28	5.2	5.4
8	4-NH$_2$	0.02	− 1.23	6.0	6.1
9	4-iPr	− 0.05	1.53	8.7	9.1
10	3-F	0.43	0.00	6.4	6.3
11	3-Cl	0.41	0.00	6.2	6.4
12	3-Br	0.44	0.00	6.4	6.3
13	3-Me	− 0.04	0.00	7.7	7.5
14	3-OMe	0.26	0.00	6.4	6.7
15	3-NO$_2$	0.67	0.00	5.4	5.7

[a] 50% inhibition of histamine (H$_1$) activity in isolated guinea pig ileum (relative to mepyramine)
[b] Eq. 18

moiety) [110].

$$pA_2 = 99.776\text{MRI} + 0.989\text{Szw} - 1.414\text{Ww} + 0.859\text{I}p_1 - 34.882$$
$$n = 11, \quad r = 0.980, \quad s = 0.228, \quad F = 37.13 . \tag{19}$$

However, in deriving this equation four compounds (Table 9, compounds 3, 6, 7, and 15) have been excluded from the study. Between these two studies, the model (Eq. 18) proposed by Singh et al. [109] is far superior when compared to the later one (Eq. 19) [110].

In another interesting study, considering the importance of asparagine, aspartic acid, threonine, and lysine in the binding of agonists/antagonists to the histamine (H$_1$) receptor, Brzezinska and co-workers [111, 112] have quantitatively investigated the antihistaminic (H$_1$) activity of substituted thiazoles and benzthiazole derivatives [113–115] (Fig. 11, Table 10) using the log P and reverse and normal phase thin layer chromatographic (TLC) parameters (Rm values) generated through impregnating the TLC plates with these selected amino acids. The idea behind impregnating the TLC plates with amino acids was to assess their contribution in the binding affinities at the receptor site and in turn use them in modeling the activity of the compounds. Accordingly,

Table 10 Thiazole and benzthiazole derivatives in modeling the anti-histaminic activity (Fig. 11) [111, 112]

No.	R	n	pA_2 [a] obsd	cald[b]	cald[c]
1	H	–	4.44	4.10	4.53
2	3-Me	–	4.00	4.18	3.83
3	3-F	–	4.53	4.79	4.62
4	3-Cl	–	4.82	5.02	4.78
5	3-Br	–	4.65	5.20	4.92
6	3-OMe	–	4.14	4.13	4.37
7	H	–	5.88	5.72	5.87
8	3-F	–	6.15	6.22	5.69
9	3-Cl	–	6.38	5.96	6.04
10	4-F	–	5.99	5.74	5.78
11	4-Br	–	5.87	6.01	6.19
12	H	–	5.98	5.78	6.10
13	Me	3	5.70	5.56	5.92
14	H	2	5.82	5.74	5.46
15	Me	2	5.60	5.25	5.60
16	$-CH_2$-Ph-4-Me	3	5.99	6.32	6.23
17	$-C_2H_4$-Ph-4-Me	2	6.08	6.14	6.12
18	$-CH_2$-Ph	2	5.77	5.90	5.74

[a] 50% inhibition of histamine (H_1) antagonist activity (isolated guinea pig ileum) relative to temelastine
[b] Eq. 20 [111]
[c] Eq. 21 [112]

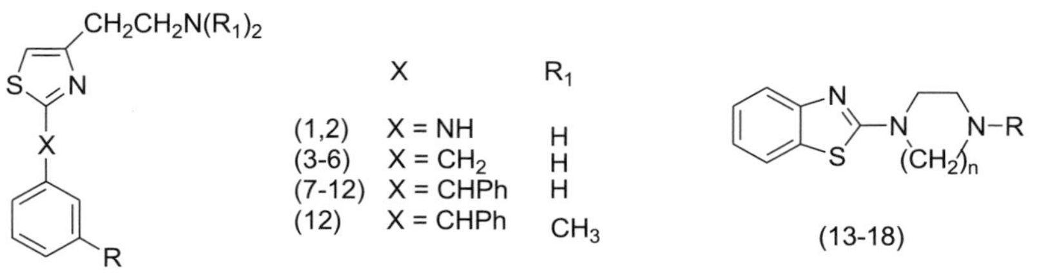

Fig. 11 Structures of thiazole and benzthiazole derivatives

the Rm values of the compounds have been measured under different chromatographic environments [111, 112]. Following are the representative QSAR equations of the antihistaminic (H_1) activity in the reverse and normal phase

TLC studies, respectively, of these compounds.

reverse phase TLC model

$$pA_2 = -1.27(0.23)S2/C + 9.59(2.60)C\text{-}S6 + 0.63(0.09)\log P + 3.96$$
$$n = 18, \quad r = 0.94, \quad s = 0.289, \quad F = 35.11 \tag{20}$$

normal phase TLC model

$$pA_2 = 1.16(0.34)S4 + 0.16(0.03)S7/C + 0.54(0.07)\log P + 4.16$$
$$n = 18, \quad r = 0.95, \quad s = 0.254, \quad F = 47.04 . \tag{21}$$

The terms S2, S4, S6, S7, and C represent, respectively, the chromatographic environment(s) formed by combinations of asparagine, lysine, asparagine + lysine, asparagine + aspartic acid + threonine and/or control (without amino acids). In these equations S2/C, S4, C-S6, and S7/C refer to the Rm values (independent descriptors) of corresponding chromatographic environments (impregnation) reflected in modeling the antihistaminic (H_1) activity. These studies have highlighted the importance of overall hydrophobicity of the compounds in deciding the activity. They have also quantified the interaction of selected amino acids with the compounds. It further suggested that the normal phase TLC parameters modeled the activity better than the reverse phase TLC parameters. This study has opened up a novel way for generating the receptor-directed interaction information in terms of amino acids for modeling the antihistaminic (H_1) activity of these compounds. One can adapt to this kind of procedure for the study of other receptors also.

3.1.5
H^+K^+-ATPase Inhibitors

Thiazolidines have been reported to be a core scaffold for antiulcer compounds. La Mattina et al. have evaluated a series of 128 compounds belonging to 4-substituted-2-guanidino thiazoles (Fig. 12) against gastric hydrogen-potassium stimulated adenosine triphosphatase (H^+K^+-ATPase), a therapeutic target of anti-ulcer drugs [116].

Goel and Madan [117] investigated the SAR of the antiulcer activity of these compounds with Wiener's topological index (Wiener number of chemical graph, W(G)) [118] and the first-order molecular connectivity index ($^1\chi$) [119] using a typical classification procedure. In the case of Wiener's

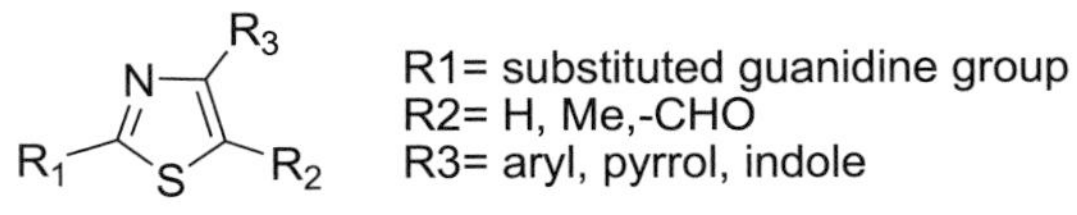

R1= substituted guanidine group
R2= H, Me,-CHO
R3= aryl, pyrrol, indole

Fig. 12 General structure of 4,5-substituted-2-guanidinothiazoles associated with H^+K^+-ATPase inhibitory activity

topological index, the compounds have been correctly classified into active and inactive groups to the order of 89%. A Wiener index value in the range of 391 to 540 was observed to be associated with the active analogues. In the case of the molecular connectivity index the overall accuracy of prediction has been reported to be 88%; and, the activity that has been found to be associated with the $^1\chi$ values is the range of 5.75 to 7.20 [117]. This approach is very handy and quick for the screening of large databases.

Borges and Takahata studied the QSAR of a different series of 4-indolyl, 2-guanidinothiazole derivatives (Fig. 13; Table 11) for their H^+K^+-ATPase inhibitory activity using 94 quantum chemical descriptors generated for this purpose [120].

They adopted partial least square (PLS), principal component regression (PCR), and back propagation neural network (BPN) methods for the derivation of the regression models. Among these approaches, the activity has been found to be best explained in the PLS analysis with five latent variables. In terms of the whole molecular structure, this has led to the suggestion that the energies of the highest occupied molecular orbitals – 9, – 8, – 7, – 5, – 3, – 2, and – 1 ($\in_{HOMO-9}$, $\in_{HOMO-8}$, $\in_{HOMO-7}$, $\in_{HOMO-5}$, $\in_{HOMO-3}$, $\in_{HOMO-2}$, $\in_{HOMO-1}$) the dipole moments along the x- and z-axis (μ_x, μ_z) and quadrupole moments yy, zz, and xz (Quad$_{yy}$, Quad$_{zz}$, and Quad$_{xz}$) are important contributors to model the activity. In terms of individual atomic centers of these analogues, the study has identified the electron densities F_3^T, F_6^T, F_4^H, F_9^H, F_4^L, F_6^L, F_7^L, F_8^L F_{11}^T, F_{13}^H, F_{14}^H, F_{12}^L, F_{13}^L, F_{14}^L, and F_{15}^L and the charges Q_2, Q_3, Q_5, Q_6, Q_{12}, and Q_{18} (the subscript numbers correspond to the atom identification number as shown in Fig. 13; the super scripts T, H, and L represent Total, HOMO, and LUMO, respectively) as important to explain the activity of the analogues. This study has explained the activity in terms of a detailed electronic profile of the compounds. However, these results are far more sophisticated and at the same time too complex to comprehend when compared to that of Goel and Madan's investigations involving Wiener's and Chi indices [117].

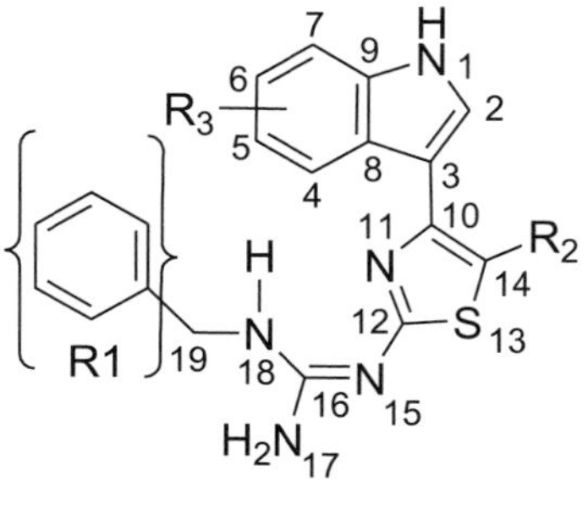

Fig. 13 General structure of 4-indolyl-2-guanidinothiazoles associated with H^+K^+-ATPase inhibitory activity

Table 11 4-Indolyl, 2-guanidinothiazole derivatives in modeling the H^+K^+-ATPase inhibitory activity (Fig. 13) [120]

No.	R_1	R_2	R_3	IC_{50}[a] obsd	Predicted[b] PLS	PCR	BPN
1	H	H	H	9.0	6.0	3.3	8.4
2	C_7H_7	H	H	1.5	1.3	3.2	1.2
3	H	Me	H	7.6	7.9	2.2	7.6
4	H	H	5-OMe	19.0	16.2	4.4	7.2
5	C_7H_7	H	5-OMe	4.3	3.5	3.7	6.4
6	H	H	5-OC$_7$H$_7$	1.2	1.3	3.9	1.4
7	C_7H_7	H	5-OC$_7$H$_7$	1.8	1.7	3.6	0.61
8	H	H	2-Me	9.0	10.6	3.1	5.2
9	C_7H_7	H	2-Me	1.7	2.2	2.90	1.1
10	H	Me	2-Me, 5-Cl	0.6	0.47	0.97	2.8
11	C_7H_7	Me	2-Me, 5-Cl	1.1	0.74	0.61	0.89
12	C_7H_7	H	4-Me	3.3	3.7	0.87	0.27
13	H	H	5-Me	1.8	2.3	2.9	2.9
14	C_7H_7	H	5-Me	0.94	1.1	2.9	0.57
15	C_7H_7	H	6-Me	1.6	1.5	3.5	2.2
16	H	H	7-Me	7.6	6.7	3.6	6.7
17	C_7H_7	H	7-Me	1.0	1.4	3.4	2.2
18	H	H	5-Cl	1.8	1.5	2.3	1.9
19	C_7H_7	H	5-Cl	0.7	0.87	0.98	0.31
20	H	Me	5-Cl	0.59	0.81	0.8	2.1
21	C_7H_7	Me	5-Cl	0.7	0.63	0.88	0.77
22	H	H	5-Br	0.96	1.1	2.6	1.0
23	C_7H_7	H	5-Br	0.7	0.52	2.3	1.1
24	H	H	5-F	7.4	9.2	4.1	9.9
25	C_7H_7	H	5-F	2.7	3.3	3.9	3.2
26	H	H	5-CO$_2$Me	6.7	5.3	2.4	0.37
27	C_7H_7	H	5-CO$_2$Me	3.1	3.4	2.1	4.2
28	C_7H_7	H	5-NHCOEt	23.8	21.7	5.9	4.0

[a] H^+K^+-ATPase inhibitory activity
[b] Predicted activity by leave one out procedure

3.1.6
Antiepileptic Agents

Thiazolidines/thiazoles in fusion with other aromatic systems are also potential bioactive scaffolds. Riluzole (Fig. 14)—a benzothiazole analogue—is known to intervene in epilepsy, a cerebral disorder of the central nervous system, via a glutamatergic mechanism [121]. The anticonvulsant activity of riluzole has been attributed to its direct action on the voltage-dependent

Table 12 Physicochemical properties and anticonvulsant activity of benzothiazolamine
derivatives (Fig. 14) [123]

No	X	45SUBST	π	MR	$-\log IC_{50}$ [a]	
					obsd	cald [b]
1	6-OCF$_3$ (Riluzole)	0	1.04	0.79	5.39	4.82
2	H	0	0.00	0.10	4.59	4.53
3	6-SMe	0	0.61	1.38	4.80	4.13
4	6-Me	0	0.56	0.56	3.82	4.54
5	6-CHMe$_2$	0	1.53	1.50	5.17	4.68
6	6-COOH	0	-0.32	0.69	3.88	–
7	6-SO$_2$Me	0	-1.63	1.35	3.30	3.22
8	6-Cl	0	0.71	0.60	5.17	4.57
9	6-OMe	0	-0.02	0.79	3.30	3.74
10	6-COPh	0	1.05	3.03	3.30	3.96
11	6-SO$_2$NH$_2$	0	-1.82	1.23	3.30	3.14
12	6-Br	0	0.86	0.89	4.85	4.50
13	6-NO$_2$	0	-0.28	0.74	3.66	4.23
14	6-CF$_3$	0	0.88	0.50	5.31	5.04
15	6-F	0	0.14	0.09	3.30	4.30
16	6-OH	0	-0.67	0.28	3.30	3.47
17	6-NH$_2$	0	-1.23	0.54	3.30	2.99
18	5-OMe	1	-0.02	0.79	3.30	3.14
19	5-Me	1	0.56	0.56	3.30	3.70
20	5-CF$_3$	1	0.88	0.50	4.00	4.02
21	5-COPh	1	1.05	3.03	3.30	3.06
22	4-OMe	1	-0.02	0.79	3.30	2.90
23	4-Me	1	0.56	0.56	3.82	3.63
24	4-Cl	1	0.71	0.60	4.58	3.68
25	4-CF$_3$	1	0.88	0.50	3.86	4.10
26	4-Ph	1	1.96	2.54	5.04	–
27	4-Me,5-F	1	0.70	0.66	3.30	3.55
28	4,5-(CHCH)$_2$	1	1.30	1.75	3.30	3.49
29	4,6-F$_2$	1	0.28	0.18	3.30	3.17
30	4,6-Cl$_2$	1	1.42	1.21	3.30	3.67
31	4,6-Me$_2$	1	1.12	1.13	3.30	3.59
32	5,6-Me$_2$	1	0.99	1.13	3.30	3.56

[a] Sodium flux (NaFl) inhibition
[b] Eq. 22

sodium channels [122]. Hays et al. have investigated the QSAR of the anticonvulsant activity of substituted 2-benzothiazolamines (Fig. 14, Table 12)
with hydrophobicity (π), molar refractivity (MR), Hammett's sigma (σ) and
Swain and Lupton's polarity ($\mathcal{F}$) and resonance ($\mathcal{R}$) constants and an indicator parameter (45SUBST) describing the importance of the 4 and/or 5

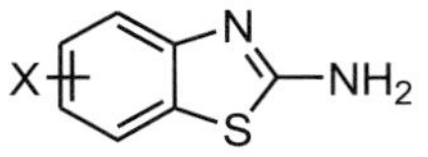

Fig. 14 General structure of benzothiazolamines associated with anticonvulsant activity

substituents [123]. This study has resulted in the following equation to explain the variation in the anticonvulsant activity of these analogues.

$$- \log \text{IC}_{50} = 0.57(0.12)\pi - 0.90(0.19)45\text{SUBST}$$
$$- 0.45(0.14)\text{MR} + 0.92(0.38)\mathcal{R} + 4.57$$
$$n = 30, \quad r = 0.794, \quad s = 0.480, \quad F = 10.80 . \tag{22}$$

On this basis it has been suggested that an increase in lipophilicity (π), decrease in size (MR), increase in resonance effect ($\mathcal{R}$), and the absence of 4 or 5 substitution of the benzothiazolamine ring system would be favorable for anticonvulsant activity. This has been presented as an example of QSAR in fused systems. In the literature more study reports are available where the thiazole moiety existed as part of the fused ring systems [77].

3.2
Metabolic Disorders

3.2.1
β_3-receptor Agonists

Thiazoles have been investigated as a potential drug scaffold in the treatment of metabolic disorders such as obesity, diabetes, cataracts, and thrombosis. They have been reported to be associated with the β_3 agonist activity (a subtype of β-adrenergic receptor) through the activation of adenyl cyclase [124, 125]. This receptor has been implicated in obesity, a condition of abnormal body weight resulting from the accumulation of extra adipose tissue due to a state of positive energy balance characterized by diminished energy expenditure when compared to the energy intake [126]. A systematic exploration involving several imidazolindinone (**a**), imidazolone (**b**), indoline (**c**), and thiazole (**d,e**) derivatives in the stimulation of adenyl cyclase revealed the advantage of the thiazole derivatives (**e**) over the rest of compounds as β_3 agonists (Fig. 15) [124, 125, 127, 128].

In this context, Hanumantharao et al. [129] have carried out a QSAR analysis of the β_3 agonist activity of the thiazole derivatives (**e**) with the molecular descriptors generated from the MOE programme [71]. This has resulted in QSAR models (Eq. 23 and 24) of these compounds in terms of hydrophobicity of VdW surface areas with different polarities ($S \log P$, $S \log P_{\text{VSA0}}$ and $S \log P_{\text{VSA8}}$), electrostatic potential energy (E_{ele}), and zero-order connectivity

Table 13 Physicochemical/electronic properties and β_3-agonist activity of thiazole derivatives (comp. **e**, Fig. 15) [129]

No.	n	R	E_{ele}	$S\log P$	$S\log P_{VSA0}$	$S\log P_{VSA8}$	pEC_{50}[a] obsd	cald[b]	Cald[c]
1	0	C_8H_{17}	− 0.04	6.87	43.40	119.68	− 1.00	− 0.92	− 0.86
2	0	$CH_3CONH(CH_2)_5$	− 0.04	4.89	61.41	63.07	− 1.70	− 1.42	− 1.78
3	0	C_2H_4Ph	− 0.05	5.75	43.40	6.47	− 1.52	− 1.69	− 1.46
4	0	CH_2-naphthyl	− 0.04	6.71	43.40	6.47	− 0.70	− 0.67	− 0.65
5	0	2-benzofuranyl	− 0.04	6.38	43.40	6.47	− 0.96	− 0.88	− 0.75
6	0	3-indolyl	− 0.04	6.12	43.40	6.47	− 0.69	− 0.81	− 0.86
7	0	3-pyridinyl	− 0.04	5.03	43.40	6.47	− 1.60	− 1.26	− 1.44
8	0	2-pyridinyl	− 0.04	5.03	43.40	6.47	− 1.34	− 1.41	− 1.52
9	0	3,4,diF-Ph	− 0.04	5.91	43.40	6.47	− 0.85	− 0.91	− 1.01
10	0	$PhcC_6H_{11}$	− 0.04	7.76	43.40	81.94	− 0.08	− 0.22	− 0.23
11	0	Ph-tBu	− 0.04	6.93	43.40	6.47	− 0.38	− 0.40	− 0.40
12	0	4-OH-Ph	− 0.04	5.34	68.78	6.47	− 1.65	− 1.91	− 1.34
13	0	4-AcNH-Ph	− 0.05	5.59	43.40	6.47	− 1.65	− 1.54	− 1.71
14	0	$4F$-$PhCH_2$	− 0.04	5.70	43.40	6.47	− 0.79	− 0.96	− 1.13
15	1	C_8H_{17}	− 0.04	6.58	61.41	138.55	− 0.99	− 1.10	− 0.99
16	1	CH_2-naphthyl	− 0.04	6.22	61.41	6.47	− 1.08	− 0.90	− 0.86

[a] β_3-agonist activity
[b] Eq. 23
[c] Eq. 24

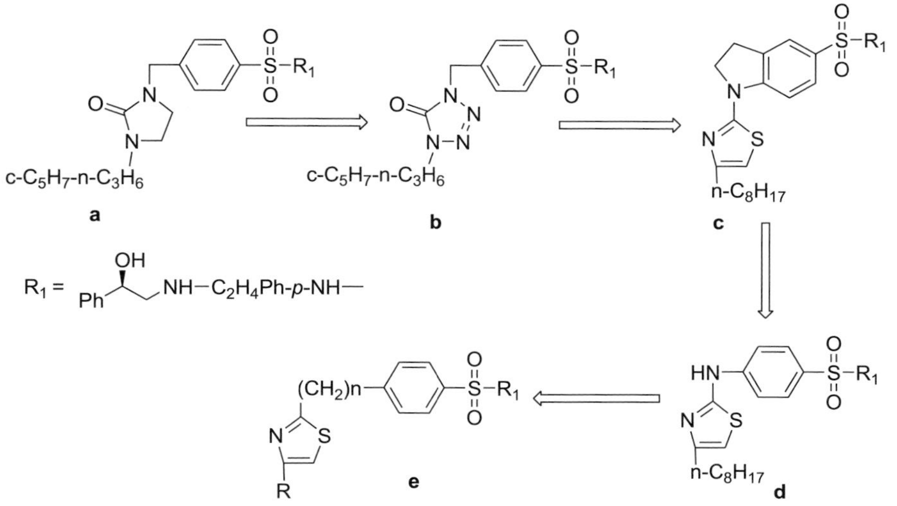

Fig. 15 Development of thiazoles (**d, e**) as β-adrenergic receptor (β_3) agonists from their predecessors (**a–c**)

index (chi0) (Table 13).

$$pEC_{50} = 0.266(0.039)chi0 + 88.998(14.954)E_{ele}$$
$$- 0.030(0.006)S\log P_{VSA0} - 3.880$$
$$n = 16, \quad r = 0.936, \quad Q^2 = 0.712, \quad s = 0.191, \quad F = 28.30 \tag{23}$$

$$pEC_{50} = 45.869(15.591)E_{ele} + 0.558(0.071)S\log P$$
$$- 0.003(0.001)S\log P_{VSA8} - 2.526$$
$$n = 16, \quad r = 0.930, \quad Q^2 = 0.798, \quad s = 0.200, \quad F = 25.36 . \tag{24}$$

These equations have led them to suggest that molecules with a high electrostatic potential energy and that are lipophilic in nature would be favorable for β_3-agonist activity.

3.2.2
Neuropeptide Y5 Receptor Antagonists

Other then the β_3 adrenergic receptor agonistic activity, some thiazole derivatives have been reported to show antiobesity activity through the neuropeptide Y5 receptor (NPY5R) [130, 131]. The antagonists of the NPY5 receptor (NPY5R) are known to decrease the food intake in animal models [132]. A virtual screening of a large database with the pharmacophore model (Fig. 16) developed from the hydrophobic and topological similarity of three

seed structures (compounds **a–c**, Fig. 16) has paved the way for the discovery
of some thiazoles as NPY5R antagonists [130, 133–135].

From this virtual screening, 31 compounds (compound **e**, Fig. 16) have
been identified to show the activity at $IC_{50} < 10\,\mu M$. Among these a thiazole
analogue, 2-(3-benzonitrile)amino-5-benzoyl-thiazole (compound **f**, Fig. 16),
has been found to be the most active compound with an IC_{50} of 40 nM for
the mouse Y5 receptor and even showed activity in a mice feeding model at

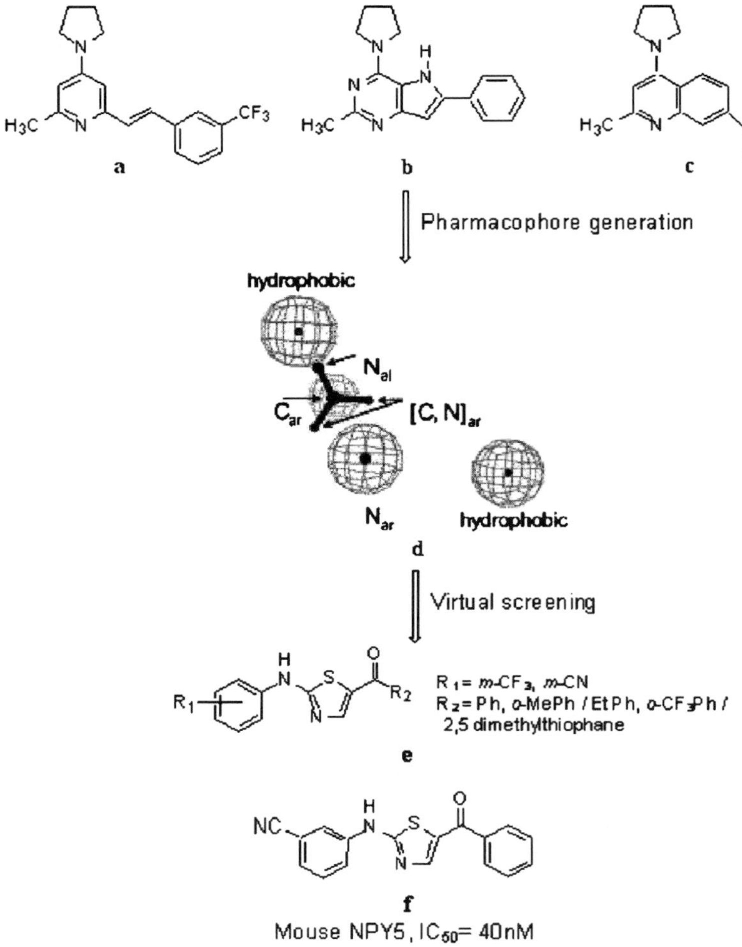

Fig. 16 Strategy adopted in the identification of thiazoles (**e**) as potential NPY5 receptor
antagonists from the virtual screening of 3D-pharmacophore (**d**) (Reprinted with permission from [130]. Copyright 2005 Elsevier Ltd.) developed from three dissimilar seed
structures (compounds **a–c**). From of this strategy compound **f** has been identified and
found to show mouse NPY5 receptor inhibition (IC_{50}) at 40 nM

10 mg/kg via intra peritoneal root [130]. The consensus between the pharmacophore patterns of the compounds **f** and **a** have been highlighted in Fig. 17. In a further SAR study, Guba et al. discovered that electron-withdrawing substituents on aryl moieties would be better for the activity. The position of choice for the substituent groups on the aryl moieties (at 2- and 5- of the thiazole), have been found to be *meta* and *ortho*, respectively [130].

In continuation of the search for more active ligands for the NPY5 receptor, Nettekoven, Guba and co-workers [131] observed that thiazole **g** derivatives showed 10 times more affinity than that of thiazole **h** derivatives for the mouse receptor (Fig. 18). This difference in the activity profile of the thiazole isomers **g** and **h** has been explained through the conformational analysis of core scaffolds (Fig. 18).

The thiazole isomers **g** and **h** differed in their minimum energy conformations (Fig. 18) due to different torsional angles between the carbonyl function and the heterocycle system. In the thiazole **g** isomer a favorable short distance interaction between the carbonyl oxygen and the thiazole sulfur ($C = O - S$ type of interaction), below the sum of the van der Waals radii (< 3.3 Å), has led to an energetically highly preferred conformation. This has found support from Iwaoka et al. [136] simulation studies on comparable molecular systems. However, in the thiazole **h** isomer the repulsive interaction between the carbonyl oxygen and the thiazole nitrogen has resulted in a high energy bent conformation. The potential energy difference between these conformer states has been found to be of the order of 10 to 7 kJ/mol. This is taken as the reason for the 10-fold increment in the activity profile of the thiazole **g** iso-

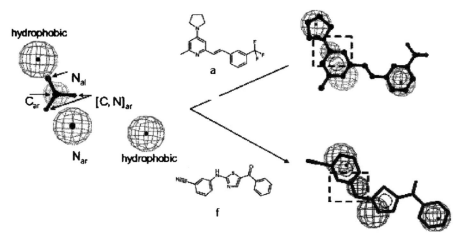

Fig. 17 Mapping of the 3D pharmacophore hypothesis onto compounds **a** and **f** (Fig. 16). The trigonal substitution pattern around C_{ar} is allowed to rotate yielding a novel scaffold in the 3D virtual screening campaign (the match is highlighted by the *dashed rectangle*). (Reprinted with permission from [130]. Copyright 2005 Elsevier Ltd.)

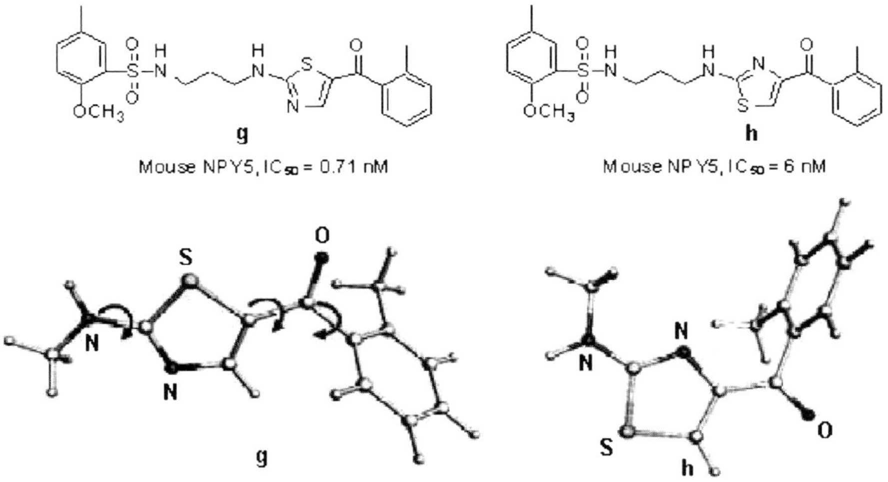

g

Mouse NPY5, IC_{50} = 0.71 nM

h

Mouse NPY5, IC_{50} = 6 nM

g

h

Fig. 18 Conformations corresponding to the minimum energies for the thiazole isomers **g** and **h**. The *arrows* represent the torsional angles sampled in both isomers. (Reprinted with permission from [131]. Copyright 2005 Elsevier Ltd.)

mer when compared to that of isomer **h**. This study has clearly demonstrated a successful structure-activity investigation involving a lead identification from the virtual screening of totally unrelated seed compounds followed by modification of the generated lead through SAR and conformational studies that finally emerged with far more active compounds compared to the initial leads.

3.2.3
PPAR Agonists

In addition to their affinity to β-adrenergic and neuropeptide Y5 receptors, thiazolidines are also known to act on peroxisome proliferator-activated receptors (PPARs). In the biological system PPARs regulate the expression of genes involved in the fatty acid and carbohydrate metabolism and in adipocyte differentiation [137]. Among the PPARs, the PPARγ subtype has received wide attention as its agonists specifically promote insulin action and glucose utilization in the peripheral tissues. Thiazolidinediones (TZDs) are among the early chemical classes reported as the agonists of PPARγ with high affinity [138, 139]. The SAR studies with the TZD analogues have suggested that an acidic head group with a central spacer group followed by a linear lipophilic tail are the primary structural requirements of PPARγ agonists. The desired linker fragment between the head group (2,4-thiazolidinedione) and the central spacer group has been suggested to be the methylene moiety. Any deviation in the linker fragment length and α,β-unsaturation in thiazolidinediones have been reported to reduce the PPARγ activity. For the lipophilic tail a variety of aryl and heteroaryl groups such as, pyridyl (**a**), oxazoyl (**b, c**), benzoxazoyls (**d**) have been found to be tolerated with the activity [138, 139, 141–143]. Different alkyl ether, alkyl ketones (**c**), olefin and some cyclic moieties have been investigated as linkers between the central spacer group and the lipophilic tail (Fig. 19). The PPARγ agonistic response of these compounds has been found to tolerate all these structural variations [140, 144].

The TDZs contain a chiral center at C-5 of the heterocyclic head group. The binding assays have indicated that the (*S*)-enantiomer of the TZD binds to the PPARγ with high affinity [145]. However, under physiological conditions it has been observed to undergo racemization [146]. These studies have provided gross structural requirements of TDZ and related ligands for the PPARγ agonistic activity. The X-ray studies of the co-crystals of TDZ analogues with different PPARs have indicated that PPARγ's ligand-binding domain (LBD) has an overall similarity with other nuclear receptors [147].

In these X-ray studies, it has been observed that the TDZ moiety binds in helices 3, 4, and 10 and to the activating function-2 by making hydrogen bonds with groups in the side chains of His-449, Tyr-473, His-323, and Ser-289. In the Y-shaped cavity, the agonist's lipophilic tail group has been found to occupy the downward position of "Y" by interacting with amino acid residues Met-348, Ile-341, and Ile-281 (Fig. 20) [148, 149]. Also, Yanagisawa and co-workers molecular modelling studies with the oximes having 5-

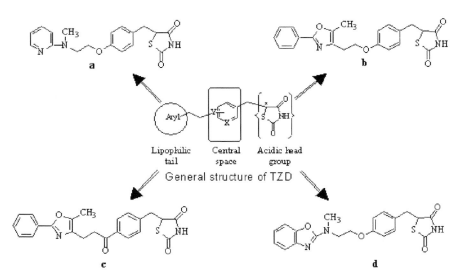

Fig. 19 Primary structural requirements of thiazolidinediones (TZDs) for PPARγ agonistic activity

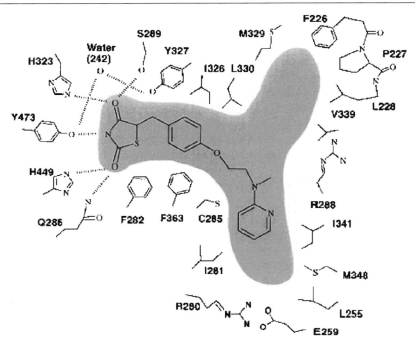

Fig. 20 Ligand (Rosiglitazone)-enzyme (PPARγ) crystal structure where the ligand is shown in the binding domain of PPARγ (Reprinted with permission from [151]. Copyright 2001 Elsevier Science Inc.)

benzyl-2,4-thiazolidinedione moieties (Fig. 21) [150] as agonists of the PPARs has shown that they occupy a similar binding domain of PPARγ [151]. All these studies have led to the suggestion that the PPARγ LBD comprises of 13 α-helices and a small four-stranded β-sheet. The helices 3, 7, and 10 have been found to form a Y-shaped cavity to accommodate the ligand scaffolds (Fig. 20).

More recent studies have confirmed that dual activation of PPARγ and PPARα subtypes would be beneficial for treating type-2 diabetes [139]. In this connection KRP-297, a TZD derivative has been found to express in vitro as well as in vivo affinity for both the PPARα and PPARγ subtypes [152]. In light of this, Desai and co-workers have investigated the PPARα/γ dual agonistic activity of a series of TZD and oxazolidinedione derivatives (comp. **a**, Fig. 22, Table 14) to assess their antidiabetic activity [153, 154].

Subsequently, Khanna et al. carried out the CoMFA and docking studies of these TZDs and oxazolidinediones in conjunction with their PPARα/γ dual activity [155]. The common CoMFA model featuring the PPARα and PPARγ binding sites has been found to best explain the affinities of these ligands for the receptors (number of significant components = 6; cross-validated

Table 14 Thiazolidinedione and oxazolidinedione derivatives in modeling (CoMFA) PPAR-α, γ and dual (d) binding affinity (comp. **a**, Fig. 22) [155]

No.	X	R_1	pIC_{50}[a] (α) obsd	perd	(γ) obsd	perd	(d) obsd	perd
1	S	OPh	7.55	7.47	7.24	7.37	14.79	14.79
2[b]	S	OPh	7.33	–	7.12	–	14.45	
3[c]	S	OPh	5.30	5.11	6.71	6.78	12.01	12.01
4	S	OPh-4-CH(Me)$_2$	5.68	5.76	6.77	6.72	12.45	12.45
5	S	OPhC(Me)$_3$	5.0	5.03	6.47	6.49	11.47	11.47
6	S	OPh-4-CH$_2$CH(Me)$_2$	5.0	5.05	6.54	6.50	11.54	11.54
7	S	OPh-4-cC$_5$H$_9$	5.70	–	6.48	–	12.18	–
8	S	OPh-Ph	5.0	5.07	6.65	6.54	11.65	11.65
9	S	OPh-4Cl	7.0	–	7.14	–	14.14	–
10	S	OPh-4F	7.55	7.25	7.11	7.15	14.66	14.66
11	S	OPh-4OMe	5.59	5.42	6.52	6.75	12.11	12.11
12	S	OPh-4OH	6.02	6.43	7.52	7.22	13.54	13.54
13	S	OPh-3,4-di-Cl$_2$	7.17	–	7.19	–	14.36	–
14	S	OPh-3-Me,4-Cl	6.79	6.92	7.25	7.13	14.04	14.04
15	S	OPh-3-Me,4-F	6.95	7.06	7.11	7.22	14.06	14.06
16	S	OcC$_6$H$_{11}$	5.85	6.10	7.43	7.27	13.28	13.28
17	S	OcC$_7$H$_{13}$	6.68	6.93	7.07	7.18	13.75	13.75
18	S	OcC$_5$H$_9$	7.38	7.09	7.20	7.26	14.58	14.58
19	S	COPh	7.05	7.04	7.72	7.66	14.77	14.77
20	S	COPh-4Cl	5.0	–	6.84	–	11.84	–
21	S	3-benzoisoxazyl	7.43	7.57	7.52	7.56	14.95	14.95
22	S	COcC$_6$H$_{11}$	6.92	6.93	7.19	7.10	14.11	14.11
23	S	CH = cC$_6$H$_{10}$	7.21	7.13	6.82	6.94	14.03	14.03
24	S	CH$_2$cC$_6$H$_{11}$	7.34	7.31	7.23	7.14	14.57	14.57
25	S	cC$_6$H$_{11}$	5.30	–	6.47	–	11.77	–
26	O	cC$_6$H$_{11}$	6.54	6.36	6.31	6.17	12.85	12.85
27	O	Ph	5.57	5.78	6.35	6.38	11.92	11.92
28	O	cC$_7$H$_{13}$	6.21	6.35	6.19	6.12	12.40	12.40
29	O	cC$_5$H$_9$	6.20	–	5.97	–	12.17	–
30	O	4-cyclohexanone	4.82	4.67	5.94	5.86	10.76	10.76
31	O	4-cyclohexanol	4.82	4.85	6.74	6.88	11.56	11.56
32	O	1,1-difluoro-4-cyclohexane	6.23	6.01	6.27	6.35	12.50	12.50
33	O	1,1-dimethyl-4-cyclohexane	5.48	5.33	5.75	5.80	11.23	11.23
34	O	4-tetrahydro-pyran	5.54	5.72	6.12	6.18	11.66	11.66

[a] PPAR α/γ and dual (d) binding affinities
[b] Without propyl group
[c] $O - (CH_2)_4 - O$ in place of $O - (CH_2)_3 - O$

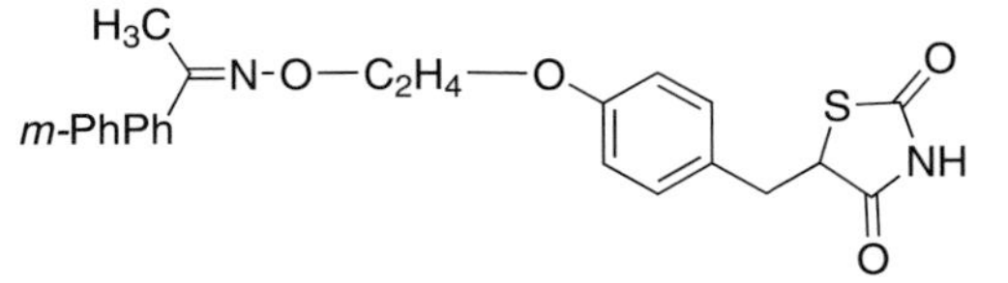

Fig. 21 2,4-thiazolidinedione carrying an oxime moiety as a PPARγ agonist

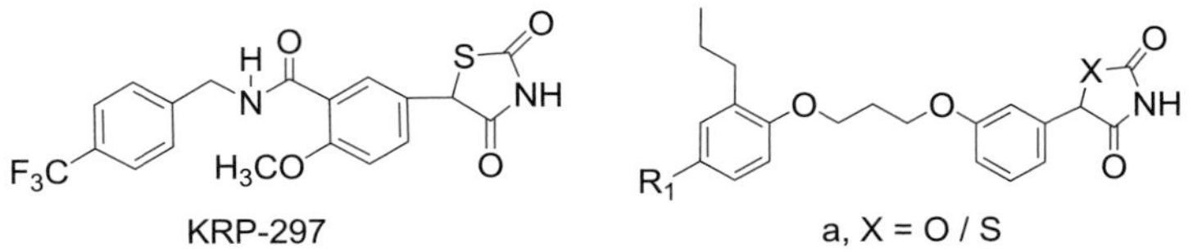

Fig. 22 Structures of KRP-297 and thiazolidinedione/oxazolidinedione derivatives associated with PPARα/γ dual agonistic activity

r^2 (r^2_{cv}) = 0.782; non cross-validated r^2 (r^2_{ncv}) = 0.985). In this, the steric and electrostatic field contributions have been found to be 0.535 and 0.465 parts respectively (comp. **a**, Fig. 22 Table 14). Apart from this, for these compounds by making use of the protein sequences of PPARγ and PPARα molecular docking studies have also been carried out in FlexX. An earlier X-ray study of PPARα and PPARγ has suggested a common Y-shaped binding pocket for both the receptors with a difference in their amino acid content that is His323 in PPARα being replaced by Tyr314 in PPARγ [156]. Against this background, it has been observed that in both the receptors, the compounds (ligands) have been found to assume a U-shape and bind to the enzyme through forming strong hydrogen bonds by involving the ligand's acidic fragment (thiazolidinedione moiety) and a set of residues (Tyr314, Tyr464, His440, and Ser280 in PPARα, and His323, Tyr473, His449, and Ser289 in PPARγ) in the active site. In light of this, the variations in the binding affinities of these compounds for the PPARα/γ-receptors have been explained by using these differences. These studies exemplify the scope of the design of ligands from targeting "a single receptor" to "multiple receptors".

3.2.4
Aldose Reductase Inhibitors

Some typical structural templates embedded with the thiazolidine frame have been reported as potent inhibitors of *aldose reductase* (AR), an enzyme in the polyol pathway responsible for the conversion of glucose to sorbitol. In this, the accumulation of sorbitol has been attributed to causing cataracts, neuropathy, and retinopathy in diabetic cases [157, 158]. The planar hydrophobic (aromatic) regions and propensity to charge transfer interactions have been

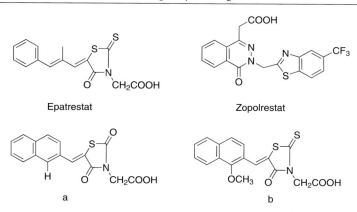

Fig. 23 Structures of epatrestat, zopolrestat, 2,4-dioxo-5-(naphthylmethylene)-3-thiazol-idineacetic acid (**a**) and its 2-thioxo analogue (**b**) as *Aldose Reductase* inhibitors

identified as some known structural requirements of the AR inhibitors [159]. Considering these features together with epatrestat (Fig. 23), a potent inhibitor of AR with a thiazolidine frame as part of its scaffold, Fresneau et al. have come up with new compounds, 2,4-dioxo-5-(naphthylmethylene)-3-thiazolidineacetic acid (compound **a**, Fig. 23) and its 2-thioxo analogue (compound **b**, Fig. 23) as AR inhibitors [160].

In a pursuit to understand the binding modes of these inhibitors with AR, a molecular modeling study has been carried out with compounds (**a**) and (**b**) and these results have been compared with the crystal structure of the complex of AR and Zopolrestat, a known AR inhibitor (Fig. 23) [160, 161]. This study has specifically identified three sites of interaction on the enzyme; site 1, an anionic pocket to accommodate a negative charge, such as a carboxylic group at the OH of Tyr48; site 2, a hydrophobic binding pocket with amino acids Trp111, Leu300, and Trp219 to accommodate aromatic moieties; and site 3 a hydrogen bonding site with SH of Cys298. This has led to the suggestion that a strong interaction at the identified sites would lead to better inhibitory activity for the compounds [160]. This study has illustrated an example of discovering novel compounds starting from the existing drugs and high active compounds.

3.2.5
Thromboxane A2 Receptor Antagonists

The emergence of thiazolidine derivatives as a structure class for platelet aggregation inhibitory activity has its roots in a marine natural product, D-cysteinolic acid (**a**) [162, 163]. The medicinal chemistry and modeling strategies have navigated D-cysteinolic acid (**a**) through various modifications to culminate in the thiazolidine analogues with different activity profiles (Fig. 24).

Among these compounds, D-cysteinolic acid (**a**) has exhibited mild platelet aggregation inhibitory activity and 2-(4-hydoxy-3-methoxyphenyl)-thiazolidine (**b**), a synthetically transformed compound of this natural product, has displayed an elevated activity, thus ushering thiazolidines into this pharmacological domain [162, 163]. In this, the elevated activity may have resulted from the restriction in the spatial disposition of nitrogen and sulfur when compared to (**a**). Also this has clearly shown the necessity of mutual proximity of these heteroatoms in the molecular skeleton for the activity. From Fig. 24, compounds **c** (3-benzoyl-2-(4-hydroxy-3-methoxyphenyl)thiazolidine), **d**, and **e** have been evaluated for their thromboxane A2 (TxA2) receptor antagonistic activity [164–166]. Pharmacologically thromboxane synthatase has a key role in the platelet aggregation and TxA2 is its potent stimulator [167]. The antagonists of TxA2 have an important place in the treatment of various circulatory disorders [168]. Also, these compounds are known to compete with the TxA2 for the same receptor site. Against this background some conformational mapping experiments between sulotroban and *o*-, *m*- and *p*-oxyacetic acid analogues of compound

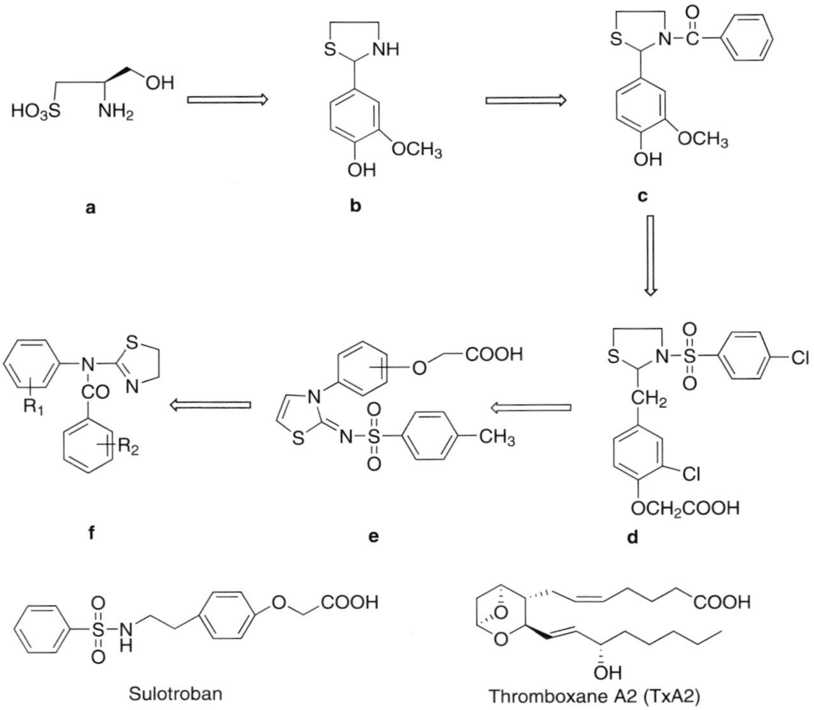

Fig. 24 D-Cysteinolic acid (**a**), sulotroban and thromboxane A2 (TxA2), in the development of different thiazoles (**b–f**) as TxA2 antagonists

e (Fig. 24) have been carried out to explore the spatial relationship between TxA2, sulotroban and the analogues of compound (e) (Fig. 24) [169]. Sulotroban (Fig. 24) is known to bind to the TxA2 receptor site in place of TxA2. For this study, in conformity with the active conformation of TxA2, the geometry of sulotroban has been considered in the form of a folded conformation [170]. This study has led Lacan et al. to suggest that the *m*-oxyacetic acid analogue of compound (e) maps the sulotroban structure space best

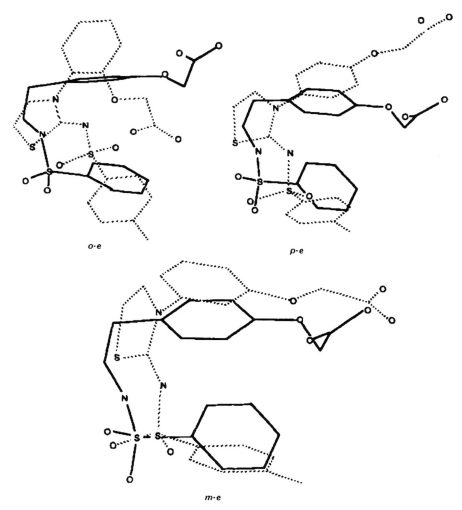

Fig. 25 Superimposition of selected minimum energy conformers of *o*-, *m*-, and *p*-phenoxyacetic acid analogues of dihydrothiazoles (*o*-e, *p*-e, and *m*-e, shown in *dotted lines*) on the folded conformation of sulotroban (shown in *solid lines*) (Reprinted with permission from [169], Copyright 1999 Elsevier Paris.)

(Fig. 25) [169, 171]. This has signified the importance of the spatial disposition of the functional groups of these thiazolidines in deciding their TxA2 antagonistic activity.

Some 2-disubstituted amino-4,5-dihydrothiazoles (compound **f**, Fig. 24) have been reported to show in vivo antithrombotic activity. A QSAR study of these thiazoles (compound **f**, Fig. 24, Table 15) with the physicochemical parameters namely Hammett's sigma (σ) constant, dipole moment (μ) and Swain and Lupton's polarity ($\mathcal{F}$) constants of R_1 and R_2 substituent groups

Table 15 2-(*N*-aryl-*N*-aroyl)amino-4,5-dihydrothiazole derivatives in modeling the antithrombotic activity (comp. **f**, Fig. 23) [172]

No.	R_1	R_2	$\log(P/100-P)$ [a] obsd	cald [b]	cald [c]
1	2-CF$_3$	H	− 0.37	− 0.37	− 0.35
2	4-F	H	− 0.22	− 0.36	− 0.34
3	2,5(OMe)$_2$	H	− 0.37	− 0.33	− 0.32
4	2-Cl,6-Me	H	− 0.37	− 0.37	− 0.36
5	2-Me	H	− 0.60	− 0.50	− 0.46
6	2-CF$_3$	4-OMe	− 0.37	− 0.25	− 0.26
7	2-Me	4-OMe	− 0.37	− 0.38	− 0.36
8	2,5(OMe)$_2$	4-OMe	− 0.18	− 0.21	− 0.22
9	4-F	4-OMe	− 0.18	− 0.29	− 0.24
10	2-Cl,6-Me	2-Me	− 0.48	− 0.39	− 0.39
11	2-CF$_3$	2-Me	− 0.22	− 0.39	− 0.38
12	4-F	2-Me	− 0.37	− 0.37	− 0.37
13	4-OMe	2-Me	− 0.48	− 0.43	− 0.41
14	4-Et	4-Cl	− 0.18	− 0.14	− 0.34
15	2-Cl,6-Me	4-Cl	− 0.30	− 0.17	− 0.24
16	4-F	4-Cl	− 0.30	− 0.19	− 0.22
17	2-Me	2-Br	− 0.37	− 0.30	− 0.34
18	H	2-Br	− 0.18	− 0.29	− 0.33
19	2-Cl,6-Me	2-Br	− 0.22	− 0.18	− 0.24
20	2,5(OMe)$_2$	4-NO$_2$	− 0.12	− 0.03	− 0.01
21	4-NO$_2$	4-NO$_2$	0.00	0.02	0.03
22	4-F	4-NO$_2$	− 0.18	− 0.06	− 0.03
23	2-Me	4-NO$_2$	− 0.14	− 0.20	− 0.15
24	2-OMe	4-NO$_2$	0.00	− 0.11	− 0.08
25	4-Cl,3-NO$_2$	4-NO$_2$	0.26	0.14	0.13

[a] P is percent protection offered, in mice, to thrombotic challenge in comparison to inodmethacin
[b] Eq. 25
[c] Eq. 26

has resulted in the following equations [172].

$$\log(P/100 - P) = 0.301(0.076)\mathscr{F} R_1 + 0.447(0.072)\mathscr{F} R_2 - 0.485$$
$$n = 25, \quad r = 0.855, \quad s = 0.097, \quad F = 29.88 \tag{25}$$

$$\log(P/100 - P) = 0.254(0.079)\mathscr{F} R_1 - 0.074(0.012)\mu R_2 - 0.449$$
$$n = 25, \quad r = 0.850, \quad s = 0.099, \quad F = 28.56 . \tag{26}$$

On the basis of these results it has been suggested that electronic factors—Hammett's sigma (σ) constant, dipole moment (μ) and Swain and Lupton's polarity ($\mathscr{F}$)—are crucial for the antithrombotic activity of these compounds. Interestingly, in all the active compounds presented in Fig. 24 the arrangement of substituents around the thiazole skeleton is different from one another. This has further widened the scope of this skeleton in the exploration of potential platelet aggregation inhibitors and thromboxane A2 receptor antagonist.

3.3
Anti-infective Agents

3.3.1
Anti-HIV Agents

Thiazolidines are also known to selectively intercept the biochemical processes of the pathogenic organisms such as bacteria, virus, and fungi, in the host species. Thiazolidines introduction into the HIV/AIDS (Human Immunodeficiency Virus/Acquired Immune Deficiency Syndrome) therapy is due to 1-aryl-1H,3H-thiazolo[3,4-a]benzimidazole (TBZ) analogues (Fig. 26). Barreca et al. have modified the TBZ by opening its "B" ring to explore the thiazolidines namely 2-(2,6-dihalo phenyl)-3-(substituted pyridin-2-yl)-thiazolidin-4-ones (Table 16) as potential anti-HIV-1 RT agents [173].

In HIV type 1, the activity of all these compounds is due to inhibition of a key viral enzyme, reverse transcriptase (RT), necessary for the catalytic transformation of single-stranded viral RNA into the double-stranded linear DNA for incorporation into the host cell chromosomes [174]. The inhibitors of HIV-1 RT fall into two main classes. The first are termed nucleoside reverse transcriptase inhibitors (NRTIs). They mimic normal substrates of RT but lack the 3-'OH group required for DNA chain elongation, and cause premature termination of the growing viral DNA strand [175]. The second are termed non-nucleoside RT inhibitors (NNRTIs) and thiazolidine analogues belong to this class. They bind allosterically to a hydrophobic pocket close to the active site and achieve highly selective suppression of HIV-1 replication with little cytotoxicity to the host cells [176, 177]. Because of this considerable research efforts have been focused on the synthesis and structure-activity relationships of large numbers of different compounds as NNRTIs [178]. The

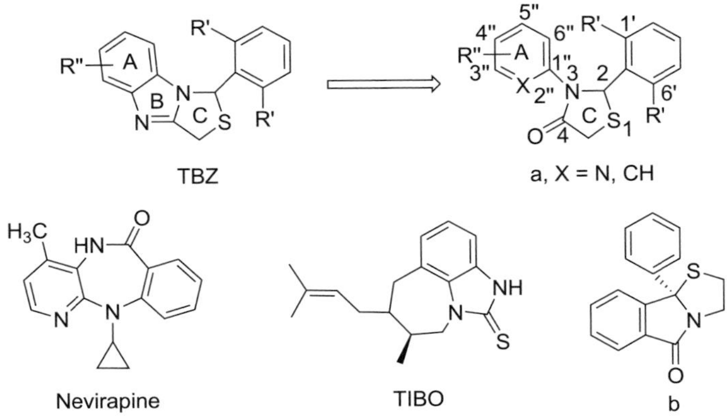

Fig. 26 Structures of TBZ, nevirapine, TIBO, (S)-(–)9b-phenyl-2,3-dihydrothiazolo[2,3-3]isoindol-5-(9bH)-one (**b**) and thiazolidin-4-ones associated with HIV-1 RT inhibitory activity

X-ray crystallographic studies of diverse NNRTIs/RT complexes have indicated that the NNRTIs, irrespective of their structure class, uniformly assume a "butterfly-like" shape in the receptor (Fig. 27) [179].

Barreca et al.'s modeling studies with 2-(2,6-dihalo phenyl)-3-(substituted pyridin-2-yl)-thiazolidin-4-ones (Fig. 26, Table 16) [173] in the GRID and AutoDock [180] programs in conjunction with nevirapine (Fig. 26) have suggested that (a) the binding mode of these thiazolidine analogues is similar to that of TBZ and other NNRTIs [173]; and (b) the aliphatic and aromatic binding regions of nevirapine match with the 2,3-diaryl moieties of thiazolidin-4-ones scaffold.

The QSAR of these compounds has been analyzed with the CP-MLR [51] procedure with physicochemical substituent constants [41] and several indicator parameters signifying their sub-structural features [181]. In this, the HIV-1 RT inhibitory activity of the compounds listed in Table 16 has been found to be correlated with the $CLOGP$, Swain and Lupton's resonance constant of the 4″ position substituent ($\mathcal{R}_{4''}$) and some indicator parameters as shown.

$$-\log EC_{50} = 0.415(0.135)CLOGP - 1.414(0.136)I_{4''} + 5.377$$
$$n = 23, \quad r = 0.921, \quad Q^2 = 0.792, \quad s = 0.310, \quad F = 55.83 \tag{27}$$

$$-\log EC_{50} = 0.404(0.145)I_{2,3} - 1.327(0.141)I_{4''} + 6.668$$
$$n = 23, \quad r = 0.916, \quad Q^2 = 0.787, \quad s = 0.319, \quad F = 52.04 \tag{28}$$

$$-\log EC_{50} = 0.439(0.141)CLOGP + 9.141(0.925)\mathcal{R}_{4''} + 5.280$$
$$n = 23, \quad r = 0.914, \quad Q^2 = 0.775, \quad s = 0.323, \quad F = 50.67 . \tag{29}$$

Table 16 2-(2,6-Dihalophenyl)-3-(substituted pyridin-2-yl)-thiazolidin-4-ones analogues in modeling the HIV-1 RT inhibitory activity (comp. **a**, X = N, Fig. 26) [181]

| No. | R'' | | | | R' | | $-\log EC_{50}$ [a] | |
	6''	3''	5''	4''	2'	6'	obsd	cald[b]
1	H	H	H	H	Cl	Cl	6.75	6.89
2	H	H	H	H	Cl	F	6.55	6.70
3	H	H	H	H	F	F	6.07	6.53
4	H	H	H	Cl	Cl	Cl	5.75	5.77
5	H	H	H	Cl	Cl	F	5.67	5.56
6	H	H	H	Cl	F	F	5.14	5.41
7	H	H	H	Br	Cl	Cl	5.82	5.83
8	H	H	H	Br	Cl	F	5.91	5.64
9	H	H	H	Br	F	F	5.31	5.47
10	H	Br	H	H	Cl	Cl	6.56	7.24
11	H	Br	H	H	Cl	F	7.19	7.06
12	H	Br	H	H	F	F	7.52	6.88
13	Me	H	H	H	Cl	Cl	–	
14[c]	Me	H	H	H	F	F	4.27	
15	H	H	Me	H	Cl	Cl	6.83	7.12
16	H	H	Me	H	Cl	F	7.00	6.94
17	H	H	Me	H	F	F	6.60	6.76
18	H	H	H	Me	Cl	Cl	5.85	5.71
19	H	H	H	Me	F	F	5.29	5.34
20	H	Me	H	H	Cl	Cl	7.36	7.12
21	H	Me	H	H	Cl	F	7.27	6.93
22	H	Me	H	H	F	F	7.08	6.76
23	H	Me	Me	H	Cl	Cl	7.04	7.35
24	H	Me	Me	H	Cl	F	7.38	7.17
25	H	Me	Me	H	F	F	7.23	6.99

[a] 50% Inhibitory concentration against cytopathic effect of HIV-1 in MT-4 cells
[b] Eq. 27
[c] Not included in the analysis

In these equations the indicator variable $I_{4''}$ has been defined for the 4''-substituent of the 3-(pyridin-2-yl) moiety (zero for hydrogen and one otherwise) of thiazolidinones (Fig. 26). Also, in these compounds the HIV-1 RT inhibitory activity has been linked to their ability to assume a butterfly-like shape in the receptor. Collectively, the substituents of 2- and 3-aryl moieties of these compounds promote this molecular shape. In this connection, the indicator parameter $I_{2,3}$ has been introduced to address the collective steric features of the 2- and 3-aryl moieties of these compounds. In this, $I_{2,3}$ has been defined to take a value of "one" if any one substituent at 2'- or 6'- of the 2-phenyl moiety or at the 3''- or 5''- positions of the 3-(pyridin-2yl) moiety is larger than hydrogen and "zero" otherwise. These equations, while favor-

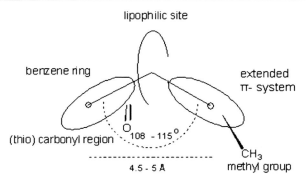

Fig. 27 The "butterfly like model" derived from the common 3D superimposition of 4,5,6,7-tetrahydro-5-methylimidazo[4,5,1-jk][1,4]benzodiazepin-2(1H)-ones (**TIBO**), Nevirapine and (S)-(–)9b-phenyl-2,3-dihydrothiazolo[2,3-3]isoindol-5-(9bH)-one (compound **b**, Fig. 25) (Reprinted with permission from [179]. Copyright 1993 American Chemical Society)

ing hydrophobic compounds, have clearly suggested that an unsubstituted 4″-position or a 4″-position substituent with a positive resonance contribution would be preferred for the HIV-1 RT inhibitory activity. The positive regression coefficient of $I_{2,3}$ has suggested the preference for substitution at one or more positions of the 2′, 6′ of 2-phenyl and/or 3″, 5″ positions of the 3-(pyridin-2yl) moiety, hence favoring the butterfly-like or an out of plane arrangement for the 2- and 3-aryl moieties. In these compounds a dual role has been assigned to the 3″-, 5″-position substituents of the 3-(pyridin-2yl) moiety (Fig. 26) in terms of providing a driving force to move the 3-pyridyl moiety to an out of plane arrangement with respect to the thiazolidinone nucleus and to fulfil hydrophobic and steric requirements at the receptor sites. The following regression equations have been derived to explain the HIV-1 RT inhibitory activity of 2-(2,6-dihalophenyl)-3-(substituted phenyl)-thiazolidin-4-ones (Fig. 26, Table 17) alone (Eq. 30) and in combination with 2-(2,6-dihalophenyl)-3-(substituted pyridin-2-yl)-thiazolidin-4-ones (Fig. 26 Table 16) (Eqs. 31 and 32) [181].

$$- \log EC_{50} = 0.454(0.100)\pi_{(2'+6')} + 0.168(0.060)MR_{3''}$$
$$- 0.014(0.005)(MR_{3''})^2 - 0.948(0.192)\mathcal{F}_{3''}$$
$$- 0.942(0.212)\mathcal{R}_{(3''+5'')} + 5.655$$
$$n = 28, \quad r = 0.867, \quad Q^2 = 0.620, \quad s = 0.247, \quad F = 13.38 \tag{30}$$

$$- \log EC_{50} = 0.356(0.080)ClogP - 0.732(0.259)\mathcal{R}_{(3''+5'')}$$
$$- 1.306(0.145)I_{4''} + 0.866(0.116)I_{2''} + 4.615$$
$$n = 51, \quad r = 0.861, \quad Q^2 = 0.695, \quad s = 0.319, \quad F = 33.12 \tag{31}$$

Table 17 2-(2,6-Dihalophenyl)-3-(substituted phenyl)-thiazolidin-4-ones analogues in modeling the HIV-1 RT inhibitory activity (comp. **a**, X = CH, Fig. 26) [181]

No.	R″		R′		$-\log EC_{50}$ [a]	
	3′	5′	2′	6′	obsd	cald [b]
26	H	H	Cl	Cl	6.40	6.46
27	H	H	F	F	5.64	5.94
28	Br	H	Cl	Cl	6.21	6.40
29	Br	H	Cl	F	6.19	6.14
30	Br	H	F	F	5.88	5.88
31	Cl	H	Cl	Cl	6.51	6.54
32	Cl	H	Cl	F	6.21	6.28
33	Cl	H	F	F	6.69	6.02
34	Me	H	Cl	Cl	7.09	6.95
35	Me	H	Cl	F	6.67	6.69
36	Me	H	F	F	6.16	6.43
37	Et	H	Cl	Cl	6.81	6.88
38	Et	H	Cl	F	6.67	6.34
39	Et	H	F	F	6.11	6.08
40	MeO	H	Cl	Cl	6.71	6.74
41	MeO	H	Cl	F	6.52	6.71
42 [c]	MeO	H	F	F	4.93	5.65
43	NO₂	H	Cl	Cl	6.14	5.95
44	NO₂	H	Cl	F	5.88	5.71
45	NO₂	H	F	F	5.26	5.45
46	Cl	Cl	Cl	Cl	6.55	6.65
47	Cl	Cl	Cl	F	6.50	6.42
48	Cl	Cl	F	F	6.01	6.17
49	F	F	Cl	Cl	6.89	7.02
50	F	F	Cl	F	6.61	6.42
51	F	F	F	F	6.27	6.16
52	Me	Me	Cl	Cl	6.72	6.72
53	Me	Me	Cl	F	7.06	6.81
54	Me	Me	F	F	6.62	6.55

[a] 50% Inhibitory concentration against cytopathic effect of HIV-1 in MT-4 cells
[b] Eq. 30
[c] Not included in the analysis

$$-\log EC_{50} = 0.363(0.081)ClogP - 0.738(0.264)\mathcal{R}_{(3''+5'')}$$
$$+ 8.424(0.960)\mathcal{R}_{4''} + 0.883(0.118)I_{2''} + 4.585$$
$$n = 51, \quad r = 0.856, \quad Q^2 = 0.684, \quad s = 0.325, \quad F = 31.69 . \tag{32}$$

All these regression equations suggested that the hydrophobic compounds provide better inhibitory activity. Moreover, they have displayed the influence of polar/inductive effects of the $3''$- and $5''$-position substituents on the activity. The positive regression coefficient of $I_{2''}$ ["one" for 3-(pyridine-2-yl) and "zero" for 3-phenyl] in Eqs. 31 and 32 has suggested the advantage of 3-pyridyls over 3-phenyls for better inhibition of HIV-1 RT. A better blend of polarity, electronic, and hydrophobic features in 3-pyridyls, compared to the corresponding phenyl analogue, has been attributed as one reason for these compounds' improved activity. Furthermore, in the PLS analysis of the total compounds shown in Tables 16 and 17, the descriptor $I_{2''}$ has emerged as the single largest contributing descriptor and thereby suggested the importance of the 3-(pyridine-2-yl) moiety when compared to the 3-phenyl moiety in HIV-1 RT inhibition. This has further suggested that the varying centers of this molecule have almost uniform influence on the activity [181].

In addition to the physicochemical domain, the topological features of selected HIV-1 RT inhibitors, 2-(2,6-dihalophenyl)-3-(substituted pyridin-2-yl)-thiazolidin-4-ones (Fig. 26, Table 16), have been analyzed [182] with the empirical, constitutional, and graph theoretical descriptors from the DRAGON software package [47]. This study has resulted in the identification of several influential descriptors to model the inhibitory activity and the equations shown below represent some models stemming from these descriptors.

$$- \log EC_{50} = - 1.874(0.228)AECC - 300.003(30.650)PW3 + 126.245$$
$$n = 24, \quad r = 0.941, \quad Q^2 = 0.855, \quad s = 0.308, \quad F = 80.89 \tag{33}$$

$$- \log EC_{50} = - 263.490(24.496)JGI4 + 8.236(0.923)GGI7 + 21.831$$
$$n = 24, \quad r = 0.944, \quad Q^2 = 0.859, \quad s = 0.301, \quad F = 85.39 \tag{34}$$

$$- \log EC_{50} = 4.462(1.314)MATS3v - 1.037(0.103)GATS8e + 9.191$$
$$n = 24, \quad r = 0.931, \quad Q^2 = 0.824, \quad s = 0.331, \quad F = 68.59 . \tag{35}$$

Equation 33 with AECC (the average eccentricity of the topological graph) and PW3 (Randic's molecular shape descriptor representing the path/walk-3 ratio) represents a model from the simple topological descriptors, Eq. 34 with JGI4 (mean topological charge index of order 4) and GGI7 (topological charge index of order 7) represents a model from the Galvez topological charge descriptors, and Eq. 35 with MATS3v (Moran autocorrelation lag three weighed by atomic volumes) and GATS8e (Geary autocorrelation lag eight weighed by electronegativity) represents a model from the 2D autocorrelation descriptors of the compounds. These equations have prompted us to suggest that less extended or compact saturated structural templates would be better for the activity. Also, the participation of Galvez topological charge descriptors have indicated the importance of fourth and seventh eigenvalues of the polynomial of the corrected adjacency matrix of the compounds' charge indices. The 2D autocorrelation descriptors have suggested the role of three and eight centered structural fragments of the compounds in the activity information. In terms of predictivity these equations have fared better than those involving physicochemical descriptors alone [182]. However, the topological models are more complex in nature when compared to physicochemical descriptors.

Roy and Leonard [183] have also presented QSAR models for the HIV-1 RT inhibitory activity of the thiazolidinones listed in Tables 16 and 17 along with some more similar analogues (Table 18) [184–187] using the hydrophobicity and molar refractivity, quantum chemical and topological and indicator parameters as descriptors. In this, the 3-pyridyls/phenyls (Tables 16 and 17) and 3-(pyrimidin-2-yls) (Table 18) [186] have become part of the dataset. Additionally, four compounds with the thiazolidin-4-thione nucleus have been included in the dataset. In Fujita-Ban [188] and mixed (Hansch and Fujita-Ban) approaches 7 to 17 descriptor models have been discovered for the cytopathicity effect (EC_{50}) and cytotoxic effect (CC_{50}) of the compounds. The following equations show the minimum descriptor models for each activity from this study.

$$
\begin{aligned}
-\log EC_{50} = {} & 0.307(0.090)S_{ssssC} - 0.079(0.056)S\_aasC \\
& + 0.837(0.291)\pi_{3''} - 1.363(0.355)E6 + 2.729(1.074)E10 \\
& - 1.293(0.389)E18 + 6.663(4.395)D1 - 1.339(0.732)I_{N_Z} \\
& - 0.490(0.203)N_{MeRR1} - 0.501(0.508)I_{MeOR3} \\
& - 1.847(0.708)I_{MeR4} - 1.171(0.700)I_{MeR6} \\
& - 2.103(0.435)I_{MeR7} - 19.634(13.581)
\end{aligned}
$$

$$
n = 66, \quad r = 0.911, \quad Q^2 = 0.504, \quad s = 0.334, \quad F = 19.60 \tag{36}
$$

$$
\begin{aligned}
-\log CC_{50} = {} & 0.051(0.048)S\_ssssC - 0.035(0.028)S\_ssO \\
& + 0.066(0.044)S\_sBr + 0.788(0.676)\pi_{2'+6'} \\
& - 0.360(0.379)\pi^2_{2'+6'} + 0.320(0.339)I_{N\text{-}W} \\
& - 0.253(0.114)I_{N\text{-}X} - 0.332(0.126)N_{Me\text{-}2'6'} \\
& - 0.362(0.339)I_{Cl\text{-}R4'} + 1.144
\end{aligned}
$$

$$
n = 83, \quad r = 0.809, \quad Q^2 = 0.544, \quad s = 0.230, \quad F = 13.6 . \tag{37}
$$

In terms of the common structural components, Roy and Leonard's [183] observations have been in good agreement with the previous QSAR and modeling studies [181, 182]. Additionally, they have suggested that the thiazolidin-4-one nucleus is a preferred one when compared to thaizolidin-4-thione for the activity. Also, it has been noted that a methyl group at the 5- and 3'-positions is detrimental to the activity, and both the 3-(pyridin-2-yl) and 3-(pyrimidin-2-yl) moieties are preferred for HIV-1 RT inhibitory activity.

In a follow-up of our modeling studies on thiazolidine-based HIV-1 RT inhibitors, we have synthesized some thiazolidin-4-ones, metathiazanones for this activity. The QSAR studies of these compounds with physicochemical and quantum chemical descriptors have highlighted the importance of PMIZ

Table 18 2,3-Disubstituted-thiazolidin-4-ones in modeling the HIV-1 RT inhibitory activity (comp. a, Fig. 26) [183] [a]

No.	R''					R'		$\log EC_{50}$ [b]		$-\log EC_{50}$ [c]	
	2''	3''	4''	5''	6''	2'	6'	obsd	cald [d]	obsd	cald [e]
1	H	Me	H	H	H	Me	Me	–	–	1.44	1.19
2	Cl	Me	H	H	H	F	CF$_3$	–	–	1.49	1.52
3	Me	Me	H	H	H	Cl	F	2.83	3.25	1.39	1.75
4	H	Me	H	H	H	MeO	MeO	2.72	2.45	0.68	0.54
5	H	H	H	H	H	F	F	–	–	1.45	1.61
6	H	H	H	H	H	Cl	Cl	–	–	1.51	1.70
7 [f]	H	Me	H	H	H	F	MeO	2.94	2.67	0.84	1.13
8	Cl	H	H	Me	H	F	CF$_3$	2.67	2.42	1.03	0.89
9	H	H	H	Me	H	MeO	MeO	2.48	2.72	0.60	0.53
10	H	Me	H	H	H	Me	Me	2.81	2.49	0.55	0.69
11	Cl	Me	H	H	H	F	CF$_3$	2.52	2.84	0.92	0.91
12	Me	Me	H	H	H	Cl	F	4.30	3.75	1.42	1.35
13	H	Me	H	H	H	MeO	MeO	2.87	3.16	0.62	0.49
14	H	Br	H	H	H	Me	Me	2.20	2.69	0.57	0.92
15 [f]	H	Br	H	H	H	F	MeO	4.47	3.62	1.00	0.80
16	Cl	Br	H	H	H	F	CF$_3$	3.18	2.93	1.06	1.13
17	Me	Br	H	H	H	Cl	F	3.46	3.91	1.68	1.52
18	H	Me	H	Me	H	Me	Me	2.36	2.38	–	–
19	H	Me	H	Me	H	F	MeO	3.70	3.39	–	–
20	Cl	Me	H	Me	H	F	CF$_3$	2.23	2.76	0.79	0.90
21	Me	Me	H	Me	H	Cl	F	3.63	3.67	1.45	1.35
22	H	Me	H	Me	H	MeO	MeO	2.95	3.14	0.51	0.71
23	H	Me	H	H	H	F	F	2.19	2.60	1.41	1.31
24	H	Br	H	H	H	F	F	3.24	2.81	1.51	1.51
25 [f]	H	H	H	Me	H	F	MeO	3.06	3.11	–	–
26 [f]	H	Me	H	H	H	F	MeO	3.54	3.46	0.78	1.70
27	H	Br	H	H	H	F	MeO	2.67	2.42	1.03	0.891
28 [f]	H	H	H	H	Cl	F	F	–	–	1.53	1.36
29 [f]	H	H	H	Me	H	Cl	F	–	–	1.47	1.22
30 [f]	H	H	H	H	Me	F	F	–	–	0.70	1.03
31	H	Me	H	H	H	F	F	2.47	2.14	1.58	1.03

[a] Compounds shown are additional to those listed in Tables 16 and 17; comp. 1–7, 3-phenyls; comp. 8–30, 3-(pyridin-2-yl); comp. 31–35, 3-(pyridin-2-yl) with methyl at C5 of thiazolidine-4-one; comp. 36–38, 3-(pyridin-3-yl); comp. 39 and 40, 3-(pyridin-4-yl); comp. 41–49, 3-(pyrimidin-2-yl)

[b] 50% Inhibitory concentration against cytopathic effect of HIV-1 in MT-4 cells

[c] Concentration required to reduce 50% viability of MT-4 cells

[d] Eq. 36

[e] Eq. 37

[f] Test set compound

Table 18 (continued)

No.	R″					R′		$\log EC_{50}$ [b]		$-\log EC_{50}$ [c]	
	2″	3″	4″	5″	6″	2′	6′	obsd	cald [d]	obsd	cald [e]
32	H	Me	H	H	H	Cl	Cl	2.42	2.62	1.05	1.35
33	H	Me	H	Me	H	F	F	2.51	2.83	1.45	1.35
34	H	Me	H	H	H	Cl	Cl	2.88	2.68	0.85	1.28
35 [f]	H	Me	H	Me	H	F	F	2.69	1.46	1.43	1.04
36	H	H	H	–	H	F	F	1.02	1.02	–	–
37	H	H	H	–	H	Cl	Cl	–	–	1.63	1.45
38 [f]	H	H	H	–	H	Cl	Cl	–	–	1.63	1.45
39	H	H	H	H	–	Cl	Cl	–	–	2.00	1.93
40	H	H	H	H	–	F	F	–	–	1.47	1.54
41	H	Me	–	H	H	F	F	3.41	3.57	–	–
42	H	Me	–	H	H	Me	Me	2.47	2.29	0.71	0.47
43	H	Me	–	H	H	F	MeO	3.27	3.46	0.50	0.76
44	Cl	Me	–	H	H	F	CF₃	3.04	2.70	0.72	0.67
45	Me	Me	–	H	H	Cl	F	3.11	3.50	0.81	1.16
46 [f]	H	Me	–	H	H	MeO	MeO	3.04	3.07	0.64	0.34
47	H	Me	–	Me	H	Cl	Cl	4.23	4.07	1.36	1.12
48	H	Me	–	H	H	Cl	Cl	3.55	3.18	1.61	1.49
49	H	Me	–	Me	H	F	F	2.19	2.58	1.35	1.21

[a] Compounds shown are additional to those listed in Tables 16 and 17; comp. 1–7, 3-phenyls; comp. 8–30, 3-(pyridin-2-yl); comp. 31–35, 3-(pyridin-2-yl) with methyl at C5 of thiazolidine-4-one; comp. 36–38, 3-(pyridin-3-yl); comp. 39 and 40, 3-(pyridin-4-yl); comp. 41–49, 3-(pyrimidin-2-yl)
[b] 50% Inhibitory concentration against cytopathic effect of HIV-1 in MT-4 cells
[c] Concentration required to reduce 50% viability of MT-4 cells
[d] Eq. 36
[e] Eq. 37
[f] Test set compound

(moment of inertia Z-component), $\log P_{(o/w)}$ and $MNDO_{dipole}$ in the HIV-1 RT inhibitory activity (comp. **a** and **b** Fig. 28, Table 19; comp. **c** Fig. 28, Table 20) [189, 190]. In these models, Eq. 38 has been derived for thiazolidin-4-ones, metathiazanones (Table 19) and Eq. 39 has been derived for thiazolidin-4-ones with the furfuryl moiety in place of pyridyl at N3 (Table 20).

$$-\log EC_{50} = -0.0018(0.0003)PMIZ + 10.702$$
$$n = 5, \quad r = 0.950, \quad Q^2 = 0.573, \quad s = 0.297, \quad F = 27.91 \tag{38}$$

$$-\log EC_{50} = 1.479(0.206)\log P_{(o/w)} + 0.649(0.114)MNDO\_dipole - 1.191$$
$$n = 15, \quad r = 0.920, \quad Q^2 = 0.742, \quad s = 0.501, \quad F = 33.60 . \tag{39}$$

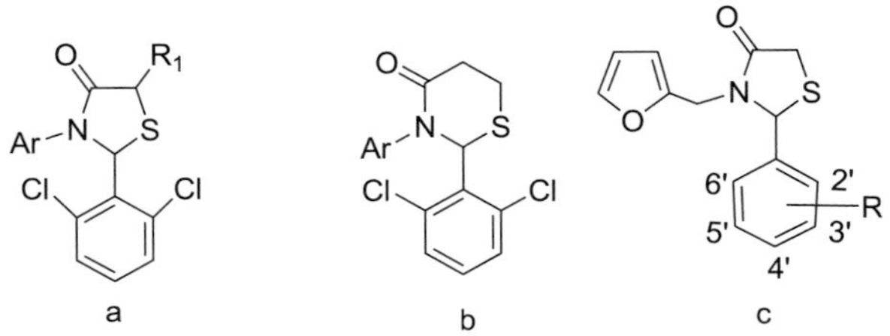

O
R1
Ar
N
S
Cl
Cl
a
Comp. 1,2, 4, 5, 7 and 8
O
Ar
N
S
Cl
Cl
b
Comp. 3 and 6
O
N
S
O
6'
5'
4'
3'
2'
R
c

Fig. 28 Thiazolidin-4-ones and related analogues explored for HIV-1 RT inhibitory activity

Table 19 Physicochemical properties and HIV-1RT inhibitory activity of thiazolidin-4-one and metathiazanone derivatives (comp. **a** and **b**, Fig. 28) [189]

No.	Ar	*R*	PMIZ	$-\log EC_{50}$ [a] obsd	cald [b]
1	pyridin-2-yl	Me	2550.89	–	6.09
2	pyridin-3-yl methyl	H	3200.88	4.81	4.92
3	pyridin-3-yl methyl	H	3284.88	5.18	4.77
4	pyridin-3-yl methyl	Me	3307.00	4.56	4.73
5	furan-2-yl methyl	H	2273.00	6.69	6.60
6	furan-2-yl methyl	H	2894.60	5.25	5.47
7	furan-2-yl methyl	Me	2963.52	–	5.35
8	pyridin-2-yl	H	2573.53	–	6.05

[a] 50% Inhibitory concentration against cytopathic effect of HIV-1 in MT-4 cells
[b] Eq. 38

In this study thiazolidin-4-ones have been found to retain the activity even with the introduction of a furfuryl moiety in place of pyridyl at N3. This has expanded the nature and scope of the chemical space around the thiazolidin-4-one nucleus (comp. **c**, Fig. 28, Table 20) [190].

3.3.2 Antifungal Agents

The antifungal activity of azoles, the emergence of opportunistic fungal infections due to AIDS, and economic considerations of agro and food industry has generated considerable momentum to discover and explore other prototypes for the chemotherapy of fungal diseases. In this context, some thiazolidines, thiazoles, and thiadiazoles have been found to display antifungal activity against different human and plant pathogenic fungi.

Table 20 Physicochemical properties and HIV-1RT inhibitory activity of thiazolidin-4-one derivatives (comp. **c** Fig. 28) [190]

No.	R					Log $P_{(o/w)}$	MNDO _dipole	$-\log EC_{50}$ [a]	
	2′	3′	4′	5′	6′			obsd	cald[b]
1	Cl	H	H	H	Cl	3.49	3.17	6.69	6.02
2	F	H	H	H	F	2.61	3.76	5.54	5.10
3	Cl	H	H	H	F	3.33	4.46	6.24	6.63
4	H	H	Cl	H	H	2.90	1.78	4.06	4.25
5	Cl	H	H	H	H	2.90	3.38	5.19	5.28
6	H	Cl	H	Cl	H	3.60	0.78	4.07	4.64
7	OMe	H	H	H	H	2.26	3.15	3.94	4.19
8	H	H	OMe	H	H	2.26	3.22	3.77	4.24
9	OMe	H	OMe	OMe	H	2.00	2.42	3.28	3.33
10	H	OMe	OMe	OMe	H	1.75	2.84	3.15	3.24
11	OMe	H	OMe	H	OMe	2.24	0.01	4.86	4.85
12	Me	H	H	H	H	2.60	2.21	4.54	4.09
13	Me	H	H	H	Me	2.90	2.21	5.11	4.53
14	F	F	F	F	F	3.06	1.04	4.33	4.01
15	1-Naphthyl					3.53	2.47	5.26	5.62

[a] 50% Inhibitory concentration against cytopathic effect of HIV-1 in MT-4 cells
[b] Eq. 39

In an exploratory study, some 2,3,4-substituted thiazolidines (Fig. 29; Table 21) have been discovered to show a broad spectrum in vitro antifungal activity against human pathogenic fungi namely *Candida albicans* (CA), *Cryptococcus neoformans* (CN), *Tricophyton mentagrophyte* (TM), and *Aspergillus fumigatus* (AF) [191]. The antifungal activities of these compounds have been analyzed in 2D- and 3D-QSAR studies to map the activity and to identify the common pharmacophore for the design of broad spectrum antifungals. The Apex-3D module of InsightII [69] has been put to use in the 3D-QSAR analysis. The following 3D-QSAR equations have emerged from molecular superimposition patterns based on three-biophoric sites of these compounds for their activity against the fungal strains CN, TM, and AF, respectively (Fig. 30).

$$-\log \text{MIC(CN)} = 0.013(0.002)\text{TREF} - 0.427(0.079)\text{IR}_{\text{m}}$$
$$+ 4.475(1.593)\pi_{\text{a}}(\text{ss1}) + 0.099(0.041)\text{REF}_{\text{a}}(\text{ss2})$$
$$- 0.095(0.026)\text{REF}_{\text{a}}(\text{ss3}) - 0.102(0.037)\text{REF}_{\text{a}}(\text{ss4})$$
$$+ 0.356(0.090)\text{I-Ring(ss5)} - 0.026$$
$$n = 50, \quad r = 0.808, \quad s = 0.197, \quad F = 11.26 \tag{40}$$

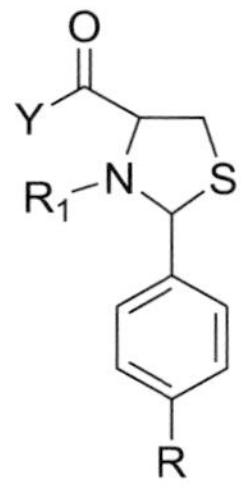

Fig. 29 General structure of 2,3,4-substituted thiazolidines associated with antifungal activity

$$- \log \text{MIC(TM)} = 0.005(0.001)\text{TREF} - 0.485(0.115)\pi_a(\text{ss1})$$
$$- 0.058(0.015)\text{REF}_a(\text{ss2}) - 0.027(0.012)\text{REF}_a(\text{ss3})$$
$$- 0.064(0.016)\text{REF}_a(\text{ss4}) + 0.080(0.027)\text{REF}_a(\text{ss5})$$
$$+ 0.452$$

$$n = 53, \quad r = 0.805, \quad s = 0.145, \quad F = 14.11 \tag{41}$$

$$- \log \text{MIC(AF)} = -0.419(0.135)\text{HA}(\text{ss1}) - 0.261(0.138)\text{HA}(\text{ss2})$$
$$- 0.362(0.090)\text{HA}(\text{ss3}) + 0.420(0.141)\text{HD}(\text{ss4})$$
$$+ 2.594(0.522)\pi_a(\text{ss5}) - 0.313(0.150)\pi_a(\text{ss6})$$
$$+ 0.130(0.034)\text{REF}_a(\text{ss7}) + 1.086$$

$$n = 36, \quad r = 0.841, \quad s = 0.119, \quad F = 9.63 . \tag{42}$$

In development of these equations, for the 3D-molecular superimpositions, the carbonyl oxygen in the vicinity of C4, S1, N3, the oxygen of t-butyloxycarbonyl or the benzoyl group attached to N3 and/or the C2-phenyl moiety of thiazolidines have provided the biophoric sites (Fig. 30). The parameters of these equations represent global attributes of the compounds (global parameters—TREF, total refractivity; IR_m, the nature of the meta-substituent of the C2-phenyl moiety) and/or conformational dependent attributes from the molecular superimposition pattern (secondary sites or ss; descriptors identified with suffix "ss"). In 3D-QSAR studies, all the secondary sites are local to each equation. They are conformation sensitive and location specific in the 3D-structure space. In these equations π_a, REF_a, I-Ring, HA and HD addressed the hydrophobicity, refractivity, aromatic ring indicator, H-acceptor and H-donor properties, respectively, of atoms or groups (secondary sites) at the predefined locations in the 3D space of corresponding models (Fig. 30). These equations have led to the suggestion that AF is sensitive to the structural and conformational features of the compounds. Figure 31 provides a composite 2D-representation of the biophoric and secondary sites pattern of the thiazolidines 3D-equations (Fig. 30; Eqs. 40 to 42).

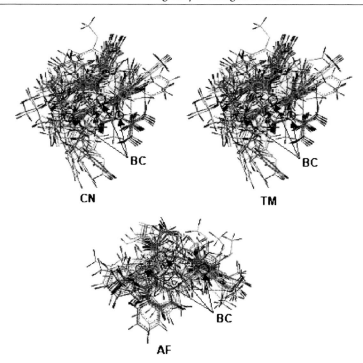

Fig. 30   Apex3D QSAR models of thiazolidines (Table 21) for their activity against *Cryptococcus neoformans* (CN) (Eq. 40), *Tricophyton mentagrophyte* (TM) (Eq. 41), and *Aspergillus fumigatus* (AF) (Eq. 42). In the molecular superimposition, *circles identified with arrows* represent the locations of the biophoric centers (BC)

A fundamental difference between 3D- and a 2D-QSAR equation is the non-existence of conformational dependent secondary sites in the latter. Hence, a direct transposition of 3D- and 2D-models is not always possible but the global properties of the chemical structures, if relevant to the activity, may show their presence in both of them. Moreover, in a broader perspective, all 2D-QSAR parameters—physicochemical as well as structural—can be considered as one or the other form of global descriptors. In light of this, to bridge the 2D- and 3D-features the following 2D-QSAR equations have been derived for the antifungal activity of 2,3,4-substituted thiazolidines (Table 21).

$$- \log \text{MIC(CN)} = 0.011(0.003)MRR_2 + 0.751(0.162)VwR_1$$
$$- 0.335(0.061)IR_m + 0.706$$
$$n = 41, \quad r = 0.800, \quad s = 0.187, \quad F = 21.86 \tag{43}$$

Table 21 Physicochemical properties and antifungal activity of 2,3,4-substituted thiazolidines (Fig. 29) [191]

No.	R^a	$R_1{}^b$	TREF	FR_p	σR_p	VwR_1	MRR_2	$-\log MIC^c$ CN obsd	TM obsd	cald^d	AF obsd	cald^e	CA obsd	cald^f
1	H	H	55.6	0.00	0.00	0.02	0.00	0.62	0.62	0.74	I		I^g	
2	Cl	H	60.4	0.41	0.23	0.02	0.00	I	0.69	0.75	I		I	
3	F	H	55.8	0.43	0.06	0.02	0.00	0.74	0.74	0.72	I		I	
4	OH	H	57.3	0.29	−0.37	0.02	0.00	I	I		I		I	
5	OMe	H	62.0	0.26	−0.27	0.02	0.00	I	I		I		I	
6	$	H	68.5	0.00	0.00	0.02	0.00	I	I		I		I	
7	£	H	75.0	0.26	−0.27	0.02	0.00	I			I		I	
8	H	¥	80.2	0.00	0.00	0.90	0.00	1.39	0.78	0.82	1.09	1.09	0.49	0.50
9	Cl	¥	85.0	0.41	0.23	0.90	0.00	1.44	0.79	0.72	0.84	0.82	0.54	0.54
10	F	¥	80.4	0.43	0.06	0.90	0.00	0.82	0.84	0.75	I		0.51	0.52
11	OH	¥	81.9	0.29	−0.37	0.90	0.00	0.81	0.82	0.72	I		0.51	0.52
12	OMe	¥	86.6	0.26	−0.27	0.90	0.00	1.43	0.81	0.73	0.83	0.83	0.53	0.53
13	$	¥	93.1	0.00	0.00	0.90	0.00	0.87	0.53	0.79	I		0.57	0.56
14	£	¥	99.5	0.26	−0.27	0.90	0.00	0.83	1.17	0.89	I		0.53	0.53
15	H	§	84.9	0.00	0.00	0.78	0.00	1.40	0.83	0.58	1.10	1.12	0.50	0.51
16	Cl	§	89.7	0.41	0.23	0.78	0.00	1.14	0.84	0.89	0.84	0.82	0.54	0.54

a $, £ Correspond to 2,5-dimethoxy and 3,4,5-trimethoxy, respectively
b ¥, § Correspond to CO_2-t-Bu and COPh, respectively
c Minimum inhibitory concentration in NCCLS protocol
d Eq. 41
e Eq. 42
f Eq. 46
g Inactive under test conditions

Table 21 (continued)

No.	R^a	R_1^b	TREF	FR_p	σR_p	VwR_1	MRR_2	CN obsd	TM obsd	cald^d	AF obsd	cald^e	CA obsd	cald^f
								\multicolumn — $-\log MIC^c$						
17	F	§	85.1	0.43	0.06	0.78	0.00	0.82	0.82	0.86	I		0.52	0.52
18	OMe	§	91.4	0.26	-0.27	0.78	0.00	1.44	0.54	0.67	0.84	0.87	0.54	0.54
19	$	§	97.8	0.00	0.00	0.78	0.00	1.17	0.57	0.70	1.17	1.15	0.57	0.57
20	£	§	104.3	0.26	-0.27	0.78	0.00	0.91	0.91	0.96	I		0.60	0.60
21	H	¥	97.5	0.00	0.00	0.90	1.03	1.18	0.88	0.92	1.18	1.08	0.58	0.58
22	H	§	102.3	0.00	0.00	0.78	1.03	1.19	0.58	0.45	0.88	0.92	0.58	0.58
23	Cl	¥	102.3	0.41	0.23	0.90	1.03	1.52	0.92	0.95	0.92	1.10	0.62	0.61
24	Cl	§	107.1	0.41	0.23	0.78	1.03	1.22	0.92	1.10	0.92	1.06	0.62	0.62
25	F	¥	97.7	0.43	0.06	0.90	1.03	I	0.90	0.92	I		0.60	0.59
26	F	§	102.5	0.43	0.06	0.78	1.03	1.51	0.60	0.53	0.90	1.07	0.60	0.60
27	OMe	¥	104.0	0.26	-0.27	0.90	1.03	1.52	0.91	0.95	0.91	1.10	0.61	0.61
28	OMe	§	108.7	0.26	-0.27	0.78	1.03	1.52	0.92	0.98	1.22	0.91	0.62	0.61
29	$	¥	110.5	0.00	0.00	0.90	1.03	0.94	0.94	0.98	I		0.64	0.64
30	$	§	115.2	0.00	0.00	0.78	1.03	0.95	1.55	1.16	I		0.65	0.64
31	£	¥	116.9	0.26	-0.27	0.90	1.03	0.97	1.27	1.00	I		0.67	0.67
32	£	§	121.7	0.26	-0.27	0.78	1.03	0.98	0.98	1.18	I		0.68	0.67

[a] $, £ Correspond to 2,5-dimethoxy and 3,4,5-trimethoxy, respectively
[b] ¥, § Correspond to CO_2-t-Bu and COPh, respectively
[c] Minimum inhibitory concentration in NCCLS protocol
[d] Eq. 41
[e] Eq. 42
[f] Eq. 46
[g] Inactive under test conditions

Table 21 (continued)

No.	R^a	$R_1{}^b$	TREF	FR_p	σR_p	VwR_1	MRR_2	$-\log MIC^c$ CN obsd	TM obsd	cald^d	AF obsd	cald^e	CA obsd	cald^f
33	H	¥	122.2	0.00	0.00	0.90	25.36	1.56	0.96	1.04	0.96	1.09	0.66	0.66
34	H	§	126.9	0.00	0.00	0.78	25.36	1.26	0.96	1.07	0.96	1.09	0.66	0.66
35	Cl	¥	127.0	0.41	0.23	0.90	25.36	0.99	0.99	1.07	0.99	1.09	I	
36	Cl	§	131.7	0.41	0.23	0.78	25.36	0.99	1.00	1.09	1.00	1.09	0.69	0.69
37	F	¥	122.4	0.43	0.06	0.90	25.36	1.28	0.98	1.03	1.28	1.07	0.68	0.67
38	F	§	127.1	0.43	0.06	0.78	25.36	1.88	0.98	1.05	1.28	1.07	0.68	0.68
39	Me	¥	128.6	0.26	-0.27	0.90	25.36	1.89	0.99	0.96	1.29	1.07	0.69	0.69
40	OMe	§	133.4	0.26	-0.27	0.78	25.36	1.29	1.29	1.08	1.29	1.07	0.69	0.69
41	$	¥	135.1	0.00	0.00	0.90	25.36	1.31	1.01	1.11	1.01	1.09	0.71	0.72
42	$	§	139.8	0.00	0.00	0.78	25.36	1.62	1.02	1.09	1.02	0.98	0.72	0.72
43	£	¥	141.5	0.26	-0.27	0.90	25.36	1.34	1.04	1.02	1.04	1.09	0.74	0.75
44	£	§	146.3	0.26	-0.27	0.78	25.36	1.34	1.04	1.17	1.04	1.09	0.74	0.76
45	H	¥	111.2	0.00	0.00	0.90	14.96	1.53	0.93	0.99	0.93	0.93	0.62	0.62
46	H	§	116.0	0.00	0.00	0.78	14.96	0.93	0.63	0.77	0.93	0.91	0.63	0.62
47	Cl	¥	116.0	0.41	0.23	0.90	14.96	1.86	0.66	0.93	1.26	1.08	0.66	0.66

a $, £ Correspond to 2,5-dimethoxy and 3,4,5-trimethoxy, respectively
b ¥, § Correspond to CO_2-t-Bu and COPh, respectively
c Minimum inhibitory concentration in NCCLS protocol
d Eq. 41
e Eq. 42
f Eq. 46
g Inactive under test conditions

Table 21 (continued)

No.	R^a	R_1^b	TREF	FR_p	σR_p	VwR_1	MRR_2	$-\log MIC^c$ CN obsd	$-\log MIC^c$ TM obsd	cald^d	AF obsd	AF cald^e	CA obsd	CA cald^f
48	Cl	§	120.8	0.41	0.23	0.78	14.96	0.96	0.95	0.74	0.66	0.65	0.66	0.66
49	F	¥	111.4	0.43	0.06	0.90	14.96	1.55	0.94	0.88	0.94	1.09	0.64	0.64
50	F	§	116.2	0.43	0.06	0.78	14.96	1.25	0.65	0.77	0.65	0.65	0.65	0.64
51	OMe	¥	117.7	0.26	−0.27	0.90	14.96	0.96	0.96	1.02	0.66	0.68	0.66	0.65
52	OMe	§	122.4	0.26	−0.27	0.78	14.96	0.96	0.98	0.88	0.66	0.65	0.66	0.66
53	$	¥	124.1	0.00	0.00	0.90	14.96	0.98	0.98	1.05	I		0.68	0.68
54	$	§	128.9	0.00	0.00	0.78	14.96	0.99	0.99	1.09	I		0.69	0.69
55	£	¥	130.6	0.26	−0.27	0.90	14.96	1.31	0.71	0.87	1.31	1.89	0.71	0.71
56	£	§	135.3	0.26	−0.27	0.78	14.96	1.31	1.01	0.79	1.31	1.32	0.71	0.72

[a] $, £ Correspond to 2,5-dimethoxy and 3,4,5-trimethoxy, respectively

[b] ¥, § Correspond to CO_2-t-Bu and COPh, respectively

[c] Minimum inhibitory concentration in NCCLS protocol

[d] Eq. 41

[e] Eq. 42

[f] Eq. 46

[g] Inactive under test conditions

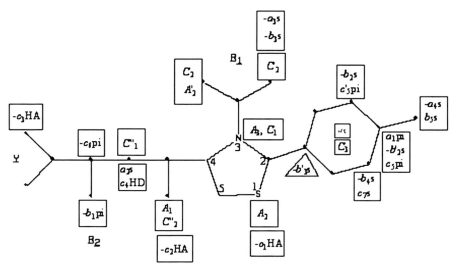

Fig. 31 A composite representation of the thiazolidines' superimposition pattern corresponding to Eqs. 40–42 in 2D-structure space. Annotations in italic capital letters *A* and *B* stand for the most probable locations of biophoric centers of Eq. 40 (also Eq. 41) and Eq. 42, respectively, and the italic small case *a*, *b* and *c* stand for secondary site locations of Eqs. 40, 41, and 42, respectively. The subscripts of annotations represent the *i* th center/site. The most probable atoms for biophoric centers: A_1 is O, A_2 is O or S, A_3 is N, C_1 is N, C_2 is O, and C_3 is ring center. Biophore distances: $A_1 - A_2$ is 4.7415 Å (sd, 0.1831), $A_2 - A_3$ is 2.4872 Å (sd, 0.1657), $A_1 - A_3$ is 3.5598 Å (sd, 0.1635), $C_1 - C_2$ is 2.2790 Å (sd, 0.0279), $C_2 - C_3$ is 5.7242 Å (sd, 0.3156), and $C_1 - C_3$ is 3.9253 Å (sd, 0.1330). The suffixes pi(π), s, HA, or HD, of *a*, *b*, and *c* indicate the nature of the secondary site, as hydrophobic, steric, hydrogen acceptor or hydrogen donor, respectively. The prefixed negative sign, if any, of *a*, *b*, and *c* indicates the sign of the *i* th site coefficient in the regression equation. A prime sign on any annotation (lower or upper case) indicates the alternative location of the corresponding site (Reprinted with permission from [191]. Copyright 2003 Wiley)

(excluded compounds: **12, 13, 19, 43, 44, 54, 56, 59** and **60**)

$$- \log \text{MIC(TM)} = 0.005(0.0007)\text{TREF} + 0.217(0.098)\text{F}R_p + 0.280$$

$$n = 47, \quad r = 0.734, \quad s = 0.113, \quad F = 25.74 \tag{44}$$

(excluded compounds: **15, 34, 38, 55, 58** and **63**)

$$- \log \text{MIC(AF)} = - 0.039(0.009)\text{MRR}_2 + 0.002(0.0004)\text{MRR}_2^2$$
$$- 2.357(1.074)\text{F}R_p + 6.048(2.787)\text{F}R_p^2$$
$$- 0.620(0.300)\sigma R_p + 0.971$$

$$n = 30, \quad r = 0.777, \quad s = 0.124, \quad F = 7.29 \tag{45}$$

(excluded compounds: **21, 29, 36, 55, 63**, and **64**)

In deriving these equations several compounds have been excluded from the dataset. In 2D-QSAR, the descriptors TREF, IR_m, MRR_2, VwR_1, FR_p, and σR_p have been found to be the parameters of choice to explain the antifungal activity of the compounds with TREF and IR_m as common variables for both 3D- and 2D-equations. With the identification of the 2D-parameters in the 3D-region, that is R_2 to the C4 region, R_1 to the N3 region, and R_p to the C2-phenyl moiety, the 2D and 3D-studies have been integrated to diagnose the activity of the compounds.

The thiazolidines' (Table 21) activity against CA has been found to be distributed in a very narrow range. Also, no 3D-molecular/conformational preferences of these compounds have been noticed in deciding this antifungal activity (Eqs. 46 and 47).

$$-\log MIC(CA) = 0.179 + 0.001(0.00001)MW$$
$$n = 48, \quad r = 0.997, \quad s = 0.005, \quad F = 7689.85 \tag{46}$$

$$-\log MIC(CA) = 0.210 + 0.004(0.00001)TREF$$
$$n = 48, \quad r = 0.976, \quad s = 0.015, \quad F = 934.85 . \tag{47}$$

These equations have indicated a strong correlation between the CA inhibitory activity and global properties namely, molecular weight (MW) and total refractivity (TREF), of these compounds. This exploration has elucidated the potential of the thiazolidines as antifungal agents. Some $1:1$ adducts of triphenyltin chloride and 2,3-disubstituted thiazolidin-4-ones have been evaluated, in vitro, against the plant pathogenic fungus, Dutch elm, *Ceratocystis ulmi* [192]. However, attempts to obtain structure-activity correlations between the activity and the descriptors of triphenyltin chloride and thiazolidin-4-one adduct have been reported to be unsuccessful [192].

Zou et al. [193] have explored the QSAR of the antifungal profile of 5-[1-aryl-1,4-dihydro-6-methylpyridazin-4-one-3-yl]-2-arylamino-1,3,4-thiadiazoles (Fig. 32, Table 22) in terms of the sum of the physicochemical properties of all the substituent groups namely, hydrophobic effects ($\sum \pi$), meta- and para- Hammett's sigma constants ($\sum \sigma$), steric effects ($\sum Es$), and the inductive effects ($\sum F$).

These compounds have been tested in vivo against wheat leaf rust, *Puccinia recondite*, at a fixed dose of 0.001 M. The following equation represents the best QSAR model for the antifungal activity ($D = \log[a/(100 - a)] - \log Mw$; where, a is the percentage inhibition and Mw is the molecular weight of the tested compound) of these compounds.

$$D = 2.802(0.450) \sum \pi - 3.275(0.534) \sum \sigma + 0.995(0.214) \sum Es$$
$$- 0.936(0.21) \sum F - 2.844$$
$$n = 16, \quad r = 0.902, \quad s = 0.281, \quad F = 11.96 . \tag{48}$$

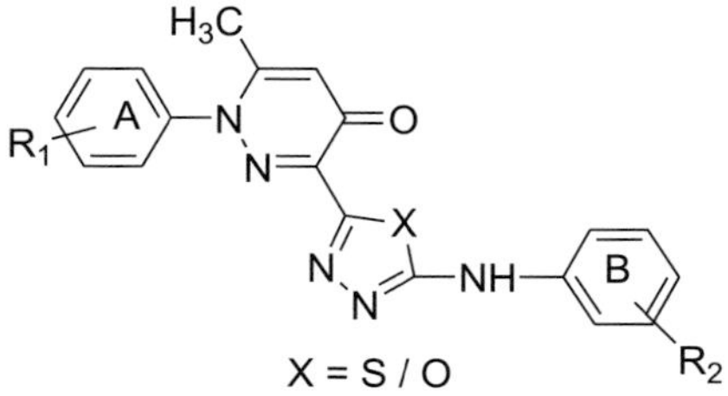

Fig. 32 General structure of 5-[1-aryl-1,4-dihydro-6-methylpyridazin-4-one-3-yl]-2-arylamino-1,3,4-thiadiazoles and oxadiazoles associated with antifungal activity

Table 22 Physicochemical properties and fungicidal activity of pyridazinonethiadiazoles (Fig. 32) [193]

	R_1	R_2	$\sum \sigma$	$\sum F$	$\sum \pi$	$\sum Es$	D^a obsd	cald^b
1	2-Cl	3-CF$_3$	0.66	0.79	1.59	− 1.21	− 2.30	− 2.49
2	2-Cl	2-F	0.29	0.07	0.85	− 1.43	− 3.57	− 2.90
3	2-Cl	H	0.23	0.41	0.71	− 0.97	− 3.16	− 2.96
4	H	2-F	0.06	− 0.34	0.14	− 0.46	− 2.58	− 2.79
5	H	3-CF$_3$	0.43	0.38	0.88	− 0.24	− 2.46	− 2.38
6	2,6-Cl$_2$	3-CF$_3$	0.89	1.19	2.30	− 2.18	− 2.70	− 2.60
7	2,6-Cl$_2$	2-F	0.52	0.47	1.56	− 2.40	− 3.01	− 3.00
8	4-Cl	3-CF$_3$	0.66	0.79	1.59	− 1.21	− 2.44	− 2.49
9	4-Cl	2-F	0.29	0.07	0.85	− 1.43	− 2.67	− 2.90
10	2,4,5-Cl$_3$	2-F	0.75	0.41	2.27	− 3.37	− 2.50	− 2.68
11	2,4,5-Cl$_3$	3-CF$_3$	1.12	0.89	3.01	− 3.15	− 2.36	− 2.04
12	2,4-di-Me	3-CF$_3$	0.09	1.61	2.00	− 2.72	− 1.69	− 1.75
13	2,4-di-Me	2-F	− 0.28	0.30	1.26	− 2.94	− 1.66	− 1.60
14	2,4-Cl$_2$	2-F	0.52	− 0.42	1.56	− 2.40	− 2.05	− 2.17
15	2,4-Cl$_2$	3-CF$_3$	0.89	0.48	2.30	− 2.18	− 1.74	− 1.93
16	2,4-Cl$_2$	H	0.46	1.20	1.42	− 1.94	− 3.23	− 3.42

^{a} $D = \log[a/(100 − a)] − \log Mw$; where, a is the percentage inhibition and Mw is the molecular weight of test compound
^{b} Eq. 48

On the basis of this equation Zou et al. have concluded that hydrophobic compounds with inductively electron-donating ortho substituents would be favorable for the activity [193]. In continuation of this, Zou and co-workers have carried out CoMFA-based 3D-QSAR analysis of these compounds together with 5-[1-aryl-1,4-dihydro-6-methylpyridazin-4-one-3-yl]-2-arylamino-1,3,4-oxadiazoles [194]. Here also the antifungal activity of these compounds has been found to be well explained by their steric and electrostatic properties. In addition to this, it has confirmed the bioisosterism

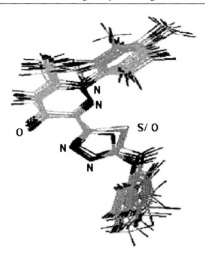

Fig. 33 Alignment and superposition models of 5-[1-aryl-1,4-dihydro-6-methylpyridazin-4-one-3-yl]-2-arylamino-1,3,4-thiadiazoles and corresponding oxadiazoles in CoMFA. (Reprinted with permission from [194]. Copyright 2002 American Chemical Society)

between the 1,3,4-thiadiazole and 1,3,4-oxadiazole analogues in expressing the antifungal activity (Fig. 33)

3.3.3
Octopamine Agonists (OA)

Several scaffolds closely related to thiazolindines have been reported to possess octopamine (OA) agonist/antagonist activity mediated through the adenylate cyclase. The OA is associated with the altered levels of cyclic AMP in the nerve tissue of insects, e.g., the migratory locust, *Locusta migratoria* L [195–199]. This has prompted the investigation of these heterocycles as safe and selective pesticides [200, 201]. Against this background Hirashima et al. investigated the QSAR of octopaminergic agonistic activity of 2-(arylimino)thiazolidine derivatives (Fig. 34, Table 23) [202] with different physicochemical parameters such as total hydrophobicity ($\sum \pi$), molar re-

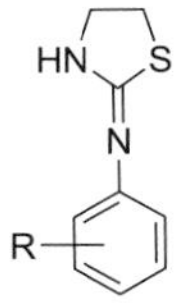

Fig. 34 General structure of 2-(arylimino)thiazolidines associated with octopaminergic agonistic activity

Table 23 Physicochemical properties and octopamine agonist activity of 2-(arylimino)thi-azolodines (Fig. 34) [203]

No.	X	$\sum \pi$	σ_P	Es^m	$logV_{max}$ [a] obsd	cald [b]
1	H	0.00	0.00	0.00	0.92	0.78
2	3-Cl	0.87	0.00	− 0.97	0.54	0.75
3	3-NO$_2$	0.23	0.00	− 2.52	0.28	0.17
4	4-F	0.12	0.06	0.00	0.75	0.83
5	4-Me	0.50	− 0.17	0.00	0.88	0.89
6	4-CN	− 0.21	0.66	0.00	0.92	0.84
7	4-EtO	0.48	− 0.24	0.00	0.69	0.87
8	2,4-Cl$_2$	1.34	0.23	0.00	1.25	1.17
9	2,4-F$_2$	0.22	0.06	0.00	0.60	0.85
10	2-Br,4-Me	1.12	− 0.17	0.00	0.89	1.05
11	2-NO$_2$, 4-Me	0.36	− 0.17	0.00	0.56	0.85
12	2-NO$_2$, 4-OMe	0.10	− 0.27	0.00	0.65	0.71
13	2-Me,4-Br	1.32	0.32	0.00	1.61	1.18
14	2-Me,4-Cl	1.16	0.23	0.00	1.49	1.13
15	2-Me,4-F	− 0.11	0.06	0.00	1.15	0.77
16	2-Me,4-I	0.01	0.18	0.00	1.38	0.82
17	2-Me,4-NO2	0.61	0.78	0.00	1.21	1.08
18	2,4-Me$_2$	0.81	− 0.17	0.00	1.02	0.97
19	2,4-(MeO)$_2$	− 0.31	− 0.27	0.00	0.53	0.66
20	2-Et, 4-Br	1.78	0.23	0.00	1.49	1.29
21	2,5-Cl$_2$	1.36	0.00	− 0.97	0.89	0.88
22	2-Cl,5-CF$_3$	1.70	0.00	− 2.40	0.66	0.59
23	2,5-F$_2$	0.43	0.00	− 0.46	1.08	0.77
24	2-Me,5-Cl	1.18	0.00	− 0.97	1.03	0.83
25	2-MeO,5-Cl	0.52	0.00	− 0.97	0.49	0.66
26	2,5-(MeO)$_2$	− 0.21	0.00	− 0.55	0.34	0.58
27	2-Cl,6-Me	0.80	0.00	0.00	0.95	0.99
28	2,6-F$_2$	0.20	0.00	0.00	0.56	0.84
29	2,6-iPr$_2$	2.30	0.00	0.00	1.38	1.39
30	3-Cl,4-Me	1.37	− 0.17	− 0.97	0.53	0.85
31	3-NO$_2$,4-Cl	1.08	0.23	− 2.52	0.56	0.44
32	3-CF$_3$,4-Cl	2.06	0.23	− 2.40	0.86	0.72
33	3,4-(MeO)$_2$	0.18	− 0.27	− 0.55	0.49	0.64
34	3,5-Cl$_2$	1.74	0.00	− 0.97	0.89	0.98
35	3,5-(Me)$_2$	1.08	0.00	− 1.24	0.57	0.74
36	3,5-(CF$_3$)$_2$	2.42	0.00	− 2.40	0.94	0.78

[a] V_{max} is the maximal efficacy of the compound against adenylate cyclase in thoracic nerve cord of *Periplantea americana* L. relative to 1 mM of octopamine
[b] Eq. 49

Table 23 (continued)

No.	X	$\sum \pi$	σ_P	Es^m	$\log V_{max}$ [a] obsd	cald [b]
37	3,5-(MeO)$_2$	0.28	0.00	– 0.55	0.60	0.71
38	2,3,4-Cl$_3$	2.21	0.23	– 0.97	0.93	1.14
39	2,4,5-Cl$_3$	2.21	2.23	– 0.97	1.13	1.49
40	2,3,5,6-F$_4$	0.86	0.00	– 0.46	1.06	0.89
41	2,3,4,5,6-F$_5$	0.98	0.06	– 0.46	0.73	0.93

[a] V_{max} is the maximal efficacy of the compound against adenylate cyclase in thoracic nerve cord of *Periplantea americana* L. relative to 1 mM of octopamine
[b] Eq. 49

fraction (MR), electronic effect sigma meta constant (σ_m), steric (Es), steric meta (Es^m), Swain and Lupton's polarity ($\mathcal{F}$) and resonance constants ($\mathcal{R}$). In these compounds the relative maximal efficacy of the compounds, log V_{max} has been shown to be correlated with the parameters $\sum \pi$, σ_P and Es^m (Eq. 49) [203].

$$\log V_{max} = 0.784(0.055) + 0.261(0.055) \sum \pi + 0.173(0.095)\sigma_P$$
$$+ 0.266(0.050)Es^m$$
$$n = 41, \quad r = 0.744, \quad s = 0.229, \quad F = 15.26 . \tag{49}$$

On the basis of this model, it has been concluded that more hydrophobic compounds having *ortho* and *para* substituents with an electron-withdrawing property and less bulky *meta*-position substituents would lead to better activity. In these compounds octopamine agonistic activity in terms of K_a, the concentration of the agonist necessary for half-maximal activation of adenylate cyclase, and its log V_{max} are mutually correlated. This has prompted us to suggest that the larger the V_{max} value, the greater would be the activity of the compounds [202].

In 2-(substituted benzylamino)-2-thiazolines (Fig. 35, Table 24) [204, 205] the octopamine agonist activity (log V_{max}) has been reported to be correlated with the parameters σ_m and Es^m as shown [206].

$$\log V_{max} = 0.663(0.125)\sigma_m + 0.118(0.036)Es^m + 1.414$$
$$n = 17, \quad r = 0.855, \quad s = 0.048, \quad F = 19.11 . \tag{50}$$

In these compounds the *meta* substituent's electronic and steric effect have been found to be important for the OA agonistic activity. Similar to the compounds of Table 24, for 2-(substituted benzylamino)-2-thiazolines (Fig. 35, Table 24) also less bulky *meta*-substituents have been reported as favorable for the activity. The positive regression coefficient of σ_m has suggested more

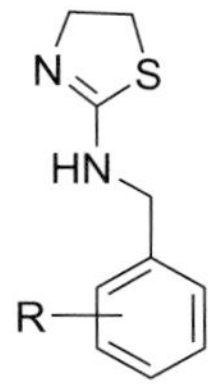

Fig. 35 General structure of 2-(substituted benzylamino)-2-thiazolines associated with octopaminergic agonistic activity

Table 24 Physicochemical properties and octopamine agonist activity of 2-(substituted benzylamino)-2-thiazolines series (Fig. 35) [206]

No.	R	σ_m	Es^m	$\log V_{max}$ [a] obsd	cald [b]
1	H	0.00	0.00	1.32	1.41
2	2-Br	0.00	0.00	1.44	1.41
3	2-Cl	0.00	0.00	1.39	1.41
4	2-F	0.00	0.00	1.46	1.41
5	2-Me	0.00	0.00	1.48	1.41
6	2-CF$_3$	0.00	0.00	1.36	1.41
7	3-F	0.34	−0.46	1.61	1.58
8	3-CF$_3$	0.43	−2.40	1.42	1.42
9	4-Cl	0.00	0.00	1.35	1.41
10	4-Me	0.00	0.00	1.41	1.41
11	4-OMe	0.00	0.00	1.45	1.41
12	2,3-Cl$_2$	0.23	−0.97	1.46	1.45
13	2,4-F$_2$	0.00	0.00	1.38	1.41
14	2-F,4-Cl	0.00	0.00	1.46	1.41
15	2,5-Cl$_2$	0.23	−0.97	1.45	1.45
16	2,5-F$_2$	0.06	−0.46	1.43	1.40
17	3,5-Cl$_2$	0.74	−1.94	1.66	1.68

[a] V_{max} is the maximal efficacy of the compound against adenylate cyclase in thoracic nerve cord of *Periplantea americana* L. relative to 1 mM of octopamine
[b] Eq. 50

electron-withdrawing substituents at this position. However, in these compounds, the K_a and V_{max} are negatively correlated. This has been attributed to the phenyl substituents of these compounds [206].

Thiazolidines and related compounds have also been indicated for their pheromone production inhibitory activity in female moths, *Helvicoverpra armigera*. Pheromone production is a chemical signaling pathway related to the octopaminergic receptor in moths for controlling their reproductive be-

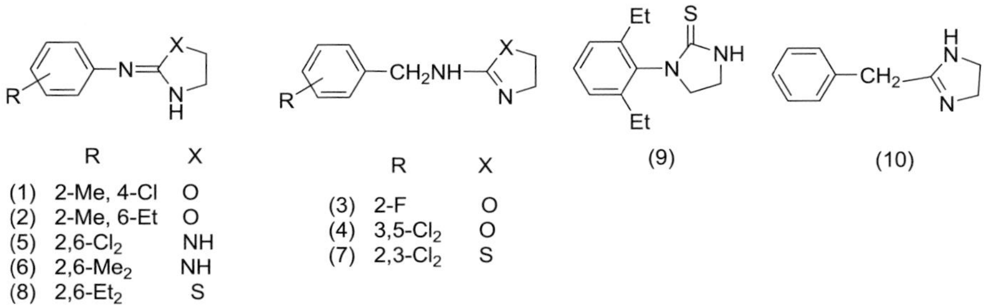

R	X
(1)	2-Me, 4-Cl	O
(2)	2-Me, 6-Et	O
(5)	2,6-Cl$_2$	NH
(6)	2,6-Me$_2$	NH
(8)	2,6-Et$_2$	S

R	X
(3)	2-F	O
(4)	3,5-Cl$_2$	O
(7)	2,3-Cl$_2$	S

Fig. 36 Different thiazoles and oxazoles associated with octopaminergic agonistic activity

Table 25 Physicochemical properties and octopamine agonist activity of thiazolidine derivatives (Fig. 36) [215, 216]

No.	LowEne	SXZF	JX	R	pK$_i$[a] obsd	cald[b]	cald[c]	cald[d]	cald[e]
1	8.85	0.69	2.18	55.60	8.21	8.20	7.95	8.02	8.32
2	63.71	0.87	2.30	60.40	5.21	5.23	5.36	5.15	5.45
3	49.98	0.78	1.86	50.80	4.00	3.99	3.95	3.82	3.96
4	– 5.69	0.88	1.89	60.20	7.16	7.19	7.16	7.22	7.17
5	52.65	0.71	2.25	57.30	8.05	8.00	8.01	8.01	7.78
6	37.67	0.80	2.29	57.80	8.41	8.43	8.84	8.46	8.29
7	73.53	0.69	1.92	66.60	4.11	4.11	4.31	4.22	3.95
8	88.28	0.67	2.33	71.50	7.21	7.25	6.89	7.22	7.11
9	106.18	0.78	2.41	74.00	5.27	5.24	5.22	5.42	5.34
10	13.91	0.86	2.00	49.10	7.11	7.11	7.06	7.19	7.38

[a] K_i is the concentration of octopamine agonist necessary for half-maximal inhibition of pheromone biosynthesis activating neuropeptide (PBAN) binding at 1 p mol/intersegment
[b] Eq. 51 [215]
[c] Eq. 52 [216]
[d] Eq. 53 [216]
[e] Eq. 54 [216]

havior [207–212]. For a set of ten compounds (Fig. 36, Table 25) [213, 214] Hirashima and co-workers have identified the following 3D-QSAR model in GFA to explain their OA agonistic activity (K_i) (Eq. 51) [215].

$$pKi = - 0.082\text{LowEne} - 14.819\text{SXZF} + 8.818\text{JX}$$
$$+ 0.004(\text{MR} - 57.7808)^2 + 3.085$$
$$n = 10, \quad r = 1.000, \quad F = 5681.367 .$$

$$(51)$$

In this equation LowEne is the energy of the most stable conformation, SXZF is the fraction of area of molecular shadow in the XZ plane over the area of the enclosing rectangle, JX is a topological parameter from Balaban's shape indices, and MR is the molecular refractivity. Hirashima and co-workers have interpreted this equation to suggest that compounds with positive JX, negative or smaller LowEne and SXZF would result in improved activity [215]. In continuation of this study they have analyzed these compounds using molecular shape analysis (MSA), molecular field analysis (MFA) and receptor surface model (RSM) [216]. In MSA a significant equation for the activity of these compounds has been discovered with information content (IC), radius of gyration (RG), and common overlap steric volume (COVS) (Eq. 52). Models from MFA have involved the potentials of the 3D-grid points generated from the atomic coordinates of the contributing models. The MFA model identified for these compounds has laid emphasis on the proton probes corresponding to lattice points H+/372, H+/298 and the methyl probe corresponding to lattice point CH3/164 (Eq. 53). The RSM analysis has highlighted the associated steric, physicochemical, and electronic properties of the grid of a hypothetical receptor site of the compounds. The RSM equation (Eq. 54) of these compounds has emerged from the electrostatic interaction energy (ELE) at grid points 137 (ELE/137) and 546 (ELE/546) and van der Waals interaction energy at point 780 (VDW/780).

$$pKi = -0.960(IC)^2 - 6.579RG - 0.055(152.765 - COSV)^2 + 39.1401$$
$$n = 10, \quad r = 0.0.992, \quad F = 119.50 \tag{52}$$

$$pKi = -0.003(H+/372)^2 + 0.063(H+/298) + 0.025CH_3/164 + 7.209$$
$$n = 10, \quad r = 0.997, \quad F = 407.92 \tag{53}$$

$$pKi = 2.763ELE/137 - 1.529x(ELE/546)^2 - 0.061VDW/780 + 8.214$$
$$n = 10, \quad r = 0.994, \quad F = 184.36. \tag{54}$$

These equations have highlighted the steric, physicochemical, and electronic requirements at the specified grid locations of the compounds and the pseudo receptor [216].

3.3.4
Trehalase Inhibitors

Also, thiazolidine derivatives have been reported to control an insect's energy metabolism through the inhibition of trehalase, an enzyme involved in the energy functions, especially in the insect-flight [217, 218]. The affinity of trehazolin (**a**) for this enzyme has motivated workers to explore 2-aryl-iminothiazolidine/oxazolidine derivatives (comp. **b**, Fig. 37, Table 26) as its inhibitors [219].

The QSAR of the trehalase inhibitory activity (pIC50) of these compounds has been investigated with dipole moment (D), lipophilicity [V_1/V_2, where V_1

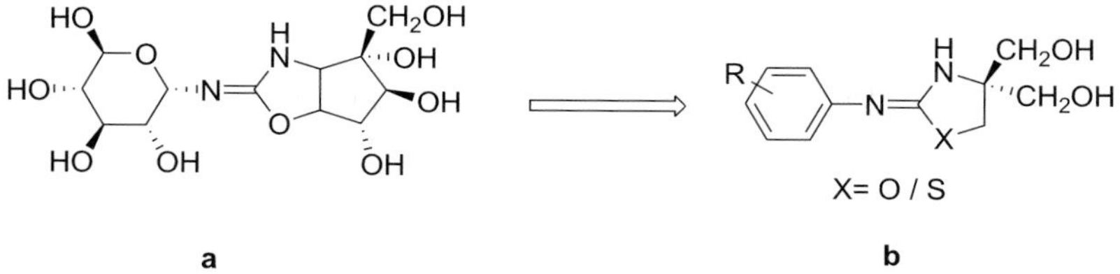

Fig. 37 Structures of trehazolin (**a**) and 2-aryliminothiazolidines/oxazolidines (**b**) associated with trehalase inhibitory activity

Table 26 Physicochemical properties and trehalase inhibitory activities of 2-arylimino-thiazolidine/oxazolidine derivatives (comp. **b**, Fig. 37) [219]

No.	R	X	d_2 (Å)^2	V_1/V_2	D	pIC$_{50}$ ^{a} obsd	cald ^{b}
1	2-F	O	4.55	3.21	0.19	4.21	4.05
2	2-F	S	4.55	3.62	0.16	3.08	2.97
3	4-F	O	4.55	2.97	0.15	3.58	3.62
4	4-F	S	4.55	3.28	0.18	3.24	2.77
5	2,4-F$_2$	O	4.55	3.14	0.16	3.05	3.53
6	2,4-F$_2$	S	4.55	3.55	0.16	2.21	2.48
7	3-Cl, 4-F	O	5.09	3.35	0.14	4.45	4.50
8	3-Cl, 4-F	S	5.10	3.70	0.14	3.62	3.60
9	2,3,4-F$_3$	O	4.77	3.38	0.16	4.07	3.81
10	2,3,4-F$_3$	S	4.77	3.63	0.15	3.15	3.34

^{a} Percent inhibition = $[1 - A/B] \times 100$, where A and B are the molarity of glucose solution with and without test compound
^{b} Eq. 55

and V_2 correspond to the non-polar surface area and polar surface area (Å^2) of the water solvation shell,respectively] and length (d_2).

$$pIC_{50} = 0.376(0.124)D + 3.174(0.624)d_2 - 2.071(0.529)V_1/V_2 - 5.109$$
$$n = 10, \quad r = 0.830, \quad s = 0.335, \quad F = 9.79 . \tag{55}$$

On the basis of this model, it has been concluded that polar compounds with higher d_2 values would be better for the activity. Moreover, it has been suggested that hydrophilic compounds would lead to improved inhibitory activity.

4
Concluding Remarks

The present review has addressed two issues namely, applications of QSAR in rational drug design and more importantly the review has highlighted the potential of the thiazolidine ring system as a biologically privileged scaffold. Over the years there has been substantial progress in QSAR methodologies from the classical to high-dimensional modeling studies leading to more realistic prediction of physicochemical properties as well as the biological activities of compounds under investigation. Through the examples cited in this review we have tried to emphasize that by rational design, biological activity inherent to the chemical entities can be accentuated culminating in therapeutically useful molecules. The studies on neuropeptide Y5 receptor (NPY5R) antagonists as antiobesity agents has clearly illustrated how the modeling studies can be focused in the design of novel and highly active compounds for the identified target. In this case, highly active NPY5R antagonists with a thiazole scaffold have been designed using modeling studies starting from totally unrelated chemical entities [130, 131]. QSAR study of the antihistaminic (H_1) activity of substituted thiazoles and benzthiazole derivatives [111, 112] using the logP and normal and reversed phase thin layer chromatographic (TLC) parameters generated through impregnating the TLC plates with the amino acids of the receptor domain provided insight into the nature and requirements of receptor domain vis-à-vis the compounds under investigation. Also, it provided a handle to deal with the much larger problems of receptor domain with simple yet reliable tools. This study has opened up a novel way of generating the receptor-directed interaction information in terms of amino acids of the active site. One can also adapt this kind of procedure for the study of other receptors. Another interesting example discussed in this review involves a simple classification of antiulcer activity of 4-substituted-2-guanidino thiazoles with topological indices namely Wiener's and molecular connectivity indices [117]. These approaches are very useful and quick for the screening of large databases. The QSAR and modeling results of thiazolidin-4-ones as HIV-1 RT inhibitors from different groups have shown consistency from different perspectives and provided scope to expand the substitution around the thiazolidinone moiety by the introduction of furfuryl in place of pyridyl at N3 [173, 181–183, 189, 190]. Through these and other examples we have tried to highlight the prospects of this heterocyclic system as well as QSAR and modeling studies in lead identification, development, and optimization in the drug discovery program.

Acknowledgements Two of the authors (VRS and MKG) gratefully acknowledge CSIR, New Delhi, India for financial support in the form of a Senior Research Fellowship. CDRI communication no. 6894.

References

1. Abraham DJ (2003) Burgers Medicinal Chemistry and Drug Discovery, 6th edn. Wiley, New Jersey
2. Frühbeis H, Klein R, Wallmeier H (1987) Angew Chem Int Ed 26:403
3. Livingstone D (1995) Data Analysis for Chemists: Applications to QSAR and Chemical Product Design. Oxford University Press, New York
4. Negwer M (1994) Organic-Chemical Drugs and Their Synonyms, 7th edn. Akademie Verlag, VCH Publishers, New York
5. Koch P, Perrotti E (1974) Tetrahedron Lett 15:2899
6. Cook AH, Heilbron I (1949) In: The chemistry of penicillin, chap 25. Princeton University Press, Princeton, p 921
7. Metzger JV (1984) Comprehensive heterocyclic chemistry: The Structure, Reactions, Synthesis and Uses of Heterocyclic compounds. In: Katrizky AR, Rees CW, Potts KT (eds) Thiazoles and their benzo derivatives. Pergamon Press, New York, p 235
8. Pesek J, Frost J (1975) Tetrahedron 31:907
9. Brich A, Harris S (1930) Biochem J 24:1080
10. Schubert J (1935) Biochem J 111:671
11. Kallen RG (1971) J Am Chem Soc 93:6236
12. Wystouch A, Lisowski M, Pedyczak A, Siemion IZ (1992) Tetrahedron Assym 3:1401
13. Patek M, Darke B, Lebel M (1995) Tetrahedron Lett 36:2227
14. Pesek JJ, Frost JH (1975) Tetrahedron 31:907
15. Dains FB, Krober OA (1939) J Am Chem Soc 61:1830
16. Damico JJ, Harman MH (1955) J Am Chem Soc 77:476
17. Bon V, Tisler M (1962) J Org Chem 27:2878
18. Rao RP (1961) J Indian Chem Soc 38:784
19. Bhargava PN, Chaurasia MR (1969) J Pharm Sci 58:896
20. Chaubey VN, Singh H (1970) Bull Chem Soc Jpn 43:2233
21. Wilson FJ, Burns R (1922) J Chem Soc 121:870
22. Bougault J, Cattelain E, Chabrier P, Quevauviller A (1949) Bull Soc Chim Fr 16:433
23. Surrey AR, Cutler RA (1954) J Am Chem Soc 76:578
24. Srivastava SK, Srivastava SL, Srivastava SD (2000) J Indian Chem Soc 77:104
25. Shrama RC, Kumar D (2000) J Indian Chem Soc 77:492
26. Baraldi PG, Simoni D, Moroder F, Manferdini S, Mucchi L, Vecchia FD (1982) J Heterocycl Chem 19:557
27. Srivastava T, Haq W, Katti SB (2002) Tetrahedron 58:7619
28. Rawal RK, Srivastava T, Haq W, Katti SB (2004) J Chem Res 368
29. Holmes C, Chinn JP, Look GC, Gordon EM, Gallop MA (1995) J Org Chem 60:7328
30. Popp FD, Rajopadhye M (1987) J Heterocycl Chem 24:261
31. Brown FC, Jones RS, Kent M (1963) Can J Chem 41:817
32. Liu H, Li Z, Anthonsen T (2000) Molecules 5:1055
33. Kato T, Ozaki T, Tamura K, Suzuki Y, Akima M, Ohi N (1999) J Med Chem 42:3134
34. Troutman HD, Long LM (1948) J Am Chem Soc 70:3436
35. Surrey AR (1948) J Am Chem Soc 70:4262
36. Omar MT, EI-Khamry AE, Sherif FA (1981) J Heterocycl Chem 18:633
37. Brown FC (1961) Chem Rev 61:463
38. Danila G (1978) Rev Chim (Bucharest) 29:820 Chem Abstr (1979) 90:72086p
39. Danila G (1978) Rev Chim (Bucharest) 29:1152 Chem Abstr (1979) 90:152037p
40. Singh SP, Parmar SS, Raman K, Stenberg VI (1981) Chem Rev 81:175

41. Hansch C, Leo A (1979) Substituent Constants for Correlation Analysis in Chemistry and Biology. Wiley-Interscience, New York
42. Stewart JJP (2004) J Mol Mode 10:155
43. CS Chem3D Ultra Cambridge Soft Corporation, Cambridge, USA
44. Katritzky AR, Lobnov V, Karelson M (1994) CODESSA (Comprehensive Descriptors for Structural and Statistical Analysis). University of Florida, Gainesville, FL, http://www.codessa-pro.com
45. Katritzky AR, Perumal S, Petrukhin R, Kleinpeter E (2001) J Chem Inf Comput Sci 41:569
46. Liu SS, Yin CS, Li ZL, Cai SX (2001) J Chem Inf Comput Sci 41:321
47. Todeschini R, Consonni V, Mauri A, Pavan M (2003) DRAGON software version 3.0. http://disat.unimib.it/chm/Dragon.htm
48. Wold S, Ruhe A, Wold H, Dunn WJ (1984) J Sci Stat Comput 5:135
49. Rogers D, Hopfinger AJ (1994) J Chem Inf Comput Sci 34:854
50. Cerius^2 Version 4.8 (2002) Life science modeling and simulation software. Accelrys, San Diego, USA
    http://www.accelrys.com/products/cerius2/
51. Prabhakar YS (2003) QSAR Comb Sci 22:583
52. Liu SS, Liu HL, Yin CS, Wang LS (2003) J Chem Inf Comput Sci 43:964
53. Gasteiger J, Zupan J (1993) Angew Chem Intl Ed Engl 32:503
54. Manallack DT, Ellis DD, Livingstone DJ (1994) J Med Chem 37:3758
55. Kövesdi I, Dominguez-Rodriguez MF, Ôrfi L, Náray-Szabó G, Varró A, Gy Papp J, Mátyus P (1999) Med Res Rev 19:249
56. Cramer RD, Patterson DE, Bunce JD (1998) J Am Chem Soc 110:5959
57. Klebe G, Abraham U, Meitzner T (1994) J Med Chem 37:4130
58. Hopfinger AJ (1980) J Am Chem Soc 102:7196
59. Tokarski JS, Hopfinger AJ (1994) J Med Chem 37:3639
60. Hopfinger AJ, Tokarsi JS (1997) In: Charifson PS (ed) Practical applications of computer-aided drug design. Marcel Dekker Inc, New York, p 105
61. Hahn M (1995) J Med Chem 38:2080
62. Hahn M, Rogers D (1995) J Med Chem 38:2091
63. Golender VE, Vorpogel ER (1993) Computer Assisted-Pharmacophore Identification. In: Kubinyi H (ed) 3D-QSAR in Drug Design: Theory, Methods and Application. ESCOM Science Publishers, Leiden, Netherlands, p 137
64. Rational Drug Design Software Catalyst, Accelrys, San Diego, CA http://www.accelrys.com/products/catalyst/
65. Cox B, Denyer JC, Binnie A (2000) Prog Med Chem 37:83
66. Walters WP, Stahl MT, Murcko MA (1998) Drug Discovery Today 3:160
67. Morris GM, Goodsell DS, Halliday RS, Huey R, Hart WE, Belew RK, Olson AJ (1988) J Comput Chem 19:1639
68. AutoDock: Automated docking of flexible ligands to macrimolecules http://www.scripps.edu/mb/olson/doc/autodock
69. Insight II, version 2.3.0, Biosym Technologies Inc, 9685, Scranton Road, San Diego, CA, USA, 1993
70. SYBYL 6.9 Tripos Inc., St. Louis, MO, USA
71. MOE: The Molecular Operating Environment, Chemical Computing Group Inc. Quebec, Canada. http://www.chemcomp.com
72. Klose N, Niedbolla K, Schwazrtz K, Bottcher I (1983) Arch Pharm 1983:316
73. Satsangi RK, Zaidi SM, Misra VS (1983) Pharmazie 38:341

74. Pignatello R, Mazzone S, Panico AM, Mazzone G, Penissi G, Castano R, Matera M, Blandino G (1991) Eur J Med Chem 26:929

75. Hadjipavlou-Litina D, Geronikaki A, Sotiropoulou E (1993) Res Commun Chem Pathol Pharmacol 79:355

76. Geronikaki A, Hadjipavlou-Litina D (1993) Pharmazie 48:948

77. Garg R, Kurup A, Mekapati SB, Hansch C (2003) Chem Rev 103:703

78. Fleischmann R, Iqbal I, Slobodin G (2002) Expert Opin Pharmacother 10:1501

79. Naito Y, Goto T, Akahoshi F, Ono S, Yoshitomi H, Okada T, Sugiyama N, Abe S, Hanada S, Hirata M, Watanabe M, Fukaya C, Yokoyama K, Fujita T (1991) Chem Pharm Bull 39:2323

80. Naito Y, Yamaura Y, Inoue Y, Fukaya C, Yokoyama K, Nakagawa Y, Fujita T (1992) Eur J Med Chem 27:645

81. Verloop A, Hoogenstraaten W, Tipkar J (1976) In: Ariens EJ (ed) Drug Design, vol. VII. Academic Press, New York, pp 165-207

82. Liu SS, Cui SH, Shi YY, Wang LS (2002) Internet Electron J Mol Des 1:610

83. Liu SS, Yin CS, Shi YY, Cai SX, Li ZL (2001) Chin J Chem 19:751

84. Liu SS, Liu LH, Shi YY, Wang LS (2002) Internet Electron J Mol Des 1:310 http://www.biochempress.com

85. Vigorita MG, Ottanà R, Monforte F, Maccari R, Monforte MT, Trovato A, Taviano MF, Miceli N, De Luca G, Alcaro S, Ortuso F (2003) Bioorg Med Chem 11:999

86. Pettipher ER, Higgs GA, Henderson B (1986) Proc Natl Acad Sci USA 83:874

87. Panico AM, Geronikaki A, Mgonzo R, Cardile V, Gentile B, Doytchinova I (2003) Bioorg Med Chem 11:2983

88. McInnes IB, Leung PB, Field M, Huang FP, Sturrock RD, Kinninm J (1996) J Exp Med 184:1519

89. Misra R, Stephan S, Chander CL (1999) Inflamm Res 48:119

90. De Nanteuil G, Portevin B, Benoist A (2001) Il Farmaco 56:107

91. Litina DH, Geronikaki A, Mgonzo R, Doytchinova I (1999) Drug Develop Res 48:53

92. Ellis GA, Blake DR (1993) Ann Rheum Dis 52:241

93. van Muijlwijk-Koezen JE, Timmerman H, Vollinga RC, von Drabbe F, Kunzel J, de Groote M, Visser S, IJzerman AP (2001) J Med Chem 44:739

94. van Tilburg EW, van der Klein PAM, de Groote M, Beukers MW, IJzerman AP (2001) Bioorg Med Chem Lett 11:2017

95. Jung KY, Kim SK, Gao ZG, Gross AS, Melman N, Jacobson KA, Kim YC (2004) Bioorg Med Chem 12:613

96. Fan M, Qin W, Mustafa SJ (2003) Am J Physiol Lung Cell Mol Physiol 284:L1012

97. Ezeamuzie CI (2001) Biochem Pharmacol 61:1551

98. Avila MY, Stone RA, Civan MM (2002) Invest Ophthalmol Vis Sci 43:3021

99. Mubagwa K, Flameng W (2001) Cardiovasc Res 52:25

100. Liang BT, Jacobson KA (1998) Proc Natl Acad Sci USA 95:6995

101. Tracey WR, Magee W, Masamune H, Kennedy SP, Knight DR, Buchholz RA, Hill R (1997) J Cardiovasc Res 33:410

102. von Lubitz DKJE (1999) Eur J Pharmacol 371:85

103. Bhattacharya P, Leonard JT, Roy K (2005) Bioorg Med Chem 13:1159

104. Bhattacharya P, Leonard JT, Roy K (2005) J Mol Model 11:516

105. Borghini A, Pietra D, Domenichellia P, Bianucci AM (2005) Bioorg Med Chem 13:5330

106. Naruto S, Motoc I, Marshall GR (1985) Eur J Med Chem 20:529

107. Borea PA, Bertolasi V, Gilli G (1986) Arzneim Forsch 36:895

108. Diurno MV, Mazzoni O, Piscopo E, Caliganao A, Giordano F, Bolognese A (1992) J Med Chem 35:2910
109. Singh P, Ojha TN, Shrama RC, Tiwari S (1994) Indian J Pharm Sci 57:162
110. Agrawal VK, Sachan S, Khadikar PV (2000) Acta Pharma 50:281
111. Brzezinska E, Koska G, Walczynski K (2003) J Chromatogr A 1007:145
112. Brzezinska E, Koska G, Klimczak A (2003) J Chromatogr A 1007:157
113. Walczynski K, Timmerman H, Zuiderveld OP, Zhang MQ, Glinka R (1999) Il Farmaco 54:533
114. Walczynski K, Guryn R, Zuiderveld OP, Zhang MQ, Timmerman H (1999) Il Farmaco 54:684
115. Walczynski K, Guryn R, Zuiderveld OP, Zhang MQ and Timmerman H (2000) Il Farmaco 55:569
116. La Mattina JL, McCarthy PA, Reiter LA, Holt WF, Yeh LA (1990) J Med Chem 39:543
117. Goel A, Madan AK (1995) J Chem Inf Comput Sci 35:504
118. Wiener H (1947) J Am Chem Soc 69:2636
119. Randić M (1965) J Am Chem Soc 97:6609
120. Borges EG, Takahata Y (2002) J Mol Struct (Theochem) 580:263
121. Mizoule J, Meldrum B, Martine M, Croucher M, Ollat C, Uzan A, Legvand JJ, Gueremy C, LeFur G (1995) Neuropharmacology 24:767
122. Benoit E, Escande D (1991) Pflugers Arch 419:603
123. Hays SJ, Rice MJ, Ortwine DF, Johnson G, Schwarz RD, Boyd DK Copeland LF, Vartanian MG, Boxer PA (1994) J Pharma Sci 83:1425
124. Mathvink RJ, Barritta AM, Candelore MR, Cascieri MA, Deng L, Tota L, Strader CD, Wyvratt MJ, Fisher MH, Weber AE (1999) Bioorg Med Chem Lett 9:1869
125. Mathvink RJ, Tolman JS, Chitty D, Candelore MR, Cascieri MA, Colwell LF, Deng L Jr, Feeney WP, Forrest MJ, Hom GJ, MacIntyre DE, Tota L, Wyvratt MJ, Fisher MH, Weber AE (2000) Bioorg Med Chem Lett 10:1971
126. Nisoli E, Tonello C, Carruba MO (2003) Curr Med Chem Central Nervous Syst Agents 3:257
127. Naylor EM, Parmee ER, Colandrea VJ, Perkins L, Brockunier L, Candelore MR, Cascieri MA, Colwell LF, Deng L, Feeney WP, Forrest MJ, Hom GJ, MacIntyre DE, Strader CD, Tota L, Wang PR, Wyvratt MJ, Fisher MH, Weber AE (1999) Bioorg Med Chem Lett 9:755
128. Shih TL, Candelore MR, Cascieri MA, Chiu SHL, Colwell LF, Deng L, Feeney WP, Forrest MJ, Hom GJ, MacIntyre DE, Miller RR, Stearns RA, Strader CD, Tota L, Wyvratt MJ, Fisher MH, Weber AE (1999) Bioorg Med Chem Lett 9:1251
129. Hanumantharao P, Sambasivarao SV, Soni LK, Gupta AK, Kaskhedikar SG (2005) Bioorg Med Chem 15:3167
130. Guba W, Neidhart W, Nettekoven M (2005) Bioorg Med Chem Lett 15:1599
131. Nettekoven M, Guba W, Neidhart W, Mattei P, Pflieger P, Roche O, Taylor S (2005) Bioorg Med Chem Lett 15:3446
132. Criscione L, Rigollier P, Batzl-Hartmann C, Rueger H, Stricker-Krongrad A, Wyss P, Brunner L, Whitebread S, Yamaguchi Y, Gerald C, Heurich RO, Walker MW, Chiesi M, Schilling W, Hofbauer KG, Levens N (1998) J Clin Invest 102:2136
133. Fukami T, Okamoto O, Fukuroda T, Kanatani A, Ihara M (1998) PCT Int Appl WO 9840356 (Banyu Pharmaceutical Co Ltd, Japan) 78
134. Norman MH, Chen N, Han N, Liu L, Hurt CR, Fotsch CH, Jenkins TJ, Moreno OA (1999) PCT Int Appl WO9940091 (Amgen Inc, USA) 469
135. Breu V, Dautzenberg F, Guerry P, Nettekoven MH, Pieger P (2002) PCT Int Appl WO2002020488 (F. Hoffmann-La Roche, Switzerland) 62

136. Iwaoka M, Takemoto S, Tomoda S (2002) J Am Chem Soc 124:10613
137. Mangelsdorf DJ, Evans RM (1995) Cell 83:841
138. Sohda T, Mizuno K, Tawada H, Sugiyama Y, Fujita T, Kawamastu Y (1982) Chem Pharm Bull 30:3563
139. Sohda T, Mizuno K, Imamiya E, Sugiyama Y, Fujita T, Kawamastu Y (1982) Chem Pharm Bull 30:3580
140. Hulin B, McCarthy PA, Gibbs EM (1996) Curr Pharm Des 2:85
141. Sohda T, Momose Y, Meguro K, Kawamastu Y, Sugiyama Y, Ikeda H (1990) Arzneim Forsch 40:37
142. Sohda T, Mizuno K, Momose Y, Ikeda H, Fujita T, Meguro K (1992) J Med Chem 35:2617
143. Yoshioka T, Fujita T, Kanai T, Aizawa Y, Kurumada T, Hasegawa K, Horikoshi H (1989) J Med Chem 32:421
144. Hulin B, Clark DA, Goldstein SW, McDermott RE, Dambek PJ, Kappeler WH, Lamphere CH, Lewis DM, Rizzi JP (1992) J Med Chem 35:1853
145. Parks DJ, Tomkinson NCO, Villeneuve MS, Blanchard SG, Willson TM (1998) Bioorg Med Chem Lett 8:3657
146. Sohda T, Mizuno K, Kawamastu Y (1984) Chem Pharm Bull 32:4460
147. Weatherman RV, Fletterick RJ, Scanlan TS (1999) Annu Rev Biochem 68:559
148. Nolte RT, Wisely GB, Westin S, Cobb JE, Lambert MH, Kurokawa R, Rosenfeld MG, Willson TM, Glass CK, Milburn MV (1998) Nature 395:137
149. Uppenberg J, Svensson C, Jaki M, Bertilsson G, Jendeberg L, Berkenstam A (1998) J Biol Chem 273:31108
150. Yanagisawa H, Takamura M, Yamada E, Fujita S, Fujiwara T, Yachi M, Isobe A, Hagisawa Y (2000) Bioorg Med Chem Lett 10:373
151. Iwata Y, Miyamoto S, Takamura M, Yanagisawa H, Kasuya A (2001) J Mol Grap Model 19:536
152. Murakami K, Tobe K, Ide T, Monchizuki T, Ohashi M, Akanuma Y, Yazaki Y, Kadowaki T (1998) Diabetes 47:1841
153. Desai RC, Han W, Metzger EJ, Bergman JP, Gratale DF, MacNaul KL, Berger JP, Doebber TW, Leung K, Moller DE, Heck JV, Sahoo SP (2003) Bioorg Med Chem Lett 13:2795
154. Desai RC, Gratale DF, Han W, Koyama H, Metzger EJ, Lombardo VK, MacNaul KL, Doebber TW, Berger JP, Leung K, Franklin R, Moller DE, Heck JV, Sahoo SP (2003) Bioorg Med Chem Lett 13:3541
155. Khanna S, Sobhia ME, Bharatam PV (2005) J Med Chem 48:3015
156. Xu HE, Lambert MH, Montana VG, Plunket KD, Moore LB, Collins JL, Oplinger JA, Kliewer SA, Gampe Jr RT, Mckee DD, Moore LB, Willson TM (2001) Proc Natl Acad Sci USA 98:13919
157. Terashima H, Hama K, Yamamoto R, Tsuboshima M, Kikkawa R, Hatanaka I, Shigeta Y (1984) J Pharmacol Exp Ther 229:226
158. Chung SSM, Ho ESM, Lam KSL, Chung SK (2003) J Am Soc Nephrol 14:233
159. Kador PF (1988) Med Res Rev 8:325
160. Fresneau P, Cussac M, Morand J, Szymonski B, Tranqui D, Leclerc G (1998) J Med Chem 41:4706
161. Wilson DK, Bohren KM, Gabbay KH, Quiocho FA (1993) Proc Natl Acad Sci USA 90:9847
162. Mimura T, Kohama Y, Kuwahara S, Yamamoto K, Komiyama Y, Satake M, Chiba Y, Miyashita K, Tanaka T, Imanishi T, Iwata C (1988) Chem Pharm Bull 36:1110

163. Sataka M, Chiba Y, Kohama Y, Yamamoto K, Okabe M, Mimura T, Imanishi T, Iwata C (1989) Experientia 45:1110
164. Sato M, Kawashima Y, Goto J, Yamane Y, Chiba Y, Jinno S, Satake M, Imanishi T, Iwata C (1994) Chem Phram Bull 42:521
165. Ezumi K, Yamakawa M, Narisada M (1990) J Med Chem 33:1117
166. Patscheke H, Stegmeier K (1984) Thromb Res 33:277
167. Arita H, Nakano T, Hanasaki K (1989) Prog Lipid Res 28:273
168. Coleman RA, Sheldrick RLG (1989) Br J Pharmacol 96:688
169. Lacan F, Verache-Lembege M, Vercauteren J, Leger JM, Masereel B, Donge JM, Nuhrich A (1999) Eur J Med Chem 34:311
170. Ezumi K, Yamakawa M, Narisada M (1990) J Med Chem 33:1117
171. Fukumoto S, Shiraishi M, Terashita ZI, Ashida Y, Inada Y (1992) J Med Chem 35:2202
172. Saxena AK, Pandey SK, Seth P, Singh MP, Dikshit M, Carpy A (2001) Bioorg Med Chem 9:2025
173. Barreca ML, Balzarini J, Chimirri A, De Clercq E, De Luca L, Holtje HD, Holtje M, Monforte AM, Monforte P, Pannecouqe C, Rao A, Zapalla M (2002) J Med Chem 45:5410
174. Jonckheere AJ, De Clercq E (2000) Med Res Rev 20:129
175. Tantillo JPC, Ding A, Jacobomolina RG, Nanni PL, Boyer SH, Hunghes R, Pawels K, Anderis PAJ, Arnold E (1994) J Mol Biol 243:369
176. De Clercq E (1993) Med Res Rev 13:229
177. Hajos G, Riedi S, Monar J, Szabo D (2000) Drugs Fut 25:47
178. Garg R, Gupta SP, Gao H, Babu MS, Debnath AK, Hansch C (1999) Chem Rev 99:3525
179. Schaefer W, Friebe WG, Leinert M, Merttens A, Poll T, Von der Saal W, Zilch H, Nuber H, Ziegler ML (1993) J Med Chem 36:726
180. Morris GM, Goodsell DS, Halliday RS, Huey R, Hart WE, Belew RK, Olson AJ (1998) J Comput Chem 19:1639
181. Prabhakar YS, Solomon VR, Rawal RK, Gupta MK, Katti SB (2004) QSAR Comb Sci 23:234
182. Prabhakar YS, Rawal RK, Gupta MK, Solomon VR, Katti SB (2005) Comb Chem High T Scr 8:431
183. Roy K, Leonard T (2005) QSAR Comb Sci 24:279
184. Rao A, Balzarini J, Carbone A, Chimirri A, De Clercq E, Monforte AM, Monforte P, Pannecouqe C, Zapalla M (2004) Il Farmaco 59:33
185. Barreca ML, Chimirri A, De Clercq E, De Luca L, Monforte AM, Monforte P, Rao A, Zapalla M (2003) Il Farmaco 58:259
186. Rao A, Carbone A, Chimirri A, De Clercq E, Monforte AM, Monforte P, Pannecouque C, Zappala M (2003) Il Farmaco 58:115
187. Barreca ML, Chimirri A, De Luca L, Monforte AM, Monforte P, Rao A, Zapalla M, Balzarini J, De Clercq E, Pannecouqe C (2001) Bioorg Med Chem Lett 11:1793
188. Fujita T, Ban T (1971) J Med Chem 14:148
189. Rawal RK, Solomon VR, Prabhakar YS, Katti SB (2005) Comb Chem High T Scr 8:439
190. Rawal RK, Prabhakar YS, Katti SB (2005) Bioorg Med Chem 13:6771
191. Prabhakar YS, Jain P, Khan ZK, Haq W, Katti SB (2003) QSAR Comb Sci 22:456
192. Eng G, Whalen D, Musingarimi P, Tierney J, De Rosa M (1998) Appl Organomet Chem 12:25
193. Zou XJ, Jin GY, Zhang ZX (2002) J Agric Food Chem 50:1451
194. Zou XJ, Lai LH, Jin GY, Zhang ZX (2002) J Agric Food Chem 50:3757

195. Roeder T, Nathanson JA (1993) Neurochem Res 18:921
196. Roeder T, Gewecke M (1990) Biochem Pharmol 39:1793
197. Roeder T (1992) Life Sci 50:21
198. Roeder T (1995) Br J Pharmacol 114:210
199. Roeder T (1990) Eur J Pharmacol 191:221
200. Jennings KR, Kuhn DG, Kukel CF, Trotto SH, Whiteney WK (1988) Pestic Biochem Physiol 30:190
201. Ismail SMM, Baines RA, Downer RGH, Dekeyser MA (1996) Pestic Sci 46:163
202. Hirashima A, Yoshii Y, Eto M (1992) Pestic Biochem Physiol 44:101
203. Hirashima A, Tomita J, Pan C, Taniguchi E, Eto M (1997) Bioorg Med Chem 5:2121
204. Hirashima A, Yoshii Y, Eto M (1992) Biosci Biotech Biochem 56:1062
205. Hirashima A, Yoshii Y, Eto M (1994) Biosci Biotech Biochem 58:1206
206. Pan C, Hirashima A, Tomita J, Kuwano E, Taniguchi E, Eto M (1997) J Sci Biol Chem 1: www.netsci-journal.com/97v1/97013/index.htm
207. Raina AK (1993) Ann Rev Entomol 38:320
208. Ma PWK, Roelofs W (1995) Insect Biochem Molec Biol 25:467
209. Fabrias G, Barrot M, Camps F (1995) Insect Biochem Molec Biol 25:655
210. Zhu J, Millar J, Loefstedt C (1995) Archs Insect Biochem Physiol 30:41
211. Jurenka RA (1996) Arch Insect Biochem Physiol 33:245
212. Rafaeli A, Gileadi C (1997) Invertebr Neurosci 3:223
213. Hirashima A, Pan C, Katafuchi Y, Taniguchi E, Eto M (1996) J Pestic Sci 21:419
214. Hirashima A, Shinkai K, Kuwano E, Taniguchi E, Eto M (1998) Biosci Biotech Biochem 62:1179
215. Rafaeli A, Gileadi C, Hirashima A (1999) Pestic Biochem Physio 65:194
216. Hirashima A, Morimoto M, Kuwano E, Eto M (1999) Bioorg Med Chem 7:2621
217. Clifford KH (1980) Eur J Biochem 106:337
218. Sacktor BS, Wormser-Shavit E (1966) J Biol Chem 241:634
219. Qian X, Liu Z, Li Z, Song G (2001) J Agric Food Chem 19:5279

Top Heterocycl Chem (2006) 4: 251–289
DOI 10.1007/7081_034
© Springer-Verlag Berlin Heidelberg 2006
Published online: 7 April 2006

QSAR Studies on Calcium Channel Blockers

Satya P. Gupta

Department of Chemistry, Birla Institute of Technology and Science, Pilani 333031, India
spg@bits-pilani.ac.in

Abstract Calcium channel blockers (CCBs) have potential therapeutic uses against several cardiovascular and non-cardiovascular diseases. For vasospastic angina, CCBs have been found to be the most effective drugs. These drugs selectively inhibit Ca^{2+} influx into heart muscles by blocking slow inward channels for Ca^{2+} or inhibit Ca^{2+} influx into vascular smooth muscles. The result is negative inotropism of smooth muscle relaxation, which is translated into hypotension. The three principal structural classes have been found to act as potent calcium channel blockers and they are phenylalkylamines, 1,4-dihydropyridines (DHPs), and benzothiazepines. Recently, a few more classes of CCBs have been studied. This article presents a comprehensive review on quantitative structure-activity relationship (QSAR) studies on all kinds of CCBs. These QSAR studies highlight the essential structural features and physicochemical properties that the compounds should possess to act as potential CCBs and vividly describe the mechanism of interaction of CCBs with the calcium channel.

Keywords Calcium channel blockers ·
Quantitative structure-activity relationship · Phenylalkylamines · 1,4-Dihydropyridines ·
Benzothiazepines · Verapamil · Nifedipine · Diltiazem · Dihydropyrimidines ·
Phenylsulfonylindolizines

Abbreviations
CCBs   calcium channel blockers
CoMFA comparative molecular field analysis
DHPs   1,4-dihydropyridines
QSAR   quantitative structure-activity relationship
SAR   structure-activity relationship

1
Introduction

Calcium is well known to play critical roles in cellular communication and regulation. The critical role of cellular Ca^{2+} was recognized early by Ringer [1], who noted its essential role in cardiac excitation-contraction coupling. Equally, the significance of the pharmacological regulation of cellular Ca^{2+} was recognized by Fleckenstein [2]. Calcium in excess is a lethal cation, however, and uncontrolled involvement of Ca^{2+} subsequent to cellular insult or injury can lead to irreversible cell destruction and death [3]. It is anticipated that the cellular movements and storage of Ca^{2+} will be subject to a variety of regulatory processes, though all of which may not assume equal importance in every cell type or during every stimulus.

The calcium channel may be viewed as a pharmacological receptor with several specific sites with which a variety of drugs may interact. However, Fleckenstein and coworkers drew attention to a chemically heterogeneous group of agents that served as electrochemical uncouplers in the heart and whose effects were mimicked by extracellular Ca^{2+} withdrawal. These fundamental properties, extended in understanding through more detailed biochemical and electrophysiological studies [4], have served to define a group of drugs with particular chemical utility in the cardiovascular diseases. Such studies led to definition of the site of action of these agents as the L class of voltage-gated Ca^{2+} channels and to localize their site of action as the α_1-transmembrane protein [5, 6]. There exists several other classes also of voltage-gated Ca^{2+} channel, such as T, P, N, etc., distinguishable from each other by biochemical, electrophysiological, and pharmacological characteristics [7].

Calcium channels are expected to possess the following general properties [6]: (1) specific binding sites for both activator and antagonist ligands, and the possible existence of an endogenous species; (2) coupling of the binding sites to the permeation and gating machinery of the channel; (3) association with regulatory guanine nucleotide binding proteins; (4) regulation by homologous influences; and (5) alteration of expression and function in disease states. Except for the discovery of an endogenous species [8], all other properties have been largely found to be associated with the L class of voltage-gated channels [9, 10].

2
Calcium Channel Blockers

The three principal structural classes of compounds have been found to bind with the L subclass calcium channel, and they are phenylalkylamines, 1,4-dihydropyridines, and the benzothiazepines, of which the protypical representatives are verapamil (1), nifedipine (2), and diltiazem (3), respectively. For the treatment of vasospastic angina, the calcium channel blockers (CCBs; also called calcium channel antagonists) have been found to be the most effective. These drugs selectively inhibit Ca^{2+} influx into the heart muscles by blocking the slow inward channels for Ca^{2+} or inhibit Ca^{2+} influx in the vascular smooth muscle. The result is negative inotropism of smooth muscle relaxation, which is translated into hypotension. Calcium channel blockers are economically and therapeutically important for the treatment of several cardiovascular diseases, such as angina, hypertension, arrhythmias, etc. (Table 1). Besides, they are also useful for the treatment of several non-cardiovascular diseases, such as asthma, dysmenorrhea, premature labor, and several miscellaneous ailments such as cancer, epilepsy, glaucoma, etc. (Table 1) [5].

Thus, the calcium channel blockers are a major therapeutic group of agents that interact preferentially with the L class of voltage-gated channel. Their varying therapeutic actions are due to mainly the existence of several sites at

1(verapamil)

2 (nifedipine)

3 (diltiazem)

Table 1 Potential therapeutic uses of calcium channel blockers [6]

Cardiovascular	Non-cardiovascular	Other
Angina	Achalasia	Aldosteronism
Atherosclerosis	Asthma	Cancer chemotherapy
Cardioplegia	Dysmenorrhea	Epilepsy
Cerebral ischemia	Eclampsia	Glaucoma
Hypertension	Esophageal spasm	Tinnitus
Congestive heart failure	Intestinal hyperacidity	Manic syndrome
Migraine	Premature labor	Vertigo
Peripheral vascular diseases	Urinary incontinence	Motion sickness
Pulmonary hypertension	Obstructive lung disease	Tourette's syndrome
Subarachnoid hemorrhage		Spinal cord injury
Tachyarrhythmias		Antimalarial drug resistance

the L channel and the ability of these drugs to have state-dependent interaction according to the membrane potential and stimulus frequency. Analysis of the structural requirements that determine affinity and voltage dependence is important for the design of potent drugs of desired selectivity.

3
Selectivity of Calcium Channel Blockers

The calcium channel blockers are a group of drugs that exhibit a major selectivity of action both between and within structural classes. This selectivity arises from a variety of factors, such as pharmacokinetics (absorption, distribution, and metabolism), mode of Ca^{2+} mobilization (voltage-gated channel, intracellular stores, or other sources), class and subclass of voltage-gated Ca^{2+} channels modulated, state-dependent interactions (frequency and voltage dependence), and pathological state (homologous and heterologous influence on channel expression and function [5]). Of these, the state-dependent interaction of drug actions at ion channels and its significance have been explained in the terms of the modulated receptor hypothesis [4, 6, 11]. The concept of state-dependent interaction is critical to the interpretation of structure-activity relationships (SARs), because it indicates that (a) drugs may bind selectively to different channel states and many have preferential access pathways to these states and (b) drugs may exhibit qualitatively different SARs according to the channel state with which they preferentially interact.

The state-dependent interactions are critically related to the molecular features of the drugs and thus the determination of the molecular features of the compounds that would favor the state dependence of the interactions can facilitate the design of the drugs with selectivity enhanced, or reduced, to their voltage-dependent interactions.

4
Molecular Features Essential for Ca^{2+} Channel Blocking Action

Among the three principal classes of compounds acting on the calcium channel, the 1,4-dihydropyridines (DHPs) have attracted particular attention since they act both as calcium channel blockers and activators. Several SAR studies are available for 1,4-dihydropyridine antagonists, but such studies on activators are limited [12, 13]. It has been proposed that the active conformation of 1,4-dihydropyridine includes a 1,4-dihydropyridine ring in a flattened boat confirmation with the 4-phenyl group orientated in a pseudoaxial confirmation [14]. An early study of Loev et al. [15] on in vivo hypotensive activity of a series of 2,6-dimethyl-3,5-dicarboethoxy-1,4-dihydropyiridines led to the definition of some basic structural requirements for the antagonist activity of DHPs:

1. Activity increases with 4-substitution in the sequence H < Me < cyclo-alkyl < heterocyclic < phenyl and substituted phenyl.
2. Substituents in the 4-phenyl ring enhance activity in the order ortho > meta ≫ para. Electron withdrawing substituents in the ortho position are optimum but any substituent in the para position reduces the activity.
3. The 1,4-dihydropyridine ring is essential. Oxidation to pyridine abolishes the activity.
4. The presence of N1–H is essential.
5. Ester groups at the C3 and C5 are optimum. Replacement by other electron withdrawing groups including – CN and – COMe leads to reduction in the activity.

The most important determinants of antagonist activity were, however, indicated to be the 4-phenyl and C3 and C5 ester substituents. These studies of Loev et al. were largely confirmed in a variety of subsequent investigations based on in vitro pharmacological and radioligand bind approaches [4].

The high activity of 4-phenyl-DHPs was in fact attributed to their conformational properties, in which the aryl ring was supposed to have perpendicular orientation relative to the 1,4-dihydropyridine ring [15]. The 1,4-dihydropyridine ring exists as a non-planar boat-shaped structure with the N1 and C4 atoms defining the stern and bowsprit positions (Fig. 1) [16]. The phenyl ring is bound to it at a pseudo axial position and approximately bisects the pyridine ring. The rotational freedom of the phenyl ring about the C4-C1′ bond is sterically restricted and the plane of the phenyl ring is forced to lie close to the N1–C4 vertical symmetry plane of the 1,4-dihydropyridine ring. This conformation of 4-phenyl-1,4-dihydropyridines is in accordance to the

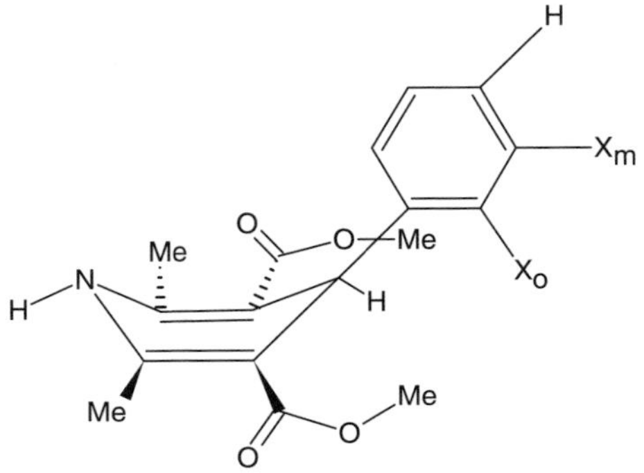

Fig. 1 The molecular geometry of DHP analogues. This geometry was adopted by Gaudio et al. to calculate some quantum mechanical parameters, which they used in their QSAR study [16]. The 1,4-dihydropyridine ring is shown to have a boat-like conformation and the phenyl ring to be bound to it at a pseudoaxial position and approximately bisecting the pyridine ring. From [16]. © 1994 John Wiley & Sons, Inc., reprinted by permission

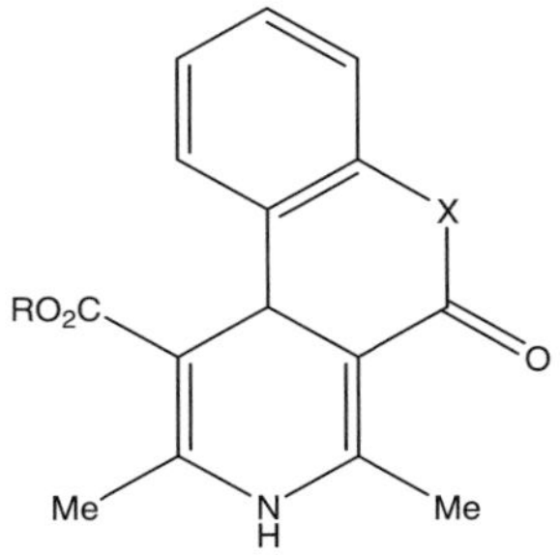

4; X = O, O–(CH$_2$)$_{1-2}$, O–(CH$_2$)$_{2-5}$–O

findings of solid-state structural studies [17, 18], which were well supported by the synthesis and activity of rigid analogues [19, 20] and by additional conformational studies [14, 21–23]. In a series of analogues, having a bridge of lactone ring between the phenyl and 1,4-dihydropyridine rings (**4**), the activity was found to increase as the rings approached the perpendicular orientation [19, 20].

The conformations of ester groups have also been found to be crucial for the activity of 1,4-dihydropyridines. Their conformations are defined with respect to the orientation of the carbonyl groups relative to the neighboring C = C bond of the 1,4-dihydropyridine ring. Both the ester groups are always found to be nearly coplanar with adjacent edges of the 1,4-dihydropyridine ring and the carbonyl group can be oriented either cis (*sp*) or trans (*ap*) to the neighboring C = C bond of the ring. Thus, the two ester groups can adopt one of the three conformations (Fig. 2): *cis–cis* (*sp,sp*), *cis–trans* (*sp,ap*) or *trans–trans* (*ap,ap*). It has been observed that ortho-substituted compounds have a

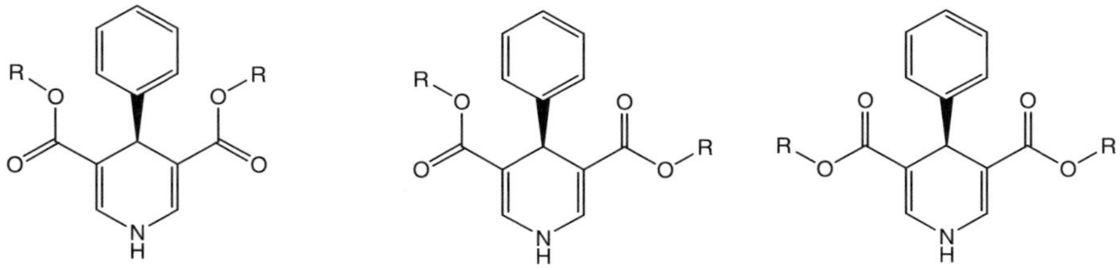

Fig. 2 Conformations of ester groups in 1,4-dihydropyridines with respect to orientation of carbonyl group relative to the neighboring $C=C$ bond of the ring; **a** *cis-cis* (*sp,sp*), **b** *cis-trans* (*sp,ap*), **c** *trans-trans* (*ap,ap*)

preference for the *sp,sp* conformation and non-ortho-substituted compounds have generally a preference for the *sp,ap* conformation. Although the relationship between these conformational preferences and the biological activity has not been well established as yet, they support the idea that there are non-equivalent binding sites adjacent to the 1,4-dihydropyridine ring, differential occupancy of which is critical to the determination of the quantitative and qualitative levels of agonist and antagonist activity.

5
QSAR Studies

QSAR (quantitative structure-activity relationship) studies provide the guidelines for making structural changes in the compounds so that drugs of higher potency can be obtained. It tries to explain the variance in biological activities of a given series of compounds in terms of physicochemical and structural properties of the molecules and thus provides a deeper insight into the mechanism of drug receptor-interactions which help tailor the drug to have the optimal interaction with the receptor.

Primarily, the QSAR studies on calcium channel blockers were related to only the 1,4-dihydropyridine class to which belongs nifedipine (2). A series of nifedipine analogues (5) as listed in Table 2 were studied by Bolger et al. [24] for their calcium channel blocking activity in terms of the molar concentration of drug (IC_{50}) leading to the 50% inhibition of the binding of [^{3}H]nitrindipine to guinea pig ileal preparation.

For this series of compounds, Mahmoudian and Richards [25] showed that the activity of ortho-substituted analogues (including the parent compound 1) had a significant correlation with Verloop's B1 parameter [26], defining the minimum width of the substituent (Eq. 1), and that the activity of meta-substituted analogues (including the parent compound 1) had a significant

5

Table 2 Bolger's data on the binding of nifedipine analogues (5) with the receptor and the physicochemical parameters [25]

Compd	X	$\log(1/IC_{50}^a)$	π	σ	B1	L
1	H	7.85	0.00	0.00	1.00	2.06
2	2-CN	9.18	– 0.33	–	1.60	4.23
3	2-NO$_2$	9.08	– 0.23	1.24	1.70	3.44
4	2-Me	8.71	0.86	– 0.13	1.52	3.00
5	2-Cl	9.78	0.76	0.68	1.80	3.52
6	2-OMe	7.87	– 0.33	0.00	1.35	3.98
7	2-F	8.44	0.00	0.54	1.35	2.06
8	3-NO$_2$	9.97	0.11	0.71	1.70	3.44
9	3-N$_3$	8.67	0.46	0.27	1.50	4.62
10	3-OMe	7.27	0.12	0.12	1.35	3.98
11	3-CN	8.68	– 0.31	0.56	1.60	4.23
12	3-Cl	9.30	0.77	0.37	1.80	3.52
13	3-F	8.49	0.22	0.34	1.35	2.06
14	3-Me	7.28	0.52	– 0.07	1.52	3.00
15	4-Cl	6.22	0.73	0.23	1.80	3.52
16	4-Me	7.18	0.63	– 0.17	1.52	3.00
17	4-F	7.46	0.15	0.06	1.35	2.06
18	4-NO$_2$	6.52	0.22	1.24	1.70	3.44

[a] The molar concentration of the drug leading to 50% inhibition of the binding of [^{3}H]nitrendipine to guinea pig ileal preparation.

correlation with the electronic parameter σ (Hammelt constant) (Eq. 2). In Eqs. 1 and 2, n is the number of data points, r is the correlation coefficient, s is the standard deviation, and F is the F-ratio between the variances of calculated and observed activities. These equations exhibited that ortho-substituents can affect the activity by their width and the meta-substituents can do so by their electron-withdrawing capabilities.

$$\log(1/IC_{50}) = 5.152 + 2.407B1_0$$
$$n = 7, \; r = 0.91, \; s = 0.32, \; F_{1,5} = 23.81 \tag{1}$$

$$\log(1/IC_{50}) = 7.543 + 3.116\sigma_m$$
$$n = 8, \; r = 0.882, \; s = 0.48, \; F_{1,6} = 21.01 \;. \tag{2}$$

For the five para-substituted compounds (including the parent one), the activity was, however, found to be correlated with Verloop's length parameter L of the substituents [25] (Eq. 3), which indicated that lengthy substituents at the para position will not be advantageous to the activity.

$$\log(1/IC_{50}) = 9.152 - 0.876L_p$$
$$n = 5, \; r = 0.94, \; s = 0.27, \; F_{1,3} = 22.27 \;. \tag{3}$$

For the whole series of Table 2, the correlation obtained was as follows:

$$\log(1/IC_{50}) = 7.430 + 2.376B1_{o,m} - 0.472L_m - 0.674L_p + 1.928\sigma_m$$
$$n = 18, \; r = 0.93, \; s = 0.43, \; F_{4,13} = 23.10 \;. \tag{4}$$

This equation suggested that the length of not only the para substituent but also of the meta substituents will be detrimental to the activity of the compounds and that the width of not only the ortho substituents but also of the meta substituents will be beneficial to the activity. However, the electronic character (electron-withdrawing) of only the meta substituents was shown to affect the activity. The hydrophobic property of substituents was found to have little effect, but when a detailed QSAR study was made by Coburn et al. [27] on a fairly large series of nifedipine analogues (Table 3) that also included the compounds of Table 2, a significant role of hydrophobic constant π of all the substituents had surfaced (Eq. 5). The IC_{50} in Eq. 5 was, however, related to the effect of compounds on tonic contractile response of longitudinal muscle strips of guinea pig ileum [27].

$$\log(1/IC_{50}) = 0.62(\pm0.09)\pi + 1.96(\pm0.29)\sigma_m - 0.44(\pm0.09)L_m -$$
$$- 1.51(\pm0.26)L_{m'} - 3.26(\pm0.33)B1_p + 14.23(\pm0.78)$$
$$n = 46, \; r = 0.90, \; s = 0.67, \; F_{5,40} = 33.93 \;. \tag{5}$$

In this equation π and σ_m have been entered for all substituents without regard to their position in the phenyl ring, but m and m' in L refer to meta positions, separately, exhibiting a marked difference in the steric effects from two meta positions. The figures with $\pm$ sign within parentheses refer to 95% confidence intervals.

For a slightly extended series of compounds than that used in deriving Eq. 4, Coburn et al. [27] found that Bolger's binding data could also depend

Table 3 Activities of nifedipine analogues (5) against muscle contraction and physico-chemical parameters [28]

Compd	X	$\log(1/IC_{50}^a)$	π	σ_m	B1	L
1	3-Br	8.89	0.86	0.39	1.95	3.83
2	$2\text{-}CF_2$	8.82	0.88	0.43	1.98	3.30
3	2-Cl	8.66	0.71	0.37	1.80	3.52
4	$3\text{-}NO_2$	8.40	− 0.28	0.71	1.70	3.44
5	$2\text{-}CH = CH_2$	8.35	0.82	0.05	1.60	4.29
6	$2\text{-}NO_2$	8.29	− 0.28	0.71	1.70	3.44
7	2-Me	8.22	0.56	− 0.07	1.52	3.00
8	2-Et	8.19	1.02	− 0.07	1.52	4.11
9	2-Br	8.12	0.86	0.39	1.95	3.83
10	2-CN	7.80	− 0.57	0.56	1.60	4.23
11	3-Cl	7.80	0.71	0.37	1.80	3.52
12	3-F	7.68	0.14	0.34	1.35	2.65
13	H	7.55	0.00	0.00	1.00	2.06
14	3-CN	7.46	− 0.57	0.56	1.60	4.23
15	3-I	7.38	1.12	0.35	2.15	4.23
16	2-F	7.37	0.14	0.34	1.35	2.65
17	2-I	7.33	1.12	0.35	2.15	4.23
18	2-OMe	7.24	− 0.02	0.12	1.35	3.98
19	$3\text{-}CF_3$	7.13	0.88	0.43	1.98	3.30
20	3-Me	6.96	0.56	− 0.07	1.52	3.00
21	3-OEt	7.96	0.38	0.10	1.35	4.92
22	3-OMe	6.72	− 0.02	0.12	1.35	3.98
23	$3\text{-}NMe_2$	6.05	0.18	− 0.15	1.50	3.53
24	3-OH	6.00	− 0.67	0.12	1.35	2.74
25	$3\text{-}NH_2$	5.70	− 1.23	− 0.16	1.50	2.93
26	3-OAc	5.22	− 0.64	0.39	1.35	4.87
27	3-OCOPh	5.20	1.46	0.21	1.70	8.15
28	$2\text{-}NH_2$	4.40	− 1.23	− 0.16	1.50	2.93
29	$3\text{-}N^+Me_3$	4.30	− 5.96	0.88	2.56	4.02
30	4-F	6.89	0.14	0.34	1.35	2.65
31	4-Br	5.40	0.86	0.39	1.95	3.83
32	4-I	4.64	1.12	0.35	2.15	4.23
33	$4\text{-}NO_2$	5.50	− 0.28	0.71	1.70	3.44
34	$4\text{-}NMe_2$	4.00	0.18	− 0.15	1.50	3.53
35	4-CN	5.46	− 0.57	0.56	2.06	4.23
36	4-Cl	5.09	0.71	0.37	1.80	3.52
37	$2,6\text{-}Cl_2$	8.72	1.42	0.74		
38	F5	8.36	0.70	1.70		
39	2-F,6-Cl	8.12	0.85	0.71		
40	$2,3\text{-}Cl_2$	7.72	1.42	0.74		
41	$2\text{-}Cl,5\text{-}NO_2$	7.52	0.43	1.08		

Table 3 continued

Compd	X	$\log(1/IC_{50}^{a})$	π	σ_m	B1	L
42	3,5-Cl$_2$	7.03	1.42	0.74		
43	2-OH,5-NO$_2$	7.00	− 0.95	0.83		
44	2,5-Me$_2$	7.00	1.12	− 0.14		
45	2,4-Cl$_2$	6.40	1.42	0.74		
46	2,4,5-Me$_3$	3.00	− 0.06	0.36		

[a] The molar concentration of the drug necessary for 50% inhibition of the contraction of guinea pig ileum induced by methylfurmethide.

upon the hydrophobicity, since they were able to derive the correlation as:

$$\log(1/IC_{50}) = 0.81(\pm0.11)\pi + 2.36(\pm0.51)\sigma_m + 0.99(\pm0.35)B1_o -$$
$$- 3.18(\pm0.49)B1_m + 9.83(\pm0.80)$$
$$n = 21, \ r = 0.95, \ s = 0.49, \ F_{4,16} = 38.91 . \tag{6}$$

Some quantum mechanical parameters were also found to be useful in accounting for the variance in calcium channel blocking activity of nifedipine analogues. For the same set of compounds as treated by Coburn et al. (Table 3), Gaudio et al. [16] had correlated the contractile response inhibition data as

$$\log(1/IC_{50}) = 0.44(\pm0.26)\pi + 1.47(\pm0.93)\sigma_m - 0.032(\pm0.011)V_w -$$
$$- 1.65((\pm0.53)B1_p - 6.5(\pm1.9)F_5^{(e)} + 0.217(\pm0.071) \ \epsilon_{rot}+$$
$$+ 17.4(\pm3.2)$$
$$n = 45, \ r = 0.95, \ s = 0.49, \ F_{6,38} = 54.69 . \tag{7}$$

In this equation, $F_5^{(e)}$ refers to the frontier electron density at the 5-position of the phenyl ring, V_w refers to the van der Walls volume of the whole molecule, and ϵ_{rot} refers to the energy barrier of the rotation of the phenyl ring. Equation 7, therefore, suggests, in addition to what one would conclude from Eq. 5, that high electron density at the 5-position and the bulk of the molecule will not be advantageous, rather a high energy barrier of the rotation of the phenyl ring will be beneficial. A high-energy barrier will mean the conformational rigidity of the phenyl ring with respect to the pyridine ring. All the quantum mechanical parameters used in Eq. 7 by Gaudio et al. were calculated using the AM1 method [28] and fully optimized geometries of the compounds.

Both Eqs. 5 and 7 show that a dominant steric effect can be produced from the para position. In fact, Gaudio et al. [16] found that the parameter V_w was relevant for only para-substituted analogues. For all mono-substituted para analogues, the correlation obtained by these authors was:

$$\log(1/\text{IC}_{50}) = 26.4(\pm 8.4) - 0.073(\pm 0.029)V_w$$
$$n = 8,\ r = 0.93,\ s = 0.46,\ F_{1,3} = 37.08. \tag{8}$$

That the bulky substituents at the para position will be detrimental to the activity was also shown by Bernstein and Wold [29] in a study on the binding affinity of a small set of compounds. However, these authors had also observed that electron-withdrawing substituents on the ring will enhance the activity.

Analogues of verapamil (1) were also studied for their calcium channel antagonist activity, and their potency for isotonic contractile response of cat capillary muscle preparation was reported [30], which was found to be correlated with the electronic constant and the molecular volume of the B-ring substituents as:

$$\log(1/\text{IC}_{50}) = 0.96\sigma + 0.63\text{MV}$$
$$n = 7,\ r = 0.994,\ s = 0.064, \tag{9}$$

which suggested that along with the electronic property, the size of the substituents will also be important for the activity of the compounds.

Calcium channel blockers bind specifically to receptor sites associated with the voltage-dependent calcium channels [31, 32]. These blockers inhibit calcium uptake [33, 34] and block smooth muscle contraction [35, 36]. All these three activities of calcium channel blockers have been found to be mutually correlated. For ten known calcium channel blockers (Table 4), Papaionnou et al. [37] derived the correlations:

$$\log \text{IC}_{50}(\text{Ca uptake}) = 0.863(\pm 0.12) \log \text{IC}_{50}(\text{binding}) - 0.538(\pm 0.87)$$
$$n = 10,\ r = 0.93,\ p < 0.0001 \tag{10}$$
$$\log \text{IC}_{50}(\text{contraction}) = 0.815(\pm 0.12) \log \text{IC}_{50}(\text{Ca uptake}) - 1.212(\pm 0.81)$$
$$n = 10,\ r = 0.925,\ p < 0.0001. \tag{11}$$

For a series of diltiazem-like calcium channel blockers (6), a comparative molecular field analysis (CoMFA) made by Corelli et al. [38] suggested that these calcium channel blockers can interact with the receptor at its negative charge site, two hydrogen-bonding sites, and three hydrophobic regions (Figs. 3 and 4). As shown in Fig. 3 (top view), the hydrophobic region 1 of the receptor surrounds the polycyclic core quite closely so that it does not accept substituents at carbons 6–8. The hydrophobic region 2 surrounds less tightly the side chain, allowing the presence of, at most, one bulky substituent. This hydrophobic region is closely adjacent to the negative charge site and a hydrogen bonding site, which interact with the protonated basic nitrogen and the lactam carbonyl oxygen, respectively, depending upon the molecules. The second hydrogen-bonding site is located in the pocket which accommodates the 4β-phenyl ring and interacts with the oxygen of the p-methoxy group.

Table 4 Some known calcium channel blockers and their different biological activities [37]

Compd	C_{50} (M) (calcium uptake) [a]	C_{50} or ED_{50} (M) (binding) [b]	IC_{50} (M) (concentration) [c]
Nifedipine	2.0×10^{-9}	9.5×10^{-10}	7.9×10^{-9}
(±)-Nimodipine	3.0×10^{-9}	1.7×10^{-9}	3.0×10^{-9}
(−)-202-791	4.0×10^{-9}	1.3×10^{-9}	3.2×10^{-8}
(±)-Verapmil	2.0×10^{-7}	(4.0×10^{-7}) [d]	1.4×10^{-7}
(±)-D-600	4.0×10^{-7}	(2.9×10^{-8}) [d]	4.4×10^{-8}
(±)-Bepridil	4.0×10^{-6}	1.3×10^{-5}	1.0×10^{-5}
(+)-Diltiazem	7.0×10^{-7}	(1.2×10^{-7}) [d]	3.8×10^{-7}
Cinnarizine	2.0×10^{-6}	2.5×10^{-6}	1.4×10^{-6}
Flunarizine	1.7×10^{-6}	1.2×10^{-6}	1.7×10^{-6}
(±)-Prenylamine	6.0×10^{-6}	3.9×10^{-7}	4.3×10^{-6}

[a] Inhibition of K^+-induced Ca^{2+}-uptake in rabbit aortic smooth muscle cells
[b] Inhibition of [^{3}H]nitrendipine binding to rat ventricular membrane preparations
[c] Inhibition of K^+-induced contraction of vascular smooth muscle
[d] ED_{50} value

6

This last interaction is supposed to be of particular importance for the affinity of CCBs.

Figure 4 presents an edge view of the compounds in the putative binding site, showing that the 4α-substituents lie almost perpendicular to the plane of the tricyclic system and occupy the hydrophobic region 3 that in turn can accept substituents as long as there is a phenyl group, but it is shaped to prevent para-substituted analogues from fitting.

For the activity of a series of **6**, however, the substituted phenyl ring at C4 and the basic side chain at C1 on the pyrrole ring were found to constitute two important pharmacophores [39] and Campiani et al. [39] also suggested that substitution on the fused phenyl ring (R_4-substituents) and the double substitution at C4 were beneficial to the activity. The substitution of the

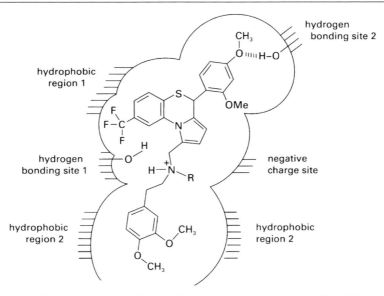

Fig. 3 A model proposed by Corelli et al. for the binding of diltiazem-like CCBs with the receptor [38]. A hypothetical compound is shown to interact with a negative charge site, two hydrogen binding sites, and two hydrophobic sites of the receptor (*top view*). Reprinted with permission from [38]. © 1997 American Chemical Society

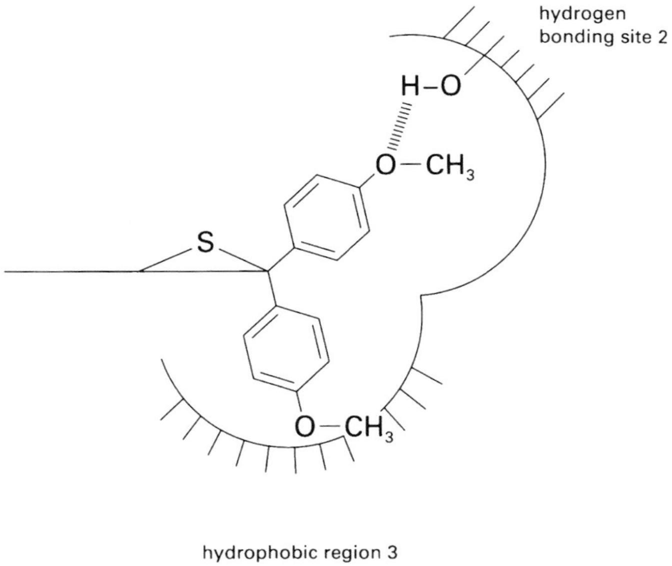

Fig. 4 An edge view of Corelli et al.'s model of the binding of diltiazem-like CCBs with the receptor [38], wherein a compound is shown to interact with hydrogen-bonding site 2 and a third hydrophobic site of the receptor. Reprinted with permission from [38]. © 1997 American Chemical Society

electron-withdrawing group in the fused phenyl ring was found to enhance the activity.

For a series of benzazepinone and benzothiazepine CCBs, Kimball et al. [40] had also identified two pharmacophores: the basic nitrogen and the phenyl methyl ethers, and proposed that the polycyclic core of these compounds could serve as a scaffold and function essentially to position the two pharmacophores in an optimal spatial situation. Accordingly, CCBs would bind to the calcium channel protein in an "inboard" binding conformation in which the side chain amine is placed over the mean plane of the molecule and in proximity to the phenyl methyl ether pharmacophore.

To explore further the mechanism of calcium channel blocking, a few series of 1,4-dihydropyrimidines (7–9) that mimic DHPs were studied for their

7

8

9

calcium channel blocking activity [41–44]. A QSAR study was made on three different series of these dihydropyrimidines, as listed in Tables 5–7, by Gupta et al. [45] to derive the following correlations.

$$\log(1/IC_{50}) = 0.733(\pm0.287)\log P + 1.207(\pm0.453)I_{R2} + 3.928(\pm1.0)$$
$$n = 16, \; r = 0.894, \; s = 0.41, \; r_{cv}^2 = 0.72, \; F_{2,13} = 25.86 \tag{12}$$

$$\log(1/IC_{50}) = 0.463(\pm0.234)MR - 0.918(\pm0.585)\pi_m(R_1) -$$
$$- 1.558(\pm0.637)\sigma_m(R_1) + 1.315(\pm0.581)I_{R3}$$
$$n = 17, \; r = 0.921, \; s = 0.34, \; r_{cv}^2 = 0.64, \; F_{4,12} = 16.76 \tag{13}$$

$$\log(1/IC_{50}) = 0.695(\pm0.230) \, {}^1\chi_{R2}^v + 1.330(\pm0.734) \, {}^1\chi_{R3}^v +$$
$$+ 0.947(\pm0.527)D_{R2} + 3.60(\pm1.467)$$
$$n = 17, \; r = 0.902, \; s = 0.35, \; r_{cv}^2 = 0.45, \; F_{3,13} = 18.83 \,. \tag{14}$$

Table 5 Analogues of 7 and their calcium entry blocking activity and physicochemical parameters

Compd	X	R_1	R_2	R_3	$\log P$	I_{R2}	$\log(1/IC_{50})$ Obsd[a]	Calcd Eq. 12
1	S	3-NO$_2$	Me	Et	2.42	1	6.89	6.91
2	S	2-NO$_2$	Me	Et	2.34	1	6.52	6.85
3	S	2-CF$_3$	Me	Et	3.44	1	7.44	7.66
4	S	2,3-Cl$_2$	Me	Et	3.88	1	7.80	7.98
5	S	3-NO$_2$	CH$_2$CH $=$ CH$_2$	Et	2.94	0	6.52	6.08
6	S	3-NO$_2$	CH$_2$(CH$_2$)3CH$_3$	Et	4.39	0	6.74	7.14
7	S	3-NO$_2$	CH$_2$C$_6$H$_5$	Et	3.50	0	6.72	6.49
8	S	3-NO$_2$	CH$_2$CH$_2$N(Me)Bn	Et	3.70	0	6.77	6.64
9	S	3-NO$_2$	CH$_2$CH$_2$N(Me)$_2$	iPr	3.23	0	5.92	6.30
10	S	3-NO$_2$	Me	Me	2.00	1	6.55	6.60
11	S	3-NO$_2$	Me	iPr	2.82	1	8.15	7.20
12	S	3-NO$_2$	Me	sBu	3.33	1	8.05	7.57
13	S	2,3-Cl$_2$	Me	iPr	4.23	1	8.22	8.23
14	S	3-NO$_2$	Me	CH$_2$CH$_2$N(Me)Bn	3.33	1	7.52	7.57
15	S	2-NO$_2$	Me	CH$_2$CH$_2$N(Me)Bn	3.24	1	5.72 [b]	7.51
16	S	2-CF$_3$	Me	CH$_2$CH$_2$N(Me)Bn	4.39	1	4.00 [b]	8.35
17	O	3-NO$_2$	Me	Et	2.15	1	6.15	6.71
18	N	3-NO$_2$	Me	Et	1.77	0	5.21	5.22

[a] Taken from [41]
[b] Not used in the derivation of Eq. 12

Table 6 Analogues of **8** and their calcium entry blocking activity and physicochemical parameters

Compd	X	R_1	R_2	R_3	$\pi_{m(R_1)}$	$\sigma_{m(R_1)}$	MR	I_{R_3}	I_X	$\log(1/IC_{50})$ Obsd[a]	Calcd Eq. 13
1	S	3-NO$_2$	COOEt	Et	−0.28	0.71	10.28	0	1	7.77	7.35
2	S	2-NO$_2$	COOEt	Et	0.00	0.00	10.28	0	1	8.51	8.19
3	S	3-Cl	COOEt	Et	0.71	0.37	10.16	0	1	7.06	6.91
4	S	2-Cl	COOEt	Et	0.00	0.00	10.16	0	1	7.77	8.14
5	S	3-CF$_3$	COOEt	Et	0.88	0.43	10.18	0	1	6.55	6.67
6	S	2-CF$_3$	COOEt	Et	0.00	0.00	10.18	0	1	7.86	8.15
7	S	2,3-Cl$_2$	COOEt	Et	0.00	0.00	10.65	0	1	8.82	8.37
8	S	2-Cl,3-NO$_2$	COOEt	Et	0.00	0.00	10.77	0	1	8.57	8.42
9	S	3-NO$_2$	COOEt	Me	−0.28	0.71	9.82	0	1	7.00	7.13
10	S	3-NO$_2$	COOEt	iPr	−0.28	0.71	10.75	1	1	8.77	8.88
11	S	3-NO$_2$	COOMe	Et	−0.28	0.71	12.45	0	1	8.44 [b]	7.14
12	S	3-NO$_2$	COOiPr	Et	−0.28	0.71	10.50	0	1	7.85	7.56
13	S	3-NO$_2$		Et	−0.28	0.71	12.45	0	1	7.96	8.35
14	S	3-NO$_2$	CONMe$_2$	Et	−0.28	0.71	10.50	0	1	6.57 [b]	7.45
15	S	3-NO$_2$	SO$_2$Ph	Et	−0.28	0.71	12.09	0	1	8.28	8.18
16	S	3-NO$_2$	CH$_2$CH$_2$CH$_3$	Et	−0.28	0.71	10.09	0	1	7.30	7.26
17	O	3-NO$_2$	COOEt	Et	−0.28	0.71	9.43	0	0	6.85	6.95
18	O	3-NO$_2$	COOEt	iPr	−0.28	0.71	9.89	1	0	8.59	8.48
19	NH	3-NO$_2$	COOEt	Et	−0.28	0.71	9.58	0	0	6.80	7.02

[a] Taken from [42]
[b] Not used in the derivation of Eq. 13

To derive these equations, $\log P$ (hydrophobic parameter), MR (molar refractivity index), and MV (molar volume) were calculated using software freely available on the internet (www.logP.com, www.daylight.com). The first-order valence molecular connectivity index $^1\chi^v$ of substituents was calculated as suggested by Kier and Hall [46, 47]. In these equations, r_{cv}^2 is cross-validated r^2 obtained by the leave-one-out jackknife procedure. Its value higher than 0.6 defines the good predictive ability of the equation. The different indicator variables in these equations were defined as follows.

$I_{R2} = 1$ for R_2 in Table 5 being a methyl group, otherwise its value is zero.

$I_{R3} = 1$ for R_3 being, in any table, an isopropyl group, otherwise its value is zero.

$I_X = 1$ or 0 for $X = S$ or 0 in any Table.

$D_{R2} = 1$ for R_2 in Table 7 being $CONH_2$ group, otherwise its value is zero.

Table 7 Analogues of **9** and their calcium entry blocking activity and physicochemical parameters

Compd	R_1	R_2	R_3	$^1\chi_{R_2}^v$	$^1\chi_{R_3}^v$	D_{R2}	log(1/IC$_{50}$) Obsd[a]	Calcd Eq. 14
1	3-NO$_2$	CONMe$_2$	iPr	1.55	1.80	0	5.50 [b]	7.16
2	3-NO$_2$	CONHMe	iPr	1.18	1.80	0	7.80 [b]	6.90
3	3-NO$_2$	CONH$_2$	iPr	0.72	1.80	1	7.92	7.53
4	3-NO$_2$	H	iPr	0.00	1.80	0	5.80	6.08
5	3-NO$_2$	CONH$_2$	Et	0.72	1.40	1	7.41	7.00
6	3-NO$_2$	CONH$_2$	Me	0.72	0.81	1	6.05	6.21
7	3-NO$_2$	CONHEt	iPr	1.74	1.80	0	7.88	7.29
8	3-NO$_2$	CONHiPr	iPr	1.86	1.80	0	7.22	7.38
9	3-NO$_2$	CONHCH$_2$Ph	iPr	3.30	1.80	0	8.52	8.38
10	3-NO$_2$	CH$_2$CH$_2$N(Me)Bn	iPr	4.16	1.80	0	8.69	8.98
11	2-NO$_2$	CONH$_2$	iPr	0.72	1.80	1	7.69	7.53
12	3-CF$_3$	CONH$_2$	iPr	0.72	1.80	1	5.77 [b]	7.53
13	2-CF$_3$	CONH$_2$	iPr	0.72	1.80	1	7.35	7.53
14	3-Cl	CONH$_2$	iPr	0.72	1.80	1	7.39	7.53
15	2-Cl	CONH$_2$	iPr	0.72	1.80	1	7.47	7.53
16	3-Br	CONH$_2$	iPr	0.72	1.80	1	7.26	7.53
17	2-Br	CONH$_2$	iPr	0.72	1.80	1	7.51	7.53
18	2,6-Cl$_2$	CONH$_2$	iPr	0.72	1.80	1	7.08	7.53
19	2,3-Cl$_2$	CONH$_2$	iPr	0.76	1.80	1	7.37	7.56
20	3-NO$_2$	CONH$_2$	iPr	0.72	1.80	1	8.07	7.53

[a] Taken from [43]
[b] Not used in the derivation of Eq. 14

In all the equations, IC_{50} stands for vasorelaxant activity of the compounds and refers to the molar concentration of the compound required to inhibit the potassium contracted rabbit aorta strips by 50%.

A very critical analysis of these equations led Gupta et al. to suggest that the most important factors that can commonly affect the activity of all the three series (Tables 5–7) are the esters group present at N1 and C6 of the pyrimidine ring. At N1, the smallest ester group (COOMe) is found to be the best and if it is replaced by an amide group, the amide group should be totally unsubstituted ($CONH_2$). Both esters and amide groups can be expected to form hydrogen bonds with the receptors, which can be sterically hindered by the presence of any bulky group in them.

At the C6 atom, the isopropyl containing ester group was suggested to be optimum. Here the ester group seems to have a steric interaction which is optimal with the isopropyl group.

In the last decade, several new classes of calcium entry blockers were studied in which phenyl sulfonylindolzine analogues had drawn more attention. Consequently, Gubin et al. [48, 49] reported two different series of these indolizine analogues: 10, in which the variations were made in the R-substituent at the 2-position of the indolizine ring and in the amine moiety (Am) of the 4-substituent of the phenyl ring; and 11, in which the indolizine ring was replaced by a variety of heterocyclic rings along with the variation in the Am moiety. These two series are listed in Tables 8 and 9, respectively. The two different assays were reported for both these series: $(IC_{50})_A$, referring to the molar concentration of the compound required to reduce [^{3}H]nitrendipine binding by 50%, and $(IC_{50})_B$, referring to the molar concentration of the compound required to block Ca^{2+} induced concentration of K^+ depolarized rat aorta by 50%. For both these activities of 10 and 11 a QSAR analysis was made by Gupta et al. [50] and the following correlations were obtained.

10

11

Table 8 Analogues of **10** with calcium entry blocking activities and the correlates

Compd	R	Am	n	π_R	σ_R	$^1\chi^v_{Am}$	I_1	I_2	$\log(1/IC_{50})_A$ Obsd[a]	Calcd Eq. 15	$\log(1/IC_{50})_B$ Obsd[a]	Calcd Eq. 16
1	CH_3	$(C_2H_5)_2N$	3	0.56	-0.17	2.05	0	0	5.28	5.66	5.47	5.71
2	CH_3	$(n\text{-}C_3H_7)_2N$	3	0.56	-0.17	3.05	0	0	6.19	6.19	6.02	6.13
3	$i\text{-}C_3H_7$	$(CH_3)_2N$	3	1.53	-0.15	0.89	0	0	6.70	6.63	6.41	6.08
4	$n\text{-}C_4H_9$	$(n\text{-}C_3H_7)_2N$	3	2.13	-0.16	3.05	0	0	6.89	7.23	6.24	6.36
5	$n\text{-}C_4H_9$	$(n\text{-}C_4H_9)_2N$	3	2.13	-0.16	4.05	0	0	7.66	7.60	6.55	6.46
6	$i\text{-}C_3H_7$	$(C_2H_5)_2N$	3	1.53	-0.15	2.05	0	0	7.29	7.45	6.64	6.96
7	$i\text{-}C_3H_7$	$(n\text{-}C_3H_7)_2N$	3	1.53	-0.15	3.05	0	0	7.39	7.99	6.99	7.39
8	$i\text{-}C_3H_7$	$(n\text{-}C_4H_9)_2N$	3	1.53	-0.15	4.05	0	0	8.68	8.36	7.67	7.49
9	$t\text{-}C_4H_9$	$(n\text{-}C_4H_9)_2N$	3	1.98	-0.20	4.05	0	0	8.02	8.30	7.28	7.18
10	$c\text{-}C_3H_5$	[see structure below]	3	1.14	-0.21	4.05	1	0	8.72	9.03	8.33	8.50
11	C_6H_{11}	$(n\text{-}C_4H_9)_2N$	3	2.51	-0.22	4.05	0	0	6.95	6.77	5.39	5.44
12	C_6H_5	$(n\text{-}C_4H_9)_2N$	3	1.96	-0.01	4.05	0	0	6.81	6.73	5.79	5.77
13	CH_3	$(n\text{-}C_4H_9)_2N$	3	0.56	-0.17	4.05	0	0	7.16	6.57	6.79	6.24
14	$n\text{-}C_3H_7$	$(n\text{-}C_4H_9)_2N$	3	1.55	-0.13	4.05	0	0	7.85	8.18	7.27	7.32
15	C_2H_5	$(n\text{-}C_3H_7)_2N$	3	1.02	-0.15	3.05	0	0	6.74	7.47	6.82	7.14
16	C_2H_5	$(n\text{-}C_4H_9)_2N$	3	1.02	-0.15	4.05	0	0	7.85	7.84	7.50	7.24
17	$c\text{-}C_3H_7$	$(n\text{-}C_4H_9)_2N$	3	1.14	-0.21	4.05	0	0	8.60	8.57	7.96	7.86
18	C_2H_5	$(n\text{-}C_4H_9)_2N$	5	1.02	-0.15	4.05	0	0	7.70	7.84	7.01	7.24

[a] Taken from [48]
[b] Not included in the derivation of Eq. 15
[c] Not included in the derivation of Eq. 16

Table 8 continued

Compd	R	Am	n	π_R	σ_R	$^1\chi^v_{Am}$	I_1	I_2	$\log(1/IC_{50})_A$ Obsd[a]	Calcd Eq. 15	$\log(1/IC_{50})_B$ Obsd[a]	Calcd Eq. 16
19	C_2H_5	$(n\text{-}C_5H_{11})_2N$	3	1.02	−0.15	5.05	0	0	8.20	8.05	7.20	7.02
20	C_2H_5	$(n\text{-}C_4H_9)_2N$	2	1.02	−0.15	4.05	0	0	7.48	7.84	7.14	7.24
21	C_2H_5	$(n\text{-}C_4H_9)_2N$	4	1.02	−0.15	4.05	0	0	8.34	7.84	7.48	7.24
22	C_2H_5		3	1.02	−0.15	2.65	0	0	6.56	7.28	6.45	7.01
23	C_2H_5	$t\text{-}C_4H_9NH$	3	1.02	−0.15	1.75	0	0	6.89	6.74	6.38	6.53
24	C_2H_5		3	1.02	−0.15	4.05	1	0	8.34	8.33	7.96	7.88
25	C_2H_5		3	1.02	−0.15	4.58	1	0	8.64	8.42	7.87	7.81
26	$i\text{-}C_3H_7$	$t\text{-}C_4H_9NH$	3	1.53	−0.15	1.75	0	0	7.76	7.26	7.38	6.78
27	$i\text{-}C_3H_7$		4	1.53	−0.15	4.08	1	0	8.35	8.82	7.84	8.13

[a] Taken from [48]
[b] Not included in the derivation of Eq. 15
[c] Not included in the derivation of Eq. 16

Table 8 continued

Compd	R	Am	n	π_R	σ_R	$^1\chi_{Am}^{v}$	I_1	I_2	$\log(1/IC_{50})_A$ Obsd[a]	$\log(1/IC_{50})_A$ Calcd Eq. 15	$\log(1/IC_{50})_B$ Obsd[a]	$\log(1/IC_{50})_B$ Calcd Eq. 16
28	$i\text{-}C_3H_7$	H_3CO-(aryl, H_3CO)-$(CH_2)_2NCH_3$	3	1.53	−0.15	4.58	1	0	9.21	8.94	8.25	8.06
29	$i\text{-}C_3H_7$	C_6H_5-$(CH_2)_2NH$	3	1.53	−0.15	2.78	0	0	8.40	7.86	7.09	7.31
30	$i\text{-}C_3H_7$	C_6H_5-CH_2NH	3	1.53	−0.15	2.62	0	0	8.35	7.78	7.64	7.24
31	$i\text{-}C_3H_7$	piperazine (N-C_6H_5, N-CH_3)	3	1.53	−0.15	4.65	0	0	7.85	8.50	6.69	7.39
32	$i\text{-}C_3H_7$	piperidine (N-C_6H_5)	3	1.53	−0.15	4.65	0	0	8.18	8.50	7.09	7.39
33	$i\text{-}C_3H_7$	$(n\text{-}C_8H_{17})_2N$	3	1.53	−0.15	8.05	0	0	8.19	8.22	4.52	4.69
34	$i\text{-}C_3H_7$	$(n\text{-}C_5H_{11})_2N$	3	1.53	−0.15	5.05	0	0	9.05	8.57	7.05	7.27
35	$i\text{-}C_3H_7$	$(n\text{-}C_4H_9)_2N$	4	1.53	−0.15	4.05	0	0	8.89	8.36	7.69	7.49

[a] Taken from [48]
[b] Not included in the derivation of Eq. 15
[c] Not included in the derivation of Eq. 16

Table 8 continued

Compd	R	Am	n	π_R	σ_R	$^1\chi^v_{Am}$	I_1	I_2	$\log(1/IC_{50})_A$ Obsd[a]	Calcd Eq. 15	$\log(1/IC_{50})_B$ Obsd[a]	Calcd Eq. 16
36	$i\text{-}C_3H_7$		3	1.53	−0.15	4.08	1	0	9.21	8.82	8.50	8.13
37	$i\text{-}C_3H_7$		4	1.53	−0.15	4.58	1	0	8.64	8.94	7.67	8.06
38	$i\text{-}C_3H_7$		3	1.53	−0.15	3.67	1	0	8.92	8.69	7.89	8.13
39	$i\text{-}C_3H_7$		3	1.53	−0.15	3.11	1	0	6.42[b]	9.62	6.40[c]	9.17
40	$i\text{-}C_3H_7$		3	1.53	−0.15	6.16	1	0	9.08	9.06	7.73	7.30

[a] Taken from [48]
[b] Not included in the derivation of Eq. 15
[c] Not included in the derivation of Eq. 16

Table 8 continued

Compd	R	Am	n	π_R	σ_R	$^1\chi^v_{Am}$	I_1	I_2	log(1/IC$_{50}$)$_A$ Obsd[a]	Calcd Eq. 15	log(1/IC$_{50}$)$_B$ Obsd[a]	Calcd Eq. 16
41	i-C$_3$H$_7$		3	1.53	-0.15	4.05	0	0	8.59	8.36	8.10	7.49
42	i-C$_3$H$_7$		3	1.53	-0.15	4.05	0	0	8.06	8.36	7.65	7.49
43	i-C$_3$H$_7$		3	1.53	-0.15	4.08	1	0	8.74	8.82	8.30	8.13
44	i-C$_3$H$_7$		3	1.53	-0.15	4.57	0	1	9.72	9.26	8.67	8.47
45	i-C$_3$H$_7$		3	1.53	-0.15	4.16	0	1	9.28	9.16	8.63	8.53

[a] Taken from [48]
[b] Not included in the derivation of Eq. 15
[c] Not included in the derivation of Eq. 16

Table 8 continued

Compd	R	Am	n	π_R	σ_R	$^1\chi_{Am}^v$	I_1	I_2	$\log(1/IC_{50})_A$ Obsd[a]	Calcd Eq. 15	$\log(1/IC_{50})_B$ Obsd[a]	Calcd Eq. 16
46	i-C$_3$H$_7$		3	1.53	-0.15	5.11	1	1	9.21	9.80	8.65	8.95
47	c-C$_3$H$_5$		3	1.14	-0.21	2.78	1	0	9.13	8.53	8.41	8.32
48	c-C$_3$H$_5$	t-C$_4$H$_9$NH	3	1.14	-0.21	1.75	0	0	7.37	7.47	7.18	7.15
49	t-C$_4$H$_9$	t-C$_4$H$_9$NH	3	1.98	-0.20	1.75	0	0	7.05	7.21	6.35	6.47

[a] Taken from [48]
[b] Not included in the derivation of Eq. 15
[c] Not included in the derivation of Eq. 16

Table 9 Analogues of **11** with calcium entry blocking activities and the correlates

Compd	R	Am	$^1\chi_X^v$	$^1\chi_{Am}^v$	I_1	I_2	D_{NI}	D_5	$\log(1/IC_{50})_A$ Obsd[a]	Calcd Eq. 17	$\log(1/IC_{50})_B$ Obsd[a]	Calcd Eq. 18
1		$(n\text{-}C_4H_9)_2N$	4.265	4.047	0	0	0	0	9.00[b]	7.31	7.88[c]	7.10
2			4.265	4.580	1	0	0	0	9.66	8.80	8.23	7.82
3		$t\text{-}C_4H_9NH$	4.265	1.750	0	0	0	0	7.92	7.30	7.48	7.11
4		$(n\text{-}C_4H_9)_2N$	4.137	4.047	0	0	0	0	7.87	7.44	7.51	7.59
5			4.137	4.580	1	0	0	0	8.59	8.94	7.73	8.30

[a] Taken from [49]

[b] Not included in the derivation of Eq. 17

[c] Not included in the derivation of Eq. 18

Table 9 continued

Compd	R	Am	$^1\chi_X^v$	$^1\chi_{Am}^v$	I_1	I_2	D_{NI}	D_5	log(1/IC$_{50}$)A Obsd[a]	Calcd Eq. 17	log(1/IC$_{50}$)B Obsd[a]	Calcd Eq. 18
6		t-C$_4$H$_9$NH	4.137	1.750	0	0	0	0	6.82	7.44	6.77 [c]	7.58
7			4.796	4.580	1	0	0	0	8.10	7.90	6.64 [c]	7.67
8			4.296	4.580	1	0	0	0	8.28	8.77	7.29	7.73
9			4.244	4.580	1	0	0	0	9.19	8.83	7.69	7.89
10			4.244	4.080	1	0	0	0	8.72	8.83	7.84	7.89

[a] Taken from [49]

[b] Not included in the derivation of Eq. 17

[c] Not included in the derivation of Eq. 18

Table 9 continued

Compd	R	Am	$^1\chi_X^v$	$^1\chi_{Am}^v$	I_1	I_2	D_{NI}	D_5	$\log(1/IC_{50})_A$ Obsd[a]	$\log(1/IC_{50})_A$ Calcd Eq. 17	$\log(1/IC_{50})_B$ Obsd[a]	$\log(1/IC_{50})_B$ Calcd Eq. 18
11			4.244	4.170	1	0	0	0	8.80	8.83	7.77	7.89
12			4.244	5.109	1	1	0	0	9.23	8.95	8.24	7.89
13			4.336	4.580	1	0	0	0	7.96	8.71	7.72	7.63
14			4.730	4.580	1	0	1	0	8.96	9.12	8.37	8.24

[a] Taken from [49]

[b] Not included in the derivation of Eq. 17

[c] Not included in the derivation of Eq. 18

Table 9 continued

Compd	R	Am	$^1\chi_X^v$	$^1\chi_{Am}^v$	I_1	I_2	D_{NI}	D_5	log(1/IC$_{50}$)$_A$ Obsd[a]	Calcd Eq. 17	log(1/IC$_{50}$)$_B$ Obsd[a]	Calcd Eq. 18
15	(structure)	$(n\text{-}C_4H_9)_2N$	4.336	4.047	0	0	0	0	6.96	7.21	7.26	6.93
16	(structure)	$(n\text{-}C_4H_9)_2N$	4.730	4.047	0	0	1	0	7.96	7.62	7.62	7.53
17	(structure)	(structure) CH_2NCH_3	4.730	4.080	1	0	1	0	9.00	9.12	8.43	8.24
18	(structure)	$t\text{-}C_4H_9NH$	4.730	1.750	0	0	1	0	6.94	7.62	7.00	7.53
19	(structure)	(structure) $(CH_2)_2NCH_3$	4.665	4.580	1	0	1	0	10.14	9.25	8.43	8.14

[a] Taken from [49]

[b] Not included in the derivation of Eq. 17

[c] Not included in the derivation of Eq. 18

Table 9 continued

Compd	R	Am	$^1\chi_X^v$	$^1\chi_{Am}^v$	I_1	I_2	D_{NI}	D_5	log(1/IC$_{50}$)$_A$ Obsd[a]	Calcd Eq. 17	log(1/IC$_{50}$)$_B$ Obsd[a]	Calcd Eq. 18
20			4.271	4.580	0	1	1	0	8.66	8.49	7.43	7.80
21			4.271	4.580	0	1	1	0	10.03 [b]	8.44	7.79	7.80
22			4.271	5.109	1	1	1	0	9.55	9.99	8.72	8.51
23			4.545	4.580	1	0	0	0	7.88	8.40	7.07	7.38

[a] Taken from [49]
[b] Not included in the derivation of Eq. 17
[c] Not included in the derivation of Eq. 18

Table 9 continued

Compd	R	Am	$^1\chi_X^v$	$^1\chi_{Am}^v$	I_1	I_2	D_{NI}	D_5	$\log(1/IC_{50})_A$ Obsd[a]	Calcd Eq. 17	$\log(1/IC_{50})_B$ Obsd[a]	Calcd Eq. 18
24		$(n\text{-}C_4H_9)_2N$	4.545	4.047	0	0	0	0	7.03	6.90	6.70	6.67
25		$t\text{-}C_4H_9NH$	4.545	1.750	0	0	0	0	6.74	6.90	6.43	6.67
26			2.641	4.580	1	0	0	1	–	–	5.77	5.96
27		$(n\text{-}C_4H_9)_2N$	2.641	4.047	0	0	0	1	–	–	5.30	5.25

[a] Taken from [49]
[b] Not included in the derivation of Eq. 17
[c] Not included in the derivation of Eq. 18

Table 9 continued

Compd	R	Am	$^1\chi_X^v$	$^1\chi_{Am}^v$	I_1	I_2	D_{NI}	D_5	log(1/IC$_{50}$)$_A$ Obsd[a]	Calcd Eq. 17	log(1/IC$_{50}$)$_B$ Obsd[a]	Calcd Eq. 18
28	(structure)	(structure) H_3CO–…–$(CH_2)_2NCH_3$, H_3CO	2.575	4.580	1	0	0	1	8.14	8.16	7.07	7.30
29	(structure)	$(n\text{-}C_4H_9)_2N$	2.575	4.047	0	0	0	1	6.73	6.66	6.92	6.59
30	(structure)	$t\text{-}C_4H_9NH$	2.575	1.750	0	0	0	1	6.61	6.66	6.63	6.59
31	(structure)	(structure) H_3CO–…–$(CH_2)_2NCH_3$, H_3CO	4.261	4.580	1	0	0	0	9.21	8.81	8.25	7.83

[a] Taken from [49]
[b] Not included in the derivation of Eq. 17
[c] Not included in the derivation of Eq. 18

Series of 10 (Table 8)

$$\log(1/IC_{50})_A = 6.577(\pm1.553)\pi_R - 2.179(\pm0.532)(\pi_R)^2 +$$
$$+ 0.949(\pm0.354) \; ^1\chi_{Am}^v - 0.081(\pm0.042)(^1\chi_{Am}^v)^2 -$$
$$- 8.496(\pm4.320)\sigma_R + 0.452(\pm0.303)I_1 + 0.768(\pm0.506)I_2 -$$
$$- 0.387(\pm1.700)$$

$$n = 48, \; r = 0.919, \; s = 0.41, \; F_{7,40} = 31.10, \; (\pi_R)_{opt} = 1.51,$$
$$(^1\chi_{Am}^v)_{opt} = 5.86 \tag{15}$$

$$\log(1/IC_{50})_B = 5.831(\pm1.217)\pi_R - 2.097(\pm0.417)(\pi_R)^2 +$$
$$+ 1.238(\pm0.278) \; ^1\chi_{Am}^v - 0.160(\pm0.033)(^1\chi_{Am}^v)^2 -$$
$$- 7.700(\pm3.315)\sigma_R + 0.647(\pm0.238)I_1 + 1.057(\pm0.396)I_2 -$$
$$- 0.067(\pm1.332)$$

$$n = 48, \; r = 0.948, \; s = 0.32, \; F_{7,40} = 50.20, \; (\pi_R)_{opt} = 1.39,$$
$$(^1\chi_{Am}^v)_{opt} = 3.87 \tag{16}$$

Series of 11 (Table 9)

$$\log(1/IC_{50})_A = 7.180(\pm4.169) \; ^1\chi_X^v - 0.993(\pm0.584)(^1\chi_X^v)^2 +$$
$$+ 1.505(\pm0.398)I_1 + 1.133(\pm0.537)D_{N1} - 5.246(\pm7.116)$$

$$n = 27, \; r = 0.90, \; s = 0.48, \; F_{4,22} = 23.39, \; (^1\chi_X^v)_{opt} = 3.62 \tag{17}$$

$$\log(1/IC_{50})_B = 114.705(\pm44.798) - 47.415(\pm20.218) \; ^1\chi_X^v +$$
$$+ 5.202(\pm2.281)(^1\chi_X^v)^2 + 0.706(\pm0.261)I_1 +$$
$$+ 0.706(\pm0362)D_{N1} - 20.518(\pm7.686)D_5$$

$$n = 28, \; r = 0.933, \; s = 0.32, \; F_{5,22} = 29.47, \; (^1\chi_X^v)_{opt} = 4.56. \tag{18}$$

Equations 15 and 16 obtained for the analogues of **10** exhibited the parallel correlations for the two activities, indicating that for both the activities the hydrophobic property and electron-donating nature of the R-substituents will be crucial. The Am moiety was shown to affect both the activities through its size delineated by the molecular connectivity index $^1\chi_{Am}^v$. In both Eqs. 15 and 16, however, π_R and $^1\chi_{Am}^v$ were shown to have parabolic correlations with the activities, each with an optimum value as shown in the equations.

The two variables I_1 and I_2 used in the above equations were the indicator parameters related to the Am moiety. The I_1 stands with a value of unity for an Am that had the methoxy groups at the 3-and 4-positions of the phenyl ring and I_2 stands with a value of unity for an Am that had the methoxy group only at the 5-position of the phenyl ring. A comparison of the coefficients of I_1 and I_2 in Eqs. 15 and 16 had suggested that the presence of the methoxy group at the 5-position would be better than at the 3,4-positions.

The parallelism between Eqs. 15 and 16 indicated that there could be good mutual relations between the two assays. This was nicely verified by Gupta et al. [50] for both the series (**10** and **11**) by deriving Eqs. 19 and 20, respectively, although such a nice parallelism was not observed to exist between Eqs. 17 and 18 obtained for the series of **11**.

$$\log(1/IC_{50})_B = 0.769(\pm0.11)\log(1/IC_{50})_A + 1.152(\pm0.891)$$
$$n = 48, \; r = 0.90, \; s = 0.37, \; F_{1,46} = 194.33 \tag{19}$$
$$\log(1/IC_{50})_B = 0.557(\pm0.107)\log(1/IC_{50})_A + 3.036(\pm0.891)$$
$$n = 27, \; r = 0.90, \; s = 0.27, \; F_{1,25} = 111.60. \tag{20}$$

For the series of **11** where there was a variation in the heterocyclic ring along with the variation in the Am moiety, the activities were shown to be primarily governed by the nature and size of the heterocyclic rings and little by the nature of the Am moiety. For the latter, only the parameter I_1 was found to be significant. However, the dependence of the activities on $^1\chi^v$ of X (the heterocyclic rings) for the two activities was not similar. While for activity A Eq. 17 exhibited a normal parabolic correlation with $^1\chi^v$, for activity B Eq. 18 exhibited the inverted parabolic correlation. This difference was attributed by Gupta et al. to the conformational changes in the receptors while interacting with the compounds.

However, the studies on the calcium channel blockers remained centered even today around the 1,4-dihydropyridine class. Since this class of compounds can also act as calcium channel activators, attention has always been drawn towards their structure-activity relationship studies. Attempts were made to differentiate in the mechanisms of their agonist and antagonist activities. On the basis of the force field and quantum mechanical calculations, Holtze and Marrer [51] discovered a unique area of the molecular potentials where Ca agonists and antagonists possess potential of opposite sign. These authors demonstrated that the molecular potential of a simple receptor site was reduced by interaction with calcium channel activators and, on the contrary, increased by interaction with calcium channel blockers. These opposite effects probably could be the basis for the opposite actions of DHP enantiomers at the potential-dependent calcium channel.

In order to explore deeper insight into the mechanism of actions of DHPs, several authors carried out molecular modeling studies on these compounds [52–54]. In a recent molecular modeling study on calcium channel blockers, nifedipine and black mamba toxin FS2 (a venom of the black mamba snake, which has been demonstrated to block the L-type calcium channel [55, 56], is a small peptide consisting of 58-74 amino acid residues and having 4–5 intramolecular disulfide bridges formed by cystein residues), Schleifer [54] observed the following:
a) Both compounds revealed pronounced hydrophobic regions parallel to aromatic and aliphatic ring systems.
b) Both compounds have two hydrogen bond acceptor spaces (*ap, sp* ester oxygens in conjugation with 2′-nitro in nifedipine).

c) At the deepest place in regard to the superposition, both molecules possess a hydrogen bond group (N1–H in nifedipine).

d) Both possess similar molecular electrostatic potentials.

e) Additional hydrophobic interactions may be postulated for bulky substituents in both the CCBs (at the *sp* ester side chain of DHPs).

Although the 3D binding site of DHPs is still not well defined, site-directed mutagenesis experiments identify hydrophobic amino acid residues in the putative trans-membrane segment IVS6 of L-type voltage-gated calcium channels as the molecular determinants for high affinity DHP binding [57].

With respect to a maximum activity possessed by DHPs, Mager et al. [52] found the following rank order of substituents parameters and positions: lipophilicity $\approx$ ortho-position > inductivity > minimum width > meta-position. A few neural network studies [58–60] on these DHPs were also made, but they were of only predictive value and could throw little light on the mechanism of their action.

A molecular modeling study on some rigid analogues of verapamil (**1**) suggested that the two actions of verapamil analogues—negative inotropic (decrease in force of cardiac muscle contraction) and negative chronotropic (decrease in rate of cardiac muscle contraction)—were because of the conformations of the molecules that differ in the orientation of their phenylethylamino groups [61].

Regarding diltiazem (**3**), some additional molecular features favorable for binding with the calcium channel and showing antagonistic effects were in-

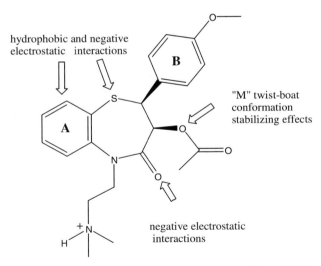

Fig. 5 Additional molecular features in diltiazem, as suggested by Schleifer and Tot [62], favorable for binding and showing antagonistic activity. From [62]. © 2000 Springer Science and Business Media. Reprinted with kind permission of Springer Science and Business Media

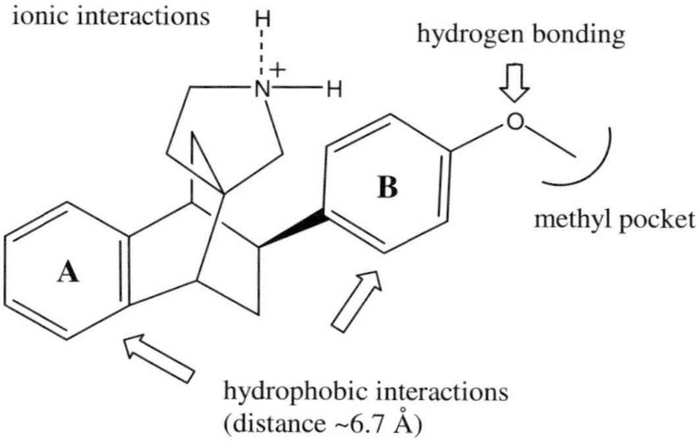

Fig. 6 The minimum requirements for binding with the diltiazem site as suggested by Schleifer and Tot [62] (shown for *spiro*-linked benzocyclo[2,2,2]octyl amine as an example). From [62]. © 2000 Springer Science and Business Media. Reprinted with kind permission of Springer Science and Business Media

dicated by Schleifer and Tot [62] as shown in Fig. 5. However, for several diltiazem mimics that did not contain sulfur, the minimum requirements for binding with the diltiazem site were suggested to be as shown in Fig. 6 for spiro-linked benzocyclo[2,2,2]octyl amine derivative, as an example.

6
Concluding Remarks

The three principal classes of compounds have been found to act as calcium channel blockers (CCBs), of which the 1,4-dihydropyridine (DHP) class has drawn the maximum attention. The various experimental and theoretical studies have delineated in detail the relationships between the structure and activity of this class of CCBs. The high activity of 4-phenyl DHPs was attributed to their conformational properties, in which the aryl ring was supposed to have perpendicular orientation relative to the DHP ring. The DHP ring exists as a non-planar boat-shaped structure with the N1 and C4 atoms defining the stern and bowsprit position. The phenyl ring is bound to it at a pseudoaxial position and approximately bisects the pyridine ring. The rotational freedom of the phenyl ring about the C4–C1′ bond is sterically restricted and the plane of the phenyl ring is forced to lie close to N1–C4 vertical symmetry plane.

The conformation of the ester groups of DHPs have also been indicated to be crucial for the activity. QSAR studies have exhibited the importance of steric factors in binding of DHPs to the calcium channel. The length and width of phenyl ring substituents are shown to govern the activity. Certain electronic properties, particularly the electron-withdrawing ability, of the meta substituents have been shown to be beneficial for the potency. A quantitative correlation has been found to exist between the conformational rigidity of the phenyl ring and the CCB activity of the DHPs and in the binding of these DHPs with the receptors the hydrophobic and hydrogen bonding interactions have been shown to be of paramount importance.

Certain other classes of compounds, e.g., dihydropyridines (**7–9**) and phenylsulfonylindolizines (**10, 11**), have also been studied for CCB activity. In these also, the hydrophobicity and steric factors have been found to play the important roles. Thus, QSAR studies have provided valuable information for the design of potent calcium channel blockers of pharmaceutical importance.

Acknowledgements The essential financial assistance for this work provided by our own organization is thankfully acknowledged.

References

1. Ringer S (1883) J Physiol (London) 4:29
2. Fleckenstein A (1983) Calcium antagonism in heart and smooth muscle: experimental facts and therapeutic prospects. Wiley, New York
3. Cheung JY, Bonventre JV, Malis CD, Leaf A (1986) N Engl J Med 14:1670
4. Janis RA, Silver P, Triggle DJ (1987) Adv Drug Res 16:309
5. Triggle DJ (1991) J Cardiov Pharmacol 18(suppl 10):S1
6. Triggle DJ (1991) Am J Hypert 4:422S
7. Bean BP (1989) Annu Rev Physiol 51:367
8. Triggle DJ (1988) Endogeneous ligands: myths and realities. In: Morad M, Nayler W, Kazda S, Schramm M (eds) The calcium channel: Structure function and implications. Springer, Berlin Heidelberg New York, pp 549–562
9. Glossman H, Streissnig J (1988) Calcium channels. Vitamins and Hormones 44:155
10. Vaghy PL, McKenna E, Itagaki K, Schwartz A (1988) Trends Pharmacol Sci 27:398
11. Hondeghem LM, Katzung BG (1985) Annu Rev Pharmacol Toxicol 24:387
12. Gupta SP (2000) Prog Drug Res 55:235
13. Gupta SP (2001) Prog Drug Res 56:121
14. Triggle DJ, Langs DA, Janis RA (1989) Med Res Rev 9:123
15. Loev B, Goodman MM, Snadon KM, Tedeschi R, Macko E (1974) J Med Chem 17:956
16. Gaudio AC, Korolkovas A, Takahata Y (1994) J Pharm Sci 83:1110
17. Triggle AM, Shefter E, Triggle DJ (1980) J Med Chem 23:1442
18. Fossheim R, Svarteng K, Mustad A, Romming C, Shefter E, Triggle DJ (1982) J Med Chem 25:126
19. Seidel W, Meyer H, Born L, Kazda S, Dompert W (1984) In: Abstracts of papers, 187th National Meeting of the American Chemical Society, St. Louis. ACS, Washington, DC
20. Baldwin JJ, Clareman DA, Lumma PK, McClure DE, Rosenthal SA, Linguist RJ, Faison EP, Kaczorowski GJ, Trumble MJ, Smith GM (1987) J Med Chem 30:690
21. Rovnyak G, Anderson N, Gougoutas J, Hedberg A, Kimbal SD, Malley M, Moreland S, Porubcan M, Pudziannowski A (1988) J Med Chem 31:936
22. Rovnyak G, Anderson N, Gougoutas J, Hedberg A, Kimbal SD, Malley M, Moreland S, Porubcan M, Pudziannowski A(1991) J Med Chem 34:2521
23. Goldman S, Stoltefuss J (1991) Angew Chem, Int Ed Engl 30:1559

24. Bolger GT, Gengo P, Klockowski R, Luchowski E, Siegel H, Janis RA, Triggle AM, Triggle DJ (1983) J Pharmacol Exp Ther 225:291
25. Mahmoudian A, Richards GW (1986) J Pharmacol 38:272
26. Verloop A, Hoogenstraaten W, Tipker J (1976) In: Ariëns EJ (ed) Drug Design, vol 7. Academic Press, New York
27. Coburn RA, Wierzba M, Suto MJ, Solo AJ, Triggle AM, Triggle DJ (1988) J Med Chem 31:2103
28. Dewar MJS, Zoebisch EG, Healy EF, Stewart JJP (1985) J Am Chem Soc 107:3902
29. Bernstein P, Wold S (1986) Quant Struct-Act Relat 5:45
30. Manhold R, Steiner R, Hass W, Kaufman R (1978) Naunyn Schmeidebergs Arch Pharmacol 302:217
31. Bolger GT, Gengo PJ, Lunchowski EM, Siegel H, Triggle DJ, Triggle RA (1982) Biochem Biophys Res Commun 104:1604
32. Murphy KMM, Snyder SH (1982) Eur J Pharmacol 77:201
33. Mayer CJ, van Breeman C, Casteels R (1972) Pflügers Arch 337:333
34. Cauvin C, Loutzenhiser R, van Breeman C (1983) Annu Rev Pharamacol Toxicol 23:373
35. Fleckenstein A (1997) Annu Rev Pharamacol Toxicol 17:149
36. Janis RA, Triggle DJ (1983) J Med Chem 26:775
37. Papaioannou S, Panzer-Knodle S, Yang PC (1987) J Pharmacol Exp Ther 241:91
38. Corelli F, Manetti F, Taft A, Campiani G, Nacci V, Botta M (1977) J Med Chem 40:125
39. Campiani G, Garofalo A, Fiorini I, Botta M, Nacci V, Taft A, Chiarini A, Budriest R, Bruni G, Romeo MR (1995) J Med Chem 38:4393
40. Kimball SD, Hunt JT, Barrish JC, Das J, Floyd DM, Lago MW, Lee VG, Spergel SH, Moreland S, Hedberg SA, Gougoutas JZ, Malley MF, Lau WF (1993) Bioorg Med Chem 1:285
41. Atwal KS, Rovnyak GC, Schwartz J, Moreland S, Hedberg A, Gougoutas JZ, Maleey MF, Floyd DM (1990) J Med Chem 33:1510
42. Atwal KS, Rovnyak GC, Kimball SD, Floyd DM, Moreland S, Swanson BN, Gougoutas JZ, Schwartz J, Smillie KM, Malley MF (1990) J Med Chem 33:2629
43. Atwal KS, Swanson BN, Unger SE, Floyd DM, Moreland S, Hedberg A, O'Reilly BC (1991) J Med Chem 34:806
44. Rovnyak GC, Atwal KS, Hedberg A, Kimball SD, Moreland S, Gougoutas JZ, O'Reilly BC, Schwartz J, Malley MF (1992) J Med Chem 35:3254
45. Gupta SP, Veerman A, Bagaria P (2004) Molec Diver 8:357
46. Kier LB, Hall LH (1976) Molecular connectivity in chemistry and drug research. Academic Press, New York
47. Kier LB, Hall LH (1983) J Pharm Sci 72:1170
48. Gubin CP, Lucchetti J, Mahaux J, Nisato D, Rosseels G, Clinet M, Polster P, Chatelain P (1992) J Med Chem 35:981
49. Gubin CP, Vogelaer HD, Inion H, Houben C, Lucchetti J, Mahaux J, Rosseels G, Peiren M, Clinet M, Polster P, Chatelain P (1993) J Med Chem 36:1425
50. Gupta SP, Mathur AN, Nagappa AN, Kumar D, Kumaran S (2003) Eur J Med Chem 38:867
51. Höltze H-D, Marrer S (1987) J Comput-Aided Mol Des 1:23
52. Mager PP, Coburn RA, Solo AJ, Triggle DJ, Rothe H (1992) Drug Des Discov 8:273
53. Belvisi L, Brossa S, Salimbeni A, Scolastico C, Todeschini R (1991) J Comput-Aided Mol Des 5:571
54. Schleifer K-J (1997) J Comput-Aided Mol Des 11:491

55. De Weille JR, Schweitz H, Macs P, Tartar A, Lazdunski M (1991) Proc Natl Acad Sci USA 88:2437
56. Teramoto N, Ogata R, Okabe K, Kamayama A, Kameyama M, Watanabe TX, Kuriyama H, Kitamura K (1996) Pflüg Arch Eur J Physiol 432:462
57. Peterson BZ, Tanada TN, Catterall WA (1996) J Biol Chem 271:5293
58. Viswanadham VN, Mueller GA, Basak SC, Weinstein JN (2001) J Chem Inf Comput Sci 41:505
59. Takahata Y, Costa MCA, Gaudio AC (2003) J Chem Inf Comput Sci 43:540
60. Hemmateenejad B, Akhond M, Miri R, Shamsipur M (2003) J Chem Inf Comput Sci 43:1328
61. Romanelli MN, Dei S, Scapecchi S, Teodoeri E, Gualtieri F, Budriesi R, Mannhold R (1994) J Comput-Aided Mol Des 8:123
62. Schleifer K-J, Tot E (2000) J Comput-Aided Mol Des 14:427

Author Index Volumes 1–4

The volume numbers are printed in italics

Subject Index

RETURN TO: CHEMISTRY LIBRARY

100 Hildebrand Hall • 510-642-3753

LOAN PERIOD	1	2		3
		1-MONTH USE		
4		5		6

ALL BOOKS MAY BE RECALLED AFTER 7 DAYS.

Renewals may be requested by phone or, using GLADIS, type **inv**
followed by your patron ID number.

DUE AS STAMPED BELOW.

NON-CIRCULATING.
UNTIL: 02-01-08

FORM NO. DD 10
1,000   7-07

UNIVERSITY OF CALIFORNIA, BERKELEY
Berkeley, California 94720–6000